Polymer Coatings

Polymer Coatings
Technologies and Applications

Edited by
Sanjay Mavinkere Rangappa,
Jyotishkumar Parameswaranpillai, and
Suchart Siengchin

CRC Press
Taylor & Francis Group
Boca Raton London New York

CRC Press is an imprint of the
Taylor & Francis Group, an **Informa** business

First edition published 2021
by CRC Press
6000 Broken Sound Parkway NW, Suite 300, Boca Raton, FL 33487-2742

and by CRC Press
2 Park Square, Milton Park, Abingdon, Oxon, OX14 4RN

© 2021 Taylor & Francis Group, LLC

CRC Press is an imprint of Taylor & Francis Group, an Informa business

No claim to original U.S. Government works

Visit the Taylor & Francis Web site at
http://www.taylorandfrancis.com

and the CRC Press Web site at
http://www.crcpress.com

ISBN: 9780367611576 (hbk)

ISBN: 978-0-429-19922-6 (ebk)

Typeset in Times
by codeMantra

Contents

Preface

Polymer coatings act as a physical barrier to protect the surface from corrosion. In our words, polymer coatings can cover and protect a multitude of surfaces, such as metals, plastics, wood, bricks, stones, glass, and flooring. Both thermoplastics and thermosetting resins can be used as coating materials. The traditional polymers such as polyvinyl alcohol, polyvinyl chloride, epoxy resins, polyurethanes, polyesters, elastomers, acrylic resins, and phenolic resins have been used as coating materials, depending on the type of applications. Techniques such as spin coating, spray coating, solvent casting, and dip coating have been used frequently for polymer coating. Traditional coatings, when subjected to harsh environments, may easily undergo decomposition. The performance of the coatings can be improved with the incorporation of active agents. Recently, stimuli-responsive polymer coatings, self-healing coatings, self-cleaning coatings, anti-fouling coatings, anti-bacterial coatings, etc., have been reported with the incorporation of active agents. The performance of the coatings can be evaluated with techniques such as salt spray test, electrochemical impedance spectroscopy, scanning vibrating electrode technique, electron microscopy, optical microscopy, fluorescence microscopic images, and CT scans.

The research on polymer coatings is thriving, and this leads to an upsurge in the number of papers, reviews, and patents published. However, only a few books have been published in the area of polymer coatings. Therefore, in the view of the recent developments in polymer coating, we believe it is befitting to judiciously edit a book on Polymer Coatings: Technologies and Applications. The objective of the proposed book is to give an update on the recent developments in polymer coatings. We hope that the present book will benefit researchers working in both academia and industry.

The book includes 18 chapters that shed light on the area of Polymer Coatings: Technologies and Applications. Chapter 1, *Introduction to Polymeric Coatings*, gives a succinct introduction to polymer coatings, its theory, and applications. Chapter 2, *Polymer Coatings Methods*, focuses on the techniques used for the preparation of polymer coating. The techniques such as spin coating, spray coating, thermal spray analysis, cold spray coating, solvent casting, drop casting, dip coating, and film casting are discussed in detail in this chapter. Chapter 3, *Polymer Coatings Based on Nanocomposites*, outlines the recent progress in nanotechnology-assisted polymer coating. Chapter 4, *Mechanism, Anti-Corrosion Protection and Components of Anti-Corrosion Polymer Coatings*, gives an overview of the anti-corrosion effect of polymer coatings. Chapter 5, *Morphology of Polymer Coatings*, gives a summary of different morphologies generated in polymer coatings. The authors emphasized that the morphologies of the generated polymer coating depend on various factors such as concentration, type of blends, processing, and phase separation. Chapter 6, *Spectroscopic Analysis of Polymer Coatings*, gives an overview of the applicability of different spectroscopic techniques for the characterization of polymer coatings. Chapter 7, *Rheology of Polymer Coatings*, outlines the rheology of paints and coating formulations. Chapter 8, *Cure Reactions of Polymer Coatings*, summarizes the recent developments in the cure reactions of various polymer coatings such as epoxy resins, polyurethanes, phenol-formaldehyde resins, polyester resins, acrylates, vinyl resins, alkyd resins, and polyester resins. Chapter 9, *Thermal and Mechanical Properties of Polymer Coatings*, gives a succinct review of the mechanical and thermal properties of different polymer-based coatings. Chapter 10, *Conducting Polymer Coatings for Corrosion Protection*, sheds light on the recent progress in the conductive polymer-based coatings. The commonly used conducting polymers for coating applications are polypyrrole and polyaniline. Chapter 11, *Stimuli-Responsive Polymer Coatings*, gives an overview of stimuli-responsive polymer coatings and their applications. Chapter 12, *Self-Healing Polymer Coatings*, discusses the application of self-healing coating to control the corrosion of the metal substrate. Chapter 13, *Hydrophobic and Hydrophilic Polymer Coatings*, emphasizes the recent progress in the hydrophilic and hydrophobic coating, their properties, theory, and applications. Chapter 14, *Antifouling, Anti-Bacterial, and Bioactive Polymer Coatings*, highlights the recent developments in anti-bacterial and anti-biofouling-based polymer coatings. Chapter 15, *Adhesion of Polymer Coatings*, discusses the relevance of adhesion in improving the performance of polymer coatings. Chapter 16, *Modeling and Simulation in Polymer Coatings*, highlights the significance of modeling and simulation as a potential solution tool for polymer coating systems. Chapter 17, *Future Challenges and Applications of Polymer Coatings*, gives an overview of the broad spectrum of applications of polymer coatings. Chapter 18, *Incorporation of CeMo Nanocontainers Loaded with Inhibitors, Water, and Chlorine Microtraps into Anticorrosion Coatings onto ZK30*, highlights the anti-corrosion activity of CeMo nanocontainers loaded with inhibitors, water, and chlorine microtraps coatings on magnesium alloy ZK30.

The editors are thankful to all the authors for their contribution. The editors also thank the editorial and publishing team for their guidance and support.

Editors

Sanjay Mavinkere Rangappa earned a BE in mechanical engineering at Visvesvaraya Technological University, Belagavi, India; an MTech in computational analysis in mechanical sciences at VTU Extension Centre, GEC, Hassan; and a PhD in mechanical engineering at Visvesvaraya Technological University, Belagavi, India. Dr Sanjay did postdoctoral work at King Mongkut's University of Technology North Bangkok, Thailand. He is a Life Member of the Indian Society for Technical Education (ISTE) and an Associate Member of Institute of Engineers (India). He is a reviewer for more than 50 international journals (Elsevier, Springer, Sage, Taylor & Francis, Wiley), book proposals, and international conferences. He has published more than 85 articles in high-quality international peer-reviewed journals, 3 editorial corners, 20+ book chapters, 1 book, and 15 books as an editor, and presented research papers at national and international conferences. His research areas include natural fiber composites, polymer composites, and advanced material technology. He is a recipient of the DAAD Academic exchange-PPP Programme (project-related personnel exchange) between Thailand and Germany at the Institute of Composite Materials, University of Kaiserslautern, Germany. He has received a Top Peer Reviewer 2019 award, Global Peer Review Awards, Powered by Publons, Web of Science Group. He is currently a research scientist at the Natural Composites Research Group Lab, Academic Enhancement Department, King Mongkut's University of Technology North Bangkok, Thailand.

Jyotishkumar Parameswaranpillai earned a PhD in polymer science and technology (chemistry) at Mahatma Gandhi University. He has research experience at various international laboratories, such as the Leibniz Institute of Polymer Research Dresden (IPF), Germany; Catholic University of Leuven, Belgium; and University of Potsdam, Germany. He has published more than 100 papers in high-quality international peer-reviewed journals on polymer nanocomposites, polymer blends and alloys, and biopolymer, and he has edited five books. He received numerous awards and recognitions, including the prestigious Kerala State Award for the Best Young Scientist 2016, INSPIRE Faculty Award 2011, and Best Researcher Award 2019 from King Mongkut's

University of Technology North Bangkok. He is currently a research professor at the Center of Innovation in Design and Engineering for Manufacturing, King Mongkut's University of Technology North Bangkok, Thailand.

Suchart Siengchin earned a DiplIng in mechanical engineering at the University of Applied Sciences, Giessen/Friedberg, Hessen, Germany; an MSc in polymer technology at the University of Applied Sciences Aalen, Baden-Wuerttemberg, Germany; an MSc in material science at the Erlangen-Nürnberg University, Bayern, Germany; and a DrIng in engineering at the Institute for Composite Materials, University of Kaiserslautern, Rheinland-Pfalz, Germany. His postdoctoral research was at Kaiserslautern University and the School of Materials Engineering, Purdue University, West Lafayette, Indiana, USA. He received the habilitation at the Chemnitz University in Sachsen, Germany. He was a lecturer of the Production and Material Engineering Department at the Sirindhorn International Thai-German Graduate School of Engineering (TGGS), King Mongkut's University of Technology North Bangkok (KMUTNB). He is a full professor at and president of KMUTNB. He received the Outstanding Researcher Award in 2010, 2012, and 2013 at KMUTNB. His research interests include polymer processing and composite material. He is Editor-in-Chief of KMUTNB International Journal of Applied Science and Technology and the author of more than 150 peer-reviewed journal articles. He has participated in presentations at more than 39 international and national conferences with respect to materials science and engineering topics.

Contributors

Sanjeev Kumar Ahuja
Department of Chemical Engineering
Thapar Institute of Engineering and
    Technology
Patiala, India

Raj Kumar Arya
Department of Chemical Engineering
Dr. B. R. Ambedkar National Institute
    of Technology
Jalandhar, India

Waisudin Badri
Claude Bernard University Lyon 1
Lyon, France

Sanghamitra Barman
Department of Chemical Engineering
Thapar Institute of Engineering and
    Technology
Patiala, India

Kian Bashandeh
J. Mike Walker '66 Department of
    Mechanical Engineering
Texas A&M University
College Station, Texas

Avinash Chandra
Department of Chemical Engineering
Thapar Institute of Engineering and
    Technology
Patiala, India

Piotr Czub
Department of Chemistry and
    Technology of Polymers
Cracow University of Technology
Kraków, Poland

Narayan Chandra Das
Rubber Technology Centre
Indian Institute of Technology
Kharagpur, India

Vaishally Dogra
Department of Mechanical Engineering
Graphic Era Hill University
Dehradun, India

Emilie Dumas
Claude Bernard University Lyon 1
Lyon, France

Abdelhamid Errachid El Salhi
Claude Bernard University Lyon 1
Lyon, France

Abdelhamid Elaissari
Claude Bernard University Lyon 1
Lyon, France

Hatem Fessi
Claude Bernard University Lyon 1
Lyon, France

M. Gagliardi
NEST
Istituto Nanoscienze-CNR
Pisa, Italy

Émilie Gagnière
Claude Bernard University Lyon 1
Lyon, France

Adem Gharsallaoui
BioDyMIA
Claude Bernard University Lyon 1
Lyon, France

Sami Ghnimi
Claude Bernard University Lyon 1
Lyon, France

Naman Jain
Mechanical Engineering Department
Meerut Institute of Engineering and
    Technology
Meerut, India

George Kordas
Sol-Gel Laboratory
Institute of Nanoscience and
    Nanotechnology (INN), NCSR
    "Demokritos"
Agia Paraskevi, Greece

G. Hemath Kumar
Department of Mechanical Engineering
Madanapalle Institute of Technology
    and Science
Chittoor, India

Noureddine Lebaz
Claude Bernard University Lyon 1
Lyon, France

Wei Liao
Claude Bernard University Lyon 1
Lyon, France

Liliana Licea-Jiménez
Centro de Investigación en Materiales
    Avanzados S.C.
CIMAV Unidad Monterrey
Apodaca, México

Denis Mangin
Claude Bernard University Lyon 1
Lyon, France

Rokbi Mansour
Department of Mechanical Engineering
Université Mohamed BOUDIAF de
    M'sila
M'sila, Algeria

Abraham Méndez-Reséndiz
Centro de Investigación en Materiales
    Avanzados S.C.
CIMAV Unidad Monterrey
Apodaca, México

Ulises Antonio Méndez-Romero
Centro de Investigación en Materiales
    Avanzados S.C.
CIMAV Unidad Monterrey
Apodaca, México

Ricardo Antonio Mendoza-Jiménez
Centro de Investigación en Materiales
    Avanzados S.C.
CIMAV Unidad Monterrey
Apodaca, México

S. Mohanty
CIPET: SARP-LARPM
Bhubaneswar, India

H. Mohit
Department of Mechanical and Process
    Engineering
King Mongkut's University of
    Technology North Bangkok
Bangkok, Thailand

Sharika T. Nair
International and Interuniversity
    Centre for Nanoscience and
    Nanotechnology
Mahatma Gandhi University
and
Department of Chemistry
St. Xavier's College Vaikom
Kottayam, India

S.K. Nayak
CIPET
Chennai, India

Emerson Escobar Nunez
Departamento de Energética y
  Mecánica
Universidad Autónoma de
  Occidente
Cali-Valle, Colombia

Fabrice Ofridam
Claude Bernard University Lyon 1
Lyon, France

Jyotishkumar Parameswaranpillai
Center of Innovation in
  Design and Engineering for
  Manufacturing
King Mongkut's University of
  Technology North Bangkok
Bangkok, Thailand

Avinash Parashar
Department of Mechanical and
  Industrial Engineering
Indian Institute of Technology
Roorkee, India

Sergio Alfonso Pérez-García
Centro de Investigación en Materiales
  Avanzados S.C.
CIMAV Unidad Monterrey
Apodaca, México

Andreas A. Polycarpou
J. Mike Walker '66 Department of
  Mechanical Engineering
Texas A&M University
College Station, Texas

Harikrishnan Pulikkalparambil
Department of Mechanical and Process
  Engineering
King Mongkut's University of
  Technology North Bangkok
Bangkok, Thailand

Sanjay Mavinkere Rangappa
King Mongkut's University of
  Technology North Bangkok
Bangkok, Thailand

Shweta Rastogi
Department of Mechanical Engineering
G B Pant University of Agriculture and
  Technology
Pantnagar, India

Manju Rawat
Department of Chemical Engineering
Thapar Institute of Engineering and
  Technology
Patiala, India

Sanjay Remanan
Rubber Technology Centre
Indian Institute of Technology
Kharagpur, India

Diego Fernando Rodíguez-Díaz
Centro de Investigación en Materiales
  Avanzados S.C.
CIMAV Unidad Monterrey
Apodaca, México

R. Ruban
Department of Mechanical Engineering
National Institute of Technology
Tiruchirappalli, India

V. Arul Mozhi Selvan
Department of Mechanical Engineering,
National Institute of Technology
Tiruchirappalli, India

Daisy Sharma
School of Chemistry and Biochemistry
Thapar Institute of Engineering and
  Technology
Patiala, India

Jyoti Sharma
School of Chemistry and Biochemistry
Thapar Institute of Engineering and
   Technology
Patiala, India

Suchart Siengchin
King Mongkut's University of
   Technology North Bangkok
Bangkok, Thailand

Anna Sienkiewicz
Department of Chemistry and
   Technology of Polymers
Cracow University of Technology
Kraków, Poland

Sandeep Kumar Singh
Indian Institute of Technology
Roorkee, India

Mohamad Tarhini
Claude Bernard University Lyon 1
Lyon, France

Arturo Román Vázquez-Velázquez
Centro de Investigación en Materiales
   Avanzados S.C.
CIMAV Unidad Monterrey
Apodaca, México

Akarsh Verma
Department of Mechanical and
   Industrial Engineering
Indian Institute of Technology
Roorkee, India

S. Verma
CIPET
Chennai, India

P. Poornima Vijayan
Department of Chemistry
Sree Narayana College for Women
   (affiliated to University of Kerala)
Kollam, India

1 Introduction to Polymeric Coatings

Manju Rawat, Daisy Sharma, Jyoti Sharma, Sanjeev Kumar Ahuja, Avinash Chandra, and Sanghamitra Barman
Thapar Institute of Engineering and Technology

Raj Kumar Arya
Dr. B.R Ambedkar National Institute of Technology

CONTENTS

1.1 INTRODUCTION

The word polymer is derived from a Greek word, where "poly" means many and "mers" means particles. So, polymer can be described as a combination of identical molecules called mers. Polymers are composed of repeat units of small compounds called monomers. Different polymers show different properties, depending upon how the monomers are linked with each other. Some polymers are hard, such as bakelite and glass; whereas some are soft, such as biopolymers, silk, and rubber. Polymers are found to have several applications and the use of both natural and man-made polymers is widespread in our society. The natural polymers include proteins and polysaccharides, and the man-made polymers include synthetic plastics and fibers.

1

Their applications range from households to industries such as transportation, aviation, and pharmaceuticals. Their extensive use is in the area of medical devices, food packaging, decoration, transportation, information technology, and so on. Polymers are represented by repeating structural units called monomer. For example, a simple and long-chain polymer polyethylene, with ethylene as a repeat unit and n as its number of repetition, is expressed as follows:

$$[-CH_2 - CH_2 -]_n$$

The dissolution of polymers is different from the low-molecular-weight compound. Long dissolution time and enhanced viscosity of their solution make them unique. To obtain homogeneous solution, one can make use of solubility parameters and predict the solubility of polymer in the solvent (Hansen, 2007). When these data are not available, trials with various solvents would be needed.

Polymeric coatings are made of polymeric materials and can be applied on a number of substrates using a variety of methods such as extrusion/dispersion and solution casting techniques. Polymeric coatings provide excellent adherence to and protection from the environment. They are so designed that they adhere well to the substrate and not peel off easily, nor degrade due to heat, moisture, salt, or chemicals. Coatings are made up with different ingredients such as solvent, additives, and thinners. Different ingredients play different roles in coatings, such as additives enhance the functional properties. Similarly, thinners reduce the viscosity of the mixture, making the final coat appear smooth and without any defect.

Coatings are mainly used in the following sectors:

- *Transportation:* The coatings provide protection from weather such as heat/light, corrosion, and scratches from accidents such as dents.
- *Household:* The coatings provide finishing to kitchen appliances, such as Teflon coating and antiheating properties to utensils.
- *Medicines:* The coatings are used to coat drugs and thereby reduce contact with viruses and bacteria.
- *Industry:* In industries, coatings provide protection to equipment such as pipelines from weathering, extreme heat/light, and corrosion; in membrane industry, coating is used for the selective separation of gases.
- *Cosmetics:* Coating is widely used in creams, glasses, nail paints, etc.
- *Energy:* In this sector, coatings act as a protective barrier against extreme heat and light, enable easy dust removal, serve as an O_2 barrier, such as in wind energy and solar energy panels, and so on.

The use of coatings have increased extensively in automobile, aerospace, aircraft, marine, membranes, magnetic media, storage devices, and food industry. They are also used in the oil and gas industry for corrosion control, encapsulation of electronic circuits, textile industry for making breathable clothing, surface engineering industry for manufacturing of multilayer coatings that are used in various transportation and infrastructure applications, implantable devices, and materials for protection from the weathering conditions.

1.2 SALIENT FEATURES OF POLYMERIC COATINGS

Polymeric coatings are prepared with or without solvents. Coatings of one polymer and one solvent are called binary coatings. Multicomponent coatings have more than one polymer/solvent. Polymeric coating chemistry is different from that of polymers. The coating includes polymer matrix or binder: the polymer phase that holds all the constituents together. In polymeric coatings, we add fillers, pigments, surfactants, plasticizers, and binders to impart hardness/flexibility, strength, weather protection, color, gloss, and reflectance. The interface plays an important role in the performance of polymeric coatings. The main types of interfaces are substrate (bottom), coating air (top), and additives (internal). Other interfaces that also play a role are solid–gas, solid–liquid, and gas–liquid interfaces (e.g., in drying paints). Flow and leveling properties greatly influence the appearance and performance of coatings (Sharma et al., 2018).

The polymer phase present on the surface and its chemistry is far different from the bulk polymers. The interesting characteristics of polymer coatings are that they are easily spread and handled. They can be applied as liquids as in the case of paints, and various additives may be added to enhance their properties.

Some other properties such as flexibility, drying, and self-healing are the ones where molecular mobility plays an important role. The advanced polymeric coatings are thermosetting in nature, despite that a large majority of polymeric materials are thermoplastics. It is easy to predict the molecular mobility of thermoplastic polymers because their entangled macromolecules are still largely independent. However, in thermosetting polymers, the network structure strongly hinders the molecular mobility.

1.3 NECESSITY OF POLYMERIC COATINGS

In our modern society, coatings are useful in several ways. Some are used for good appearance, while others are highly functional, embedded in devices or used to protect the other materials by covering them. Coatings, thus, play a vital role in our life, whether as mere protective layers or as a functional part of intricate machinery (Sharma et al., 2019a). The decorative function includes gloss and diffuse reflection, aesthetic appearance, transparency, and cleanability. The protective function includes the protection of substrate from external aspects such as humidity, light, fungi, air, bacteria, chemicals, dirt, corrosion, impact, and mechanical abrasions. Hence, the polymeric coatings are the delicate mix of polymers with different types of additives to enhance the functional properties.

1.4 CURRENT SCENARIO

Currently, the functional polymeric coatings are gaining more attention. Over the last decades, more research is focused on manufacturing of functional coatings that show faster drying rate, self-healing, easy to clean, antibacterial, antifouling, low water permeability, and less moisture retention. Functional coating properties are surface phenomena. These properties cannot be attained easily by conventional methods and techniques of synthesis. These properties are achieved by adding additives in small amounts in a polymer solution, without altering the properties of substrate, and by using new technologies such as free-radical polymerization, graft polymerization, and microemulsion polymerization for the synthesis of new binders. The technology that can be used for unique functional coatings are organic–inorganic hybrid, self-assembly, and nanotechnology; enhanced properties of current coatings such as corrosion protection, ultraviolet (UV) and heat resistance, and antiaging can be attained by using modern techniques. The smart coatings offer multifunctional properties like optical coatings, conducting coatings, self-cleaning coatings, light-controlled coatings, and so on. These are not only in commercial demand but also fuel interest for academic research.

1.4.1 Research Perspective

Drying of polymeric coatings implies removal of the solvent from the coatings. There are different methods that can be used for the drying of coatings. It can be done in ambient temperature or through thermal drying, condensation drying, and so on. Drying of polymeric coating depends on both heat and mass transfer phenomena. It is done in various stages. The drying of coating determines the final product, which can be either defective if proper drying is not done or good with uniform coating provided proper drying process has been followed. The coating should have minimum residual solvent. The minimization of residual solvent and reduction of drying time is the interest point of researchers nowadays. Researchers are making efforts to obtain defect-free polymeric coatings, with least residual solvent in it (Prashar et al., 2019; Sharma et al., 2018; Sharma et al., 2019a, 2019b, 2019c)

The solvent in which polymer blend is prepared is another point of interest. It should be cheap, environment-friendly, and easy to recover. The solvent should not cause defects to coatings such as solvent popping, sagging, and blistering. The mathematical binary (Saure et al., 1998; Vrentas and Duda, 1977a, 1977b; Vrentas and Vrentas, 1994; Zielinski and Hanley, 1999), ternary (Arya, 2014; Arya et al., 2016; Arya and Vinjamur, 2013), and quaternary (Price Jr and Cairncross, 2000; Siebel et al., 2016) diffusion models help reduce the defects caused by solvents. Vrentas and Duda (1977a, 1977b) free volume theory in conjunction with Flory–Huggins theory explains the polymer solution thermodynamics and diffusion of solvent in the polymer.

Green coatings have also attracted researchers nowadays, because of the toxicity of organic solvent. On industrial level, organic solvents are used in huge amount. These solvents are toxic in nature and cause harm to environment. Water-based coatings, known as green coatings, are cost-effective and environment-friendly. The major challenge with water-based coating is its long drying time, which is greater than that required by the organic solvent-based coatings, because of their low volatility. The long drying time cost higher amount of energy, as the process is slow. Researchers aim at optimization of drying time to obtain uniform coatings with least residual solvent (Price Jr and Cairncross, 2000).

Polymeric coatings, value-added with additives, have become major attraction in various industries such as cosmetics, food, packaging, medicines, and automobiles. The reason is incorporation of functional properties into polymeric coating by additives. Additives increase the processability, reduce the cost, and change the physical and chemical properties of polymeric coatings. Ultimately, the significant advantage of using functional polymeric coatings is that these thin layers are required in tens to hundreds of micrometers to enhance the properties of material such as strength, flexibility, color, UV absorbance, fire retardance, and impact, to reduce the energy consumption with better quality of finish product as well as impact on the environment; and add to the economics and sustainability of materials and devices (Fabbri and Messori, 2017). As already noted, these coatings are of current research interest.

1.5 SOLUBILITY CHARACTERISTICS

Polymeric coatings can be made from one or more polymers dissolved in one or more solvents. The selection of solvents is done by using Hansen solubility parameters (Hansen, 2007). The data for various polymers and solvents are given in polymer handbook by Barton (2018). In case, the data are not available, the solubility of polymer can also be seen on the basis of linearity, thermoplasticity, glass transition temperature, cross-linking degrees, crystallinity, molecular weight, molecular cohesion, polarity, tacticity, conformation, and intermolecular bonding. On the basis of tacticity, polymers structure can be classified as isotactic, syndiotactic, and atactic. These polymer molecules have same chemical structure but exhibit very different properties. The main reason for this is their entirely different configuration and geometrical structure. The atactic polymers are nonsymmetric in structure and their molecular attractive forces are weak (Abu-Sharkh, 2004). Hence, they show low melting point and good solubility, for example, polyvinyl chloride formed by free-radical mechanisms. However, syndiotactic polymers such as polystyrene formed by metallocene catalysis polymerization and isotactic polymers such as polypropylene formed by Ziegler–Natta catalysis generally show high melting point and less solubility. This is due the symmetry and closely packed structure because of strong forces between the molecules (Abu-Sharkh, 2004).

The solubility is also affected by geometry of the polymer. The combination of *cis–trans* geometry obtained after polymerization of the polymer affects its solubility. For example, 1,6-heptadiynes obtained by cyclopolymerization shows that in comparison to *cis–trans* configuration, all *trans* configurations are less soluble (Lyman, 1961). The flexibility and rigidity of polymer affect its solubility in the solvent. The more the chain flexibility of the polymer, the more its degree of rotation and freedom, thus making it easily soluble such as polystyrene. If the polymer shows more rigidity in its structure, such as polyamides or cellulose esters, its movement is hindered because of the bulky side groups or perhaps strong attractive forces, which make it less soluble (Bicerano, 2002).

The amorphous polymers take less time to dissolve in solvent. In crystalline polymers, the polymers are orderly folded and hence are resistant to dissolution. Similarly, the linear chain polymers are attached with each other closely because of the attractive forces between the molecules, which make them less dissolvable. The branched or highly cross-linked polymers are difficult to dissolve because of their branching between the molecules (Bicerano, 2002).

The intermolecular bonding also affects the solubility of a polymer. The presence of high intermolecular forces such as hydrogen bonding between the molecules makes them less soluble in solvent, as molecules are not able to interact with the solvent. Hence, the external forces such as heat and stress are applied to break the intermolecular forces between the polymer molecules to make them soluble in the solvent. If in a particular case, neither the structure information work nor the solubility parameter data are available, then a number of trials should be performed for the dissolution of polymer in various solvents (Bicerano, 2002).

1.5.1 SOLUBILITY PARAMETERS

Hildebrand and Scott in 1950 were first to introduce the solubility parameters. It is the numerical value that indicates the degree of interaction between the material and solubility behavior of solvents. For convenience, the square root of the cohesive energy density is divided into three parameters : δ_P, δ_H, and δ_D, where P is the dipolar interaction, H is the hydrogen bonding interaction, and D is the nonpolar interaction. Cohesive energy is defined as the amount of energy needed to eliminate the intermolecular interaction forces between the molecules. In the condensed state, it is defined as the internal energy of compound divided by their molar volume, as seen in Eq. (1.1). This amount of energy is required to reduce the interaction between the molecules so that they get separated from each other and are surrounded by solvent molecules:

$$\delta_t = \left(\text{ced}\right)^{1/2} = \left(\frac{\Delta E_v}{V}\right)^{1/2} \tag{1.1}$$

$$\Delta E_v = \Delta H_v - RT \tag{1.2}$$

where δ_t is the Hildebrand solubility parameter (HSP), ced is the cohesive energy density, ΔE_v is the latent heat of vaporization, V is the molar volume, R is the universal gas constant, and T is the absolute temperature. The SI units for solubility parameters are $\text{MPa}^{1/2}$.

The assignment of HSP to polymer and solvent allows prediction of behavior that can improve or maintain stability, performance, and environmental acceptability as well as reduce costs.

The compatibility relation between two materials involved in polymer solution is calculated with the help of relative energy density (RED) number. RED is calculated by difference in HSP between materials involved. The RED number range is between 0 and 1.0; 0 number shows the perfect match of the two materials, whereas 1.0 reflects the limit of compatibility. The RED number less than 1.0 indicates some compatibility, whereas higher number shows no compatibility between the two materials (Birdi, 2008; Hansen and Beerbower, 1971). The three-dimensional solubility model is given by Hansen and Beerbower (1971) and is shown in Figure 1.1.

The model can help to predict the general behavior of polymer in solvent. The three-dimensional model is used to plot polymer solubility. The three axes show the component parameters δ_P, δ_H, and δ_D and R_0 indicates the interaction radius. For the dissolution of polymer in solvent, the solubility parameters δ_P, δ_H, and δ_D of the solvent should lie within the sphere. If the solubility parameters lie near the parameter

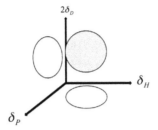

FIGURE 1.1 Solubility parameter diagram. (Redrawn from Hansen and Beerbower, 1971.)

of the sphere, minimal solubility can be observed, and completely mismatching would result when solubility parameters lie outside the sphere. The distance between solvent and center of solubility sphere can be calculated by using Eqs. (1.3) and (1.4):

$$R_a = \left[4\left(\delta_{D1} - \delta_{D2}\right)^2 + \left(\delta_{P1} - \delta_{P2}\right)^2 + \left(\delta_{H1} - \delta_{H2}\right)^2 \right]^{1/2} \tag{1.3}$$

$$\text{RED} = \frac{R_a}{R_0} \tag{1.4}$$

where R_a is the distance between solvent and the center of solubility sphere, R_0 is the distance from the center; δ_{D1}, δ_{P1}, and δ_{H1} are dispersion, polar, and hydrogen bonding solubility parameters, respectively, for the solvent; and δ_{D2}, δ_{P2}, δ_{H2} are dispersion, polar, and hydrogen bonding solubility parameters, respectively, for polymer.

For the dissolution of polymer, its complete compatibility with solvent is required. Sometimes limited solubility is also acceptable, but that leads to the phase separation problem. If there is an additive along with polymer to be dissolved in a solvent, then again full compatibility and solubility should be checked. The limited solubility of additives leads to the problem such as if some additive is soluble at elevated temperatures and if temperature is lowered, then they are not able to dissolve and hence appears like dust-type material on the surface of the polymer. It is possible to have two additives mixture in a polymer solution. Their solubility depends upon their HSP values. If their HSP values come close to the HSP value of polymer, then dissolution is possible. For the dissolution of a polymer in a solvent, the polymer solution thermodynamics is involved as described next.

1.5.2 POLYMER SOLUTION THERMODYNAMICS

According to the thermodynamics laws, the free energy of dissolution should be negative for a feasible solution. The free energy of mixing is given by

$$\Delta G_{\text{mix}} = \Delta H_{\text{mix}} - T\Delta S_{\text{mix}} \tag{1.5}$$

where T is the absolute temperature and ΔH_{mix} and ΔS_{mix} are enthalpy and entropy of mixing, respectively. The ΔS_{mix} term is always favorable for dissolution because entropy increases during dissolution. So long ΔH_{mix} is negative, ΔG_{mix} would result in a negative value. Hence, dissolution is possible. Now, consider a case of positive ΔH_{mix}.

The dissolution process is initially feasible if ΔH_{mix} is less than $T\Delta S_{mix}$. However, on continual addition of the polymer to the solution, the solution slowly becomes concentrated and proceeds to give less value of $T\Delta S_{mix}$, which eventually becomes equal to ΔH_{mix}. Finally, ΔG_{mix} becomes zero and equilibrium is attained. In the equilibrium, the polymer molecules above a particular molecular weight remain in the aggregated state, while the polymer molecules below that molecular weight become solvated (Gowariker et al., 1986).

According to Flory–Huggins theory, the entropy of mixing for polymer solution is given by

$$\Delta S_{mix} = -k\left(N_1 \ln \phi_1 + N_2 \ln \phi_{12}\right) \tag{1.6}$$

where k is the Boltzmann constant; subscripts 1 and 2 stand for the solvent and polymer, respectively; ϕ_1 and ϕ_2 are the volume fractions of solvent and polymer, respectively; and N_1 and N_2 are the number of molecules of solvent and polymer, respectively.

The volume fractions for solvent and polymer are given by Eqs. (1.7) and (1.8):

$$\phi_1 = \frac{N}{\left(N_1 + nN_2\right)} \tag{1.7}$$

$$\phi_2 = \frac{nN_2}{\left(N_1 + nN_2\right)} \tag{1.8}$$

where N_1 is the number of solvent molecules, N_2 is the number of polymer molecules, and n is the number of segments in polymer molecule.

ΔH_{mix} for polymer solution can be written as

$$\Delta H_{mix} = kT\chi N_1 \phi_2 \tag{1.9}$$

where k is the Boltzmann constant, T is the temperature, χ is the Flory–Huggins interaction parameter, N is the number of solvent molecules, and ϕ_2 is the volume fraction of polymer. χ is also called Flory–Huggins chi parameter, which can be used to check the solubility of polymer and solvent. It is a dimensionless quantity. The thermodynamics property depends on χ, for example, solubility and swelling. A small value of χ shows the best solvent for polymers. It can be divided into enthalpic and entropic terms:

$$\chi = \chi_H + \chi_S \tag{1.10}$$

where χ_H is the enthalpic component and χ_S is the entropic component. The entropic parameter has value between 0.3 and 0.4 for nonpolar systems. Usually the value is taken as 0.35. The χ is calculated by using the Bristow and Watson correlation (Eqs. 1.11–1.13) (Bristow and Watson, 1958):

$$\chi_H = \frac{V_{mix}}{RT}\left(\delta_1 - \delta_2\right)^2 \tag{1.11}$$

$$\chi = 0.35 + \frac{V_{mix}}{RT}\left(\delta_1 - \delta_2\right) \tag{1.12}$$

$$V_{mix} = x_1 V_1 + x_2 V_2 \tag{1.13}$$

where χ_H is the enthalpic part of chi parameter; χ is the interaction parameter; V_{mix} is the molar volume of the mixture of solvent and polymer, respectively, given in Eq. (1.13); δ_1 and δ_2 are solubility parameters of solvent and polymer, respectively; x_1 and x_2 are mole fractions of solvent and polymer, respectively; R is the universal gas constant; and T is the temperature in Kelvin.

Finally, for polymer solutions, the relation between ΔG_{mix} and χ is given by placing Eqs. (1.6) and (1.9) into Eq. (1.5):

$$\Delta G_{mix} = kT\left(\chi N_1 \phi_2 + N_1 \ln \phi_1 + N_2 \ln \phi_2\right) \tag{1.14}$$

1.6 ROLE OF DIFFERENT INGREDIENTS IN COATINGS

There are several ingredients in polymeric coatings. The role of each ingredient is as follows.

1.6.1 SOLVENT

Solvent is required to prepare polymeric coatings. The solvent plays an important role. It is the most important raw material used in coatings. Solvents can be organic or inorganic. Solvents strongly affect the formation of film, coating appearance, and processability. It dissolves various components into it during formation of coatings. For dissolution, the solubility parameters play key role as already described. Hence, for coatings preparation, they affect drying, viscosity, appearance, and so on of the film. In case of organic solvent, other factors that play role are their boiling point, evaporation rate, toxicity, and reactivity.

The solvent can cause different defects in the film such as blistering, bubbling, sagging, agglomeration, solvent lifting, and solvent popping. Figure 1.2 shows the defects in the polymeric coatings (Fitzsimons and Parry). Blistering might be because of the osmotic gradients caused by the use of soluble pigments, salts, retained solvents, and corrosion products. It might be because of nonosmotic blistering caused by thermal gradients produced by cold wall, compressive stress, and so on (Marrion, 2004; Chiantore and Lazzari, 2001; Gutoff and Cohen, 2006). It can be prevented by analyzing the atmospheric environment before applying the coatings, by ensuring the coating system with solubility of salt, and by analyzing the surface on which the coating is to be applied.

Bubbling might occur because of the high temperature while applying the coating and the air trapped in-between coatings and during mixing of ingredients. It is also observed while applying the coating of antifouling agents on nonuniform surface. It can be prevented by the use of proper coating equipment or application such as spray coating and mist coating, by avoiding the application of coating at high temperatures, by using proper equipment for mixing the ingredients and defoaming agents to avoid

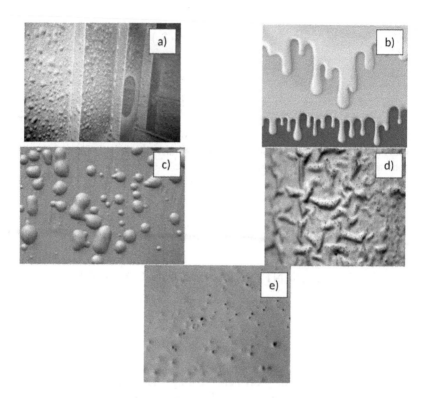

FIGURE 1.2 Defects in coating. (a) Blistering. (b) Bubbling. (c) Sagging. (d) Solvent lifting. (e) Solvent popping. (Requested for permission from Fitzsimons and Parry.)

air entrainment, and by using the data sheet to adjust the viscosity of solution by adding the suitable thinner (Marrion, 2004; Chiantore and Lazzari, 2001; Gutoff and Cohen, 2006).

Sagging can be caused due to excess multilayer coating, excessive use of thinning agents, lack of curing agents, and wrong formulation of paints. It can be prevented by using good coating techniques with correct formulation of products (Gutoff and Cohen, 2006).When the applied layers are not compatible with each other during coating, solvent lifting is caused. It can be possible when the fresh layer has strong solvent blend, which reacts with the previous undercoat layer that has not dried properly or has weak solvent. The overcoating is another cause of solvent lifting. It can be prevented by taking care of the number of times coating is applied and by checking the compatibility between the layers (Gutoff and Cohen, 2006).

Solvent popping might be because of the incorrect mixing of the solvent blend, nonuniform surface (basically porous), and high surrounding temperature. It can be checked by ensuring mixing of polymer and solvent, by using good or advance techniques for application, and by maintaining the surrounding atmosphere temperature (Gutoff and Cohen, 2006).

The coating morphology and drying rate are influenced by the type of solvent used, surface tension, additives added, viscosity of blend, different types of drying operation, and so on. Coatings are made on continuous web followed by drying in multizone dryers being operated at different drying conditions and time in each zone to reduce the solvent to optimum value without inducing any defects like blister, cracking, and so on. The primary aim of drying is to remove the solvent and produce defect-free coatings. It is obvious that the way the solvent evaporates controls the final film properties. In the industry, different types of dryers are being used to produce defect-free coatings. Before designing the real industrial dryer, it is always advisable to perform the simulation study in order to minimize the failures and hence the wasting of resources.

The drying of film is done in two stages. The first stage is solvent vapor pressure control. The resistance is produced by the layer of air above the coating surface. The second stage is diffusion control. It depends on how the solvent diffuses into the thin film to reach the surface. Various authors work on the drying of solution-casted polymeric film. The factors such as the heat and mass transfer description, availability of surface area, and equipment used for drying affect the final film appearance. They also affect the type of film defects (Alsoy, 2001; Alsoy and Duda, 1998; Okazaki et al., 1974).

Earlier, organic solvents were used in paints and coatings. But they contain hazardous pollutants that cause cancer, health problems, environmental pollution, birth defects [organo-psycho syndrome (OPS)], and so on. The hazardous solvents are listed under the hazardous air pollutants list (Spicer et al., 2002). The properties of various solvents have been compiled by Wypych and Pionteck (2016).

Finally, the type of solvent and ingredients influence the polymer miscibility and solubility. The solubility can further be validated with Hansen solubility parameters. The solvent evaporation is affected by various factors that influence the final film properties. The optimized evaporation rate can formulate the desired coatings that can meet the requirements. This can be worked out by testing the resulting formulations experimentally to confirm the predicted results.

1.6.2 Additives

Additives are the materials that are added in small amounts to the polymer matrix without affecting its physical properties. The physical and chemical properties of the coatings can be altered by adding just 0.001% and ≤5% amount of additives into it. It enhances the properties of a polymer.

Incorporation of additive into polymeric coatings makes them useful for multifunctional purposes such as automobile coatings, packaging, constructions, and electronics. The additives enhance the bulk and surface properties of polymers, as they facilitate their processability, drying, and flexibility. The additives addition can alter the physical, chemical, thermal, and morphological properties of polymers (Ambrogi et al., 2017). The coatings prepared with the additives like surfactants and plasticizers enhance their surface and bulk properties. The additives should be soluble in solvent–polymer solution.

Polymeric coatings are affected by external factors. They cannot withstand high heat, light, temperature, and weathering conditions such as UV exposure, humidity, temperature variation due to day/light, and humidity. For example, a very simple polymer film of polyvinyl chloride cracks when exposed to sunlight for a long time. Coatings life gets reduced and deteriorated; lose gloss, shine, longevity, and functional properties; and become corroded when exposed to high heat or light. This leads to mechanical failure and huge amount of loss to industries.

Increasing life, stability, and functional properties of polymeric coating is the main concern nowadays. Additives not only increase the life of coatings but also add functional properties to them. There are a variety of additives available on the market such as flame retardants (FR), UV absorbers, antioxidants, heat stabilizers, surfactants, plasticizers, colorants, pigments, coupling agents, impact modifiers, and light stabilizers. (Ambrogi et al., 2017).

FR additives are added to protect the costly plastic material from possible fire hazards. The role of these additives are that they get involved in the chemistry and physics of the combustion process and produce free radicals to stop the combustion process (Al-Malaika et al., 2017). FRs can be classified into three categories: phosphorous base, for example, ammonium polyphosphates used in epoxy, polyethylene, and so on; halogen-base, for example, chlorinates used in polyvinyl chloride, pipes, cables, and so on; brominates used in cellulosic polymers, polystyrene, and so on; and metal hydrate FRs, for example, aluminum trihydroxides used in polyvinyl chloride, poly propylene, and so on, and magnesium hydroxides used in natural rubbers (Al-Malaika et al., 2017).

UV absorbers provide stability to the polymers under UV light. They absorb the UV light and reverse the chemical reactions that occur when the polymer is left in the UV light. The main role is to reduce the aging, cracks, fading of colors, when coatings are subjected to outdoor applications. Carbon black and rutile titanium are basically used as UV absorbers because of their effective work in the wavelength range of 300–400 nm.

The antioxidant additives slow down the natural aging of the polymer, which is due to thermal or photooxidation. They are classified into peroxide decomposers and kinetic chain-breaking decomposers. Heat stabilizers additive stabilizes the polymers during processing and application when subjected to high heat. They prevent the degradation of polymers. Surfactants such as sodium dodecyl sulfate or fluoro-surfactants are added to polymer solution to reduce the surface tension and increase the drying time.

Plasticizers are used to increase the flexibility, compatibility, and durability and to reduce the tensile strength of polymeric coatings. Incorporation of plasticizers reduces the glass transition temperature of polymer (Vlacha et al., 2016). It enhances the mechanical properties, lowers the water uptake ratio, moisture content, water vapor permeability, and tensile strength, and increases the elongation at break (Balqis et al., 2017; Matet et al., 2013).

Surfactants are another class of additives. They can improve the drying rate of polymeric coatings. They are used to increase the wettability and reduce the surface tension. Surfactant chains are used to bind polymer and solvent, and sometimes for phase separation. They are mainly classified into anionic, cationic, nonionic, and polymeric. Their compatibility with polymeric solution decides their application.

The small concentration of surfactant can change the behavior of polymeric coating. They reduces the surface tension by creating the gradient between edges and center of the film (Kajiya et al., 2009; Sharma et al., 2018; Sharma et al., 2019a), due to which the solvent moves from the center to the edges and polymer moves from the edges to the center. This effect is named as Marangoni effect (Kajiya et al., 2009).

Other additives like adhesion promoters improve the adhesion of coatings to the substrate. An impact modifier improves the strength and reduces the brittleness of the coatings. Pigments and colorants impart colors to the paint and coatings (Christie, 1998).

1.6.3 PIGMENTS

Pigments are the additives that are used for the coloration of coatings. They impart color to paints, plastics, food, cosmetics, and so on based on the principle of absorption and reflection of wavelengths of visible light. They require high shear to dissolve, so that particles aggregation could not take place in the blend. Pigments are classified into two main groups: organic and inorganic pigments. The organic pigments have particle size in the range of 0.05 and 0.5 µm, whereas inorganic pigments have particle size in the range of 0.1–2.0 µm. Inorganic pigments are large in size and easily dispersible in resins. They provide better heat and light resistance, opacity, and chemical resistance than organic pigments. Examples of inorganic pigments are lead chromate, chromium oxide, titanium oxide, iron oxide, and so on (Day, 1990). Organic pigments produce more vibrant colors when added to plastics and cover less surface area. They need more shear than inorganic pigments to disperse in a blend. Some examples of organic pigments are carbon black, perylenes, phthalocyanine, and so on (Allen et al., 2002).

The basic aim of adding pigments to coatings is to provide colors to them, either for decoration purpose or for adding functional properties. Their role is much broader than imparting optical properties to paints or coatings. Some functional properties such as UV resistance, heat resistance, chemical resistance, mechanical reinforcement, and reduction of polymer degradation are also provided by pigments (Allen et al., 2002; Shao et al., 2009). On the other side, the side effect of incorporating pigments is the environmental issue. The ingredients of pigments, whether organic or inorganic such as titanium dioxide, cause pollution to the environment. It should be taken care of that the water-based solvent is used with less amount of binder. It should also be take care of that the organic solvent-based coatings with harmful ingredients like titanium dioxide and chromium dioxide should be avoided for the safety issue of the environment (Shao et al., 2009).

1.7 COATINGS INDUSTRY DEMAND

The market of coatings is huge. According to the "Industrial Coatings: A Global Market Overview" report, the industrial coatings demand is globally estimated at US$73.8 billion in 2016, and it is about to touch US$105.5 billion by 2022, increasing at the rate of 6.1% compounded annually. It has been shown that between 2016 and 2022, the largest consumer of coatings is Asia Pacific with nearly 47% followed by Europe with nearly 21%.

According to the "Paints and Coatings Market by Resin Type: Global Market Size, Share, Development, Growth, and Demand Forecast, 2013–2023" report, the increasing demand of automobile, multifunctional, electronics, and industrial coatings is the driving force for market growth. On the basis of the resin type, the coatings and paint market can by categorized as epoxy, polyvinyl chloride, nylon, polyester, spendix, acrylic, polyester, and so on.

On the basis of formulation, the paints and coatings market is categorized into water-based, solvent-based, powder coatings, and UV coatings. Nowadays industry is paying more attention toward water-based coatings, named as green coatings. The organic solvents cause damage to nature and end up being toxic. However, the water-based coatings do not cause pollution and are cost-effective.

For quite some time, the demand for protective coatings against weathering such as light, heat, impact, and so on has been increasing at 7.3% compounded annually. These coatings improve the processability, withstand environmental changes, and reduce cure cost and energy costs. Hence, these characteristics have increased demand in different wood industries, aerospace, automobile, energy sector, and shipping.

The market demand of painting and coating industries has been increasing because of the increasing demand of automobile, consumer goods, and manufacturing machines. Earlier the growth rate was 5.5%, which is now expected to increase from 7.2% to 8.3%. Moreover, government and private sectors of construction industry have joined in India and China, who are ready to spend in residential and commercial construction applications. This leads to boom the market demand of paint and coatings.

The environmental issue of zero or low VOC content in paints and coatings can be solved by using the nanotechnology, which plays a significant role in the industry. Nanoparticles will provide the smart function with superior characteristics to the coatings. They will be introduced to the coating matrix in various formulations such as in the form of ceramic or metals. The integrated product of polymer matrix and nanoparticles will increase the growth of paints and coating market.

Hence, with increasing population, the coatings demand have also increased, which ultimately boosts the economy. The major challenges ahead are the high-performance products and environment-friendly, cheap, easy-to-apply, reproducible, durable, and sustainable coatings.

1.8 CONCLUSION

The main role of coatings is in decoration, protection, and providing functionality. The different fields have different coatings requirements. For example, in adhesion, the required properties are wettability and scratch resistance. For outdoor applications, the coatings are required that can withstand the changing environmental conditions. In electrical industries, conductive coatings are required. In food industry, the coatings are required that do not allow water to permeate and absorb humidity. In membrane industry, porous coatings are required for selectivity of gases. The next-generation coatings would be multifunctional coatings that can perform two to three functions together. The next chapter presents techniques for the coating preparation.

REFERENCES

Abu-Sharkh, B.F., 2004. Influence of tacticity on solubility of propene monomer in isotactic and syndiotactic polypropylene. *Polymer*, 45(18): 6383–6389, https://doi.org/10.1016/j.polymer.2004.06.058.

Allen, N.S. Edge, M., Ortega, A., Liauw, C.M., Stratton, J. and McIntyre, R.B., 2002. Behaviour of nanoparticle (ultrafine) titanium dioxide pigments and stabilisers on the photooxidative stability of water based acrylic and isocyanate based acrylic coatings. *Polymer Degradation and Stability*, 78(3): 467–478, https://doi.org/10.1016/S0141–3910(02)00189-1.

Al-Malaika, S., Axtell, F., Rothon, R. and Gilbert, M., 2017. *Additives for Plastics, Brydson's Plastics Materials*. Elsevier, pp. 127–168, https://doi.org/10.1016/B978–0–323–35824–8.00007-4.

Alsoy, S., 2001. Predicting drying in multiple-zone ovens. *Industrial & Engineering Chemistry Research*, 40(14): 2995–3001, https://doi.org/10.1021/ie000751+.

Alsoy, S. and Duda, J., 1998. Drying of solvent coated polymer films. *Drying Technology*, 16(1–2): 15–44, https://doi.org/10.1080/07373939808917390.

Ambrogi, V., Carfagna, C., Cerruti, P. and Marturano, V., 2017. *Additives in Polymers, Modification of Polymer Properties*. Elsevier, pp. 87–108, https://doi.org/10.1016/B978-0-323-44353-1.00004-X.

Arya, R.K., 2014. Measurement of concentration profiles in thin film binary polymer-solvent coatings using confocal Raman spectroscopy: Free volume model validation. *Drying Technology*, 32(8): 992–1002, http://dx.doi.org/10.1080/07373937.2014.880714.

Arya, R.K., Tewari, K. and Shukla, S., 2016. Non-Fickian drying of binary polymeric coatings: Depth profiling study using confocal Raman spectroscopy. *Progress in Organic Coatings*, 95: 8–19, http://dx.doi.org/10.1016/j.porgcoat.2016.02.004.

Arya, R.K. and Vinjamur, M., 2013. Measurement of concentration profiles using confocal Raman spectroscopy in multicomponent polymeric coatings—model validation. *Journal of Applied Polymer Science*, 128(6): 3906–3918, https://doi.//10.1002/app.38589.

Balqis, A.I., Khaizura, M.N., Russly, A. and Hanani, Z.N., 2017. Effects of plasticizers on the physicochemical properties of kappa-carrageenan films extracted from Eucheuma cottonii. *International Journal of Biological Macromolecules*, 103: 721–732, http://dx.doi.org/doi:10.1016/j.ijbiomac.2017.05.105.

Barton, A.M., 2018. *Handbook of Poylmer-Liquid Interaction Parameters and Solubility Parameters*. Routledge, https://doi.org/10.1201/9780203752616.

Bicerano, J., 2002. *Prediction of Polymer Properties*. Third Edition, Revised and Expanded. Plastics Engineering, New York, USA, https://www.scribd.com/document/373155406/Jozef-Bicerano-Prediction-of-Polymer-Properties.

Birdi, K., 2008. *Surface and Colloid Chemistry, Handbook of Surface and Colloid Chemistry*. Boca Raton, FL: CRC Press, pp. 10–52, https://www.mobt3ath.com/uplode/book/book-36965.pdf.

Bristow, G. and Watson, W., 1958. Cohesive energy densities of polymers. Part 1.—Cohesive energy densities of rubbers by swelling measurements. *Transactions of the Faraday Society*, 54: 1731–1741, 1731–1741, https://doi.org/10.1039/TF9585401731.

Chiantore, O. and Lazzari, M., 2001. Photo-oxidative stability of paraloid acrylic protective polymers. *Polymer*, 42(1): 17–27,https://doi.org/10.1016/S0032-3861(00)00327-X.

Christie, R.M., 1998. *Pigments for plastics, Plastics Additives*. Springer, pp. 485–498, https://doi.//10.1007/978-94-011-5862-6_53.

Day, R., 1990. The role of titanium dioxide pigments in the degradation and stabilisation of polymers in the plastics industry. *Polymer Degradation and Stability*, 29(1): 73–92, https://doi.org/10.1016/0141-3910(90)90023-Z.

Fabbri, P. and Messori, M., 2017. Surface modification of polymers: Chemical, physical, and biological routes. *Modification of Polymer Properties*. Elsevier, pp. 109–130, https://doi.org/10.1016/B978-0-323-44353-1.00005-1.

Fitzsimons, B. and Parry, T., Coating Failure & Defects, *5B. Fitz's Atlas 2 and ASM Handbook*, Corrosion Pedia, https://www.ppcoatings.co.uk/wp-content/uploads/2016/06/Coating-Failure-Defects.pdf.

Gowariker, V.R., Viswanathan, N. and Sreedhar, J., 1986. *Polymer Science*. New Age International, https://doi.org/10.1002/aic.690331233.

Gutoff, E.B. and Cohen, E.D., 2006. *Coating and Drying Defects: Troubleshooting Operating Problems*. John Wiley & Sons, https://download.e-bookshelf.de/download/0000/5681/69/L-G-0000568169-0002356315.pdf.

Hansen, C.M., 2007. *Hansen Solubility Parameters: A User's Handbook*. Boca Raton, FL: CRC Press, https://doi.org/10.1201/9781420006834.

Hansen, C.M. and Beerbower, A., 1971. Solubility parameters. *Kirk-Othmer Encyclopedia of Chemical Technology*, 2(2): 889–910, https://doi.org/10.1520/MNL12218M.

Kajiya, T., Kobayashi, W., Okuzono, T. and Doi, M., 2009. Controlling the drying and film formation processes of polymer solution droplets with addition of small amount of surfactants. *The Journal of Physical Chemistry B*, 113(47): 15460–15466, https://doi.//10.1021/jp9077757.

Lyman, D., 1961. Polyurethanes. II. Effect of cis-trans isomerism on properties of polyurethanes. *Journal of Polymer Science*, 55(162): 507–514, https://doi.org/10.1002/pol.1961.1205516209.

Marrion, A., 2004. *The Chemistry and Physics of Coatings*. Royal Society of Chemistry, https://www.mobt3ath.com/uplode/book/book-23801.pdf.

Matet, M., Heuzey, M.-C., Pollet, E., Ajji, A. and Averous, L., 2013. Innovative thermoplastic chitosan obtained by thermo-mechanical mixing with polyol plasticizers. *Carbohydrate Polymers*, 95(1): 241–251, http://dx.doi.org/10.1016/j.carbpol.2013.02.052.

Okazaki, M., Shioda, K., Masuda, K., And Toei, R., 1974. Drying mechanism of coated film of polymer solution. *Journal of Chemical Engineering of Japan*, 7(2): 99–105, https://doi.org/10.1252/jcej.7.99.

Prashar, A., Sharma, J., Ahuja, S., Chandra, A. and Arya, R.K., 2019. Quaternary polymeric coating as an alternative to minimize the use of costly or toxic solvents: Poly (styrene)-ethylbenzene-toluene-p-xylene system. *Progress in Organic Coatings*, 127: 319–329, https://doi.org/10.1016/j.porgcoat.2018.11.023.

Price Jr, P.E. and Cairncross, R.A., 2000. Optimization of single-zone drying of polymer solution coatings using mathematical modeling. *Journal of Applied Polymer Science*, 78(1): 149–165, https://doi.org/10.1002/1097-4628(20001003)78:1<149::AID-APP190>3. 0. CO; 2-Y.

Saure, R., Wagner, G. and Schlünder, E.-U., 1998. Drying of solvent-borne polymeric coatings: I. Modeling the drying process. *Surface and Coatings Technology*, 99(3): 253–256, https://doi.//10.1016/S0257-8972(97)00564-1.

Shao, Y., Jia, C., Meng, G., Zhang, T. and Wang, F., 2009. The role of a zinc phosphate pigment in the corrosion of scratched epoxy-coated steel. *Corrosion Science*, 51(2): 371–379, https://doi.//10.1016/j.corsci.2008.11.015.

Sharma, D., Sharma, J., Arya, R.K., Ahuja, S. and Agnihotri, S., 2018. Surfactant enhanced drying of waterbased poly (vinyl alcohol) coatings. *Progress in Organic Coatings*, 125: 443–452, https://doi.org/10.1016/j.porgcoat.2018.06.013.

Sharma, I., Sharma, J., Ahuja, S. and Arya, R.K., 2019a. Optimization of sodium dodecyl sulphate loading in poly (vinyl alcohol)-water coatings. *Progress in Organic Coatings*, 127: 401–407, https://doi.org/10.1016/j.porgcoat.2018.11.037.

Sharma, J., Ahuja, S. and Arya, R.K., 2019b. Effect of molecular weight on residual solvent and other parameters in polymer-polymer-solvent coatings: Poly (Styrene)-poly (methyl methacrylate)-ethylbenzene system. *Progress in Organic Coatings*, 134: 119–125, https://doi.org/10.1016/j.porgcoat.2019.05.002.

Sharma, J., Ahuja, S. and Arya, R.K., 2019c. Experimental designing of polymer-polymer-solvent coatings: Poly (styrene)-poly (ethylene glycol)-chlorobenzene coating. *Progress in Organic Coatings*, 128: 181–195, https://doi.org/10.1016/j.porgcoat.2018.11.036.

Siebel, D., Scharfer, P. and Schabel, W., 2016. Prediction of diffusion in a ternary solvent–solvent–polymer blend by means of binary diffusion data: Comparison of experimental data and simulative results. *Journal of Applied Polymer Science*, 133(36), https://doi.org/10.1002/app.43899.

Spicer, C.W., Gordon, S.M., Kelly, T.J., Holdren, M.W. and Mukund, R., 2002. *Hazardous Air Pollutant Handbook: Measurements, Properties, and Fate in Ambient Air*. Boca Raton, FL: CRC Press.

Vlacha, M., Giannakas, A., Katapodis, P., Stamatis, H., Ladavos, A. and Barkoula, N.M., 2016. On the efficiency of oleic acid as plasticizer of chitosan/clay nanocomposites and its role on thermo-mechanical, barrier and antimicrobial properties–Comparison with glycerol. *Food Hydrocolloids*, 57: 10–19, http://dx.doi.org/10.1016/j.foodhyd.2016.01.003.

Vrentas, J. and Duda, J., 1977a. Diffusion in polymer–solvent systems. II. A predictive theory for the dependence of diffusion coefficients on temperature, concentration, and molecular weight. *Journal of Polymer Science: Polymer Physics Edition*, 15(3): 417–439, https://doi.//10.1002/pol.1977.180150302.

Vrentas, J. and Duda, J., 1977b. Diffusion in polymer—solvent systems. I. Reexamination of the free-volume theory. *Journal of Polymer Science: Polymer Physics Edition*, 15(3): 403–416, https://doi.//10.1002/pol.1977.180150303.

Vrentas, J. and Vrentas, C.M., 1994. Drying of solvent-coated polymer films. *Journal of Polymer Science Part B: Polymer Physics*, 32(1): 187–194, https://doi.org/10.1002/polb.1994.090320122.

Wypych, G. and Pionteck, J., 2016. *Handbook of Antistatics*. Elsevier, https://www.elsevier.com/books/handbook-of-polymers/wypych/978-1-895198-92-8.

Zielinski, J.M. and Hanley, B.F., 1999. Practical friction-based approach to modeling multicomponent diffusion. *AIChE Journal*, 45(1): 1–12, https://doi.org/10.1002/aic.690450102

2 Polymer Coatings Methods

Daisy Sharma, Manju Rawat, Jyoti Sharma, Sanjeev Kumar Ahuja, Avinash Chandra, and Sanghamitra Barman
Thapar Institute of Engineering and Technology

Raj Kumar Arya
Dr. B.R Ambedkar National Institute of Technology

CONTENTS

2.1 INTRODUCTION

In Chapter 1, we discussed the basics of polymer coatings: What are polymer coatings made of? How are they different from polymers? Besides, we discussed the importance of polymer coatings in our daily life as well as in industries. In addition, the different additives incorporated in the polymer coatings and their vital roles were also studied. In this chapter, we will discuss various methods used in the preparation of polymeric coatings. Some of the methods are traditional, like solvent casting, spin coating, thermal spraying, dip casting, and drop casting. Nowadays, new methods are being developed every day to make finer and flawless polymeric coatings that are also economical and environment-friendly. We will also discuss various theories that play a vital role in the formation of polymeric coatings.

2.2   TECHNIQUES FOR THE PREPARATION
OF POLYMERIC COATINGS

There are several techniques that are being used for the preparation of polymeric coatings, such as solvent casting, spin coating, thermal spraying, and dip casting, etc. Many researchers are experimenting with these techniques and introducing new factors in them to make better polymer coatings on a small scale.

2.3   SPIN COATING TECHNIQUE

Spin coating process is widely used to prepare thin polymer coatings on a flat surface. Emslie et al. (1958) gave the first theoretical model of the spin coating method. The authors assumed the fluid to be of Newtonian behavior (i.e., the shear rate is independent of the viscosity), and many other factors were neglected such as Coriolis force, vertical diffusivity, viscosity, and gravitational gradients. Thus, regardless of the assumptions, a basic model of the spin coating technique was provided by the authors. However, based on the model given by Emslie et al. (1958), researchers later found significant results in the spin coating process by taking into account the parameters that were disregarded (Sukanek, 1985; Chang et al., 2005; Flack et al., 1984; Hall et al., 1998; Lawrence, 1988; Meyerhofer, 1978). This process is widely used in various fields like microelectronics, sensors, packaging, corrosion protection, and biomedical purposes (Kausar, 2018). Schematic of the spin coating technique is shown in Figure 2.1.

The process involves spreading of a solvent on a flat surface and then rotating the surface so as to obtain a uniform thin coating. The rotation speed is adjustable according

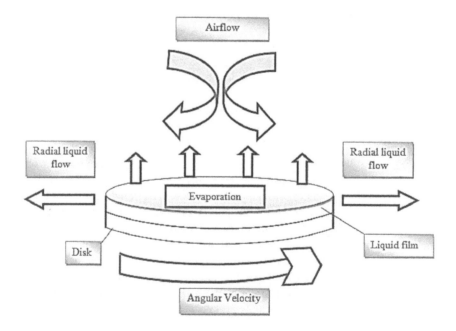

FIGURE 2.1   Schematic of the spin coating process.

to the desired thickness of the coating (Hall et al., 1998). When the rotator moves, centrifugal force plays a part and the solution is spread onto the whole surface, forcing the extra solution to move off from the edges and producing a uniform flat thin coating. The film-thinning process is considered to occur in two different stages. First, the film thinning results from the radial outflow of the liquid, which is rapidly ejected from the surface. In this initial stage, solvent evaporation is presumed to stay constant. The second stage is solvent evaporation; as the solvent evaporates, polymer concentration of the film starts to increase. After the evaporation of most of the remaining solvent, a solid uniform polymer coating is formed resulting from the concentration gradient.

2.3.1 FACTORS AFFECTING SPIN COATING OF POLYMER FILMS

Spin coating is a fairly complex process, as there are a number of parameters involved that affect the formation of a polymer film. Acrivos et al. (1960) show that for non-Newtonian liquids, uniformity itself is not guaranteed Bornside et al. (1989) consider the effect of diffusivity, solution viscosity, and variation in composition. Schwartz and Roy (2004) consider viscous, centrifugal, Coriolis, and finite contact angle effects. Hellstrom (2007) analyzed spin coating on a basic level to find why it often produces uniform coatings. Film thickness is dependent on spin speed, and the thickness of the film is usually predicted easily. Defects and nonuniformity arise from neglected factors. Higher spin velocity will lead to thinner films, and higher concentration of solvent will lead to thicker films. However, some factors have very little or no effects on thickness of a film, such as the solution amount deposited initially on the disk, the rate at which the solution is deposited, rotational acceleration, and the total spin time (Norrman et al., 2005). Hall et al. (1998) observed that diluted solutions mostly tend to form a uniform polymer film, but in higher concentration solutions film surface tend to display some degree of waviness. Higher concentration solvents may show non-Newtonian behavior, the solution may obstruct the damping of surface waves before the solution becomes immobile.

The choice of solvent is crucial for spin coating technique. Higher volatile solvent results in thicker films at a given initial viscosity and initial concentration. Also, the volatility of the solvent can be so high that it will result in nonuniformity as the chilling effect becomes dominant. Furthermore, greater polymer/solvent compatibility will result in more uniform films, i.e., resultant polymer film surface will have less topographical variation. Using a multicomponent solvent or saturating the environment above the rotating disk can prevent or reduce the skinning process, which causes defects in the polymer film. Furthermore, there are other components that should be taken into account while preparing polymer films via spin coating method: temperature, relative humidity, airflow velocity, and thermal surroundings for the solvent evaporation. Thus, for a given spin coating apparatus, it is important to obtain a certain set of operational conditions to perform the experiment successfully for formation of films.

2.4 SPRAY COATING TECHNIQUE

Spray coating technique is considered to be the most effective for coating of nonplanar and 3D surfaces. This method has great potential for large-scale production because the material wastage is less or none at all. Spray coatings are used in various fields, such as in coating of solar cells, piezoelectrics, pyroelectrics, corrosion prevention, aerospace, and automobile industries. There are various methods of spray coating: cold spray (CS) coating (Moridi et al., 2014), thermal spray coating (Mostaghimi et al., 2003), etc. Spray coating technique is shown to be better than spin coating technique; even though the spin coating method produces fine thin films, it is limited to flat or planar surfaces. On the other hand, the spray coating method has advantage of covering any kind of surface, planar or nonplanar. Ichiki et al. worked on the formation of PZT (lead zirconate titanate) films through spray coating and found that this method can be used for homogeneous film formation even on curved surfaces. Further, spray coating enables to form coatings in a larger area, which is an advantage over other polymeric coating methods. The coatings can be formed in both dry and wet environmental conditions (Ichiki et al., 2004).

2.4.1 THERMAL SPRAY PROCESS

There are a few processes that majorly constitute thermal spray technique, such as plasma, combustion spraying, wire arc, including high-velocity oxyfuel (HVOF), and conventional flame processes. In thermal spray coating technique (Petrovicova and Schadler, 2002), the feedstock material is usually in powdered form. The size of thermal spray powder typically ranges from 45 to 180 μm. Powders with narrow particle range are used to control large temperature differences in heating of the individual particles. The particles can be inserted both externally and internally. The flame or jet generated by the torch has a temperature and velocity profile controlled by gas flow and gas heating parameters. The degree of particle heating is determined by jet temperature and velocity, along with dwell time (the time through which a particle resides in the jet). The coating thickness ranges from 0.05 to 0.65 mm (Bao and Gawne, 1991; Bao et al., 1997; Bao et al., 1994; Bao et al., 1995; Rickerby, 1996). Thermal processes variables and parameters are shown in Figure 2.2 (Petrovicova and Schadler, 2002), and the schematic of the coating deposition in thermal spray coatings is shown in Figure 2.3 (Petrovicova and Schadler, 2002).

The particles impacting the substrate become flat, disk-like shapes called "splats," which spread on the substrate and fill interstices on the roughened surface. The coating forms through successive impact of viscous or molten particles on the substrate or the previously deposited particles. The characteristic layered splat structure is usually observed through the cross section of the thermal spray coating. The microstructure of the final coating and its properties depend on the intricate conditions resulting from the particle solidification process, particle splatting degree, and also the particle thermal history, which solidifies at the same location (Petrovicova and Schadler, 2002).

The properties of thermally spray polymer coatings depend on the essential material properties and their thermal history. The optimization of many of these factors has been done by several authors. Dykhuizen (1994) reviewed the impact and solidification of melted thermal spray droplets. He suggested that impacts and solidification processes have been less studied due to their difficulty of measuring, as the process occurs very fast over a small area. However, empirical evidence (Attinger et al.,

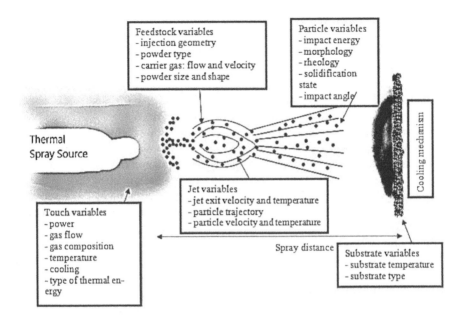

FIGURE 2.2 General thermal spray process variables and parameters. (Reprinted with permission from Petrovicova and Schadler, 2002.)

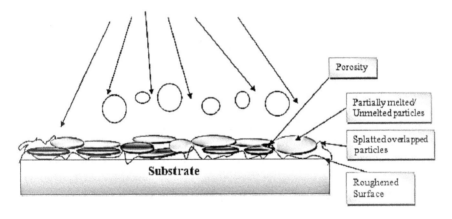

FIGURE 2.3 Coating deposition schematic, depicting microstructural features normally found in thermal spray coating. (Reprinted with permission from Petrovicova and Schadler, 2002.)

2000; Aziz and Chandra, 2000; Bhola and Chandra, 1999) suggests that parameters like impact velocity, liquid superheat, and liquid viscosity control the solidification rate and the splat size. The parameters directly influence the coating properties, such as the amount of oxides that are included in the coating. Also, if the substrate is colder, it can significantly change the shape of the droplet from a disk like to a star; however, the reason behind it is not yet clear. Mostaghimi et al. (2003) have developed a thermal spray coating model to predict microstructure of coatings, e.g., surface roughness, porosity, thickness, and residual stress as a function of spraying parameters.

In controlling the final parameters of thermally sprayed coating, particle temperature (Aziz and Ismail, 2015; Petrovicova and Schadler, 2002) upon impact is one of the main parameters. Researchers (Brogan, 2000; Brogan and Berndt, 1997) have identified lower and higher temperature thresholds of 160°C and 216°C, respectively, for sufficient particle merging in spray-formed EMAA coatings. The proposed model can be used for other polymer coatings too. The lower threshold temperature is required for the coalescence of polymer particles to achieve maximum density. Temperature below this threshold can lead to the formation of porous coatings. At 160°C, there is weak interface as the particles do not get sufficient time to diffuse over the splat boundaries. However, at the threshold temperature of 216°C, polymer coating has been found to have higher tensile strength, toughness, and elongation to break. However, if the coating is sprayed over at 216°C, there is a release of gaseous products, which ultimately lead to decrease in hardness, toughness, and strength.

Spray distance is also one of the crucial factors that have a great impact on the morphology of polymer coating. Many researches have been conducted to optimize and analyze the substrate-to-nozzle distance. Few researchers (Saitoh et al., 2012; Susanna et al., 2011; Vak et al., 2007) have identified similar patterns and three regions while spraying. First, if the spray distance is short, the particles remain wet; second, if the spray distance is long, the particles become dry; third, in the intermediate region, the particles tend to form a coating with uniform thickness. If the spray distances are longer (Petrovicova and Schadler, 2002), it causes an increase in the dwell time and partial solidification of particles. Other factors include particle size that controls the heating of the particle. Smaller particle size will improve the homogeneity and degree of particle melting. Further, the substrate needs to be properly degreased, cleaned, and dry. The surface is usually roughened by grit blasting or machining. In addition to substrate surface preparation, the surface can also be heated to achieve improved coating adhesions.

The thermal spray technique is considered a potential method for new applications for polymer coatings. The growth in this area will help to get a better understanding of processing parameters that are interrelated for coating structure and properties.

2.4.2 Cold Spray (CS) Coating Technique

CS technique is one of the spray techniques that is currently presumed to be better than the thermal spray technique as it utilizes kinetic energy instead of thermal energy for coating deposition. It helps to avoid tensile residual stress, undesired chemical reactions, and oxidation. It has come out as a promising process for the deposition of nanostructured material without altering its microstructure. CS technique can be used for various feedstock materials such as metal, ceramic and metallic matrix composite, and also for polymers.

It is a process in which solid powder particles are accelerated over the sonic velocity through a de Laval nozzle with a convergent–divergent geometry. Particles speed range between 300 and 1200 m/s. There are two types of CS techniques: high-pressure cold spray (HPCS) and low-pressure cold spray (LPCS). In HPCS, the material is injected prior to the spray nozzle throat from a high-pressure gas supply (Alkhimov et al., 1994) and in LPCS the powder is injected in the diverging section of the spray nozzle from a low-pressure gas supply (Kashirin et al., 2002). There are fewer studies of polymer coating made by CS technique. Conventional CS needs a few alterations for the deposition of polymer coatings. The polymer particles need lower velocity than metals; hence, a cylindrical nozzle is used in the place of conventional convergent–divergent nozzle, also known as de Laval nozzle. A diffuser is also placed near the nozzle exit into the carrier gas flow (Alhulaifi et al., 2012). In case of polymer deposition on metal substrates, experimental results show that there are marked differences in the initiation phase and development phase of the polymer deposition. The difference between polymer and metal depositions is the critical velocity. Since the beginning of the CS technology, experiments have been done using metallic materials only. However, finding more reliable methods will expand the poly coating applications (Moridi et al., 2014).

2.5 SOLVENT CASTING TECHNIQUE

Solvent casting technique is one of the oldest as well as the versatile and easy techniques used for the preparation of polymer coating; its schematic is shown in Figure 2.4. Since the beginning of the field of polymer chemistry, solvent casting technique has been used. This method has several advantages, for example, it can be used easily in laboratories for small-scale purposes.

In this technique, the polymer is dissolved in one or more volatile solvents (organic or water) to get a homogeneous solution with a low viscosity. The solution may either spread on a substrate or cast into a mold. The substrate is either static or dynamic in the batch or continuous process (Siemann, 2005). Another crucial advantage of this method is that an additive can be added easily and the properties of polymer coating can be easily identified.

The main advantage of this technique is that in solvent cast films, a liquid on a surface is dried without introducing any external factor such as thermal or mechanical stress. Also, this method offers a variety of additive options to the polymer. The additive can be added to any polymer–solvent solution, which then will be mixed together to form a homogeneous solution. Polymer films formed through solvent casting method usually have thicknesses that are homogeneously distributed. This method also produces clear films of higher optical clarity. There is almost no or low haze in

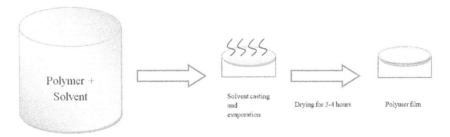

Polymer + Solvent

Solvent casting and evaporation

Drying for 3-4 hours

Polymer film

FIGURE 2.4 Solution coating technique.

the final polymer coating. The films formed are flat and have isotropic orientation. It is possible to produce high-temperature–resistant films by this method provided that the raw materials are soluble but nonmelting.

This method, however, also has few disadvantages. The film formation is highly dependent on the solvent volatility. If the solution is less volatile, it will take more time to evaporate and hence the drying of the polymer film is also delayed. Also, the less-volatile solvents will tend to leave more residual solvent in the formed film at the end. Solvent recovery in this method requires high energy and consequently high cost. Despite all disadvantages, the polymer films formed using solvent casting has high qualities as compared to those formed by any other methods (Siemann, 2005).

2.6 DROP CASTING METHOD

This method is used for the formation of small coatings on small surfaces. It requires a very small amount of solvent. Multiple droplets formed using this method provide a distinctive environment to control the shrinkage direction and the evaporation rate of the droplets (Liu et al., 2014). In this method, solution is poured onto the substrate as drops and is allowed to dry without any spreading (Riera-Galindo et al., 2018). When the droplets are projected onto the substrate, the liquid first spreads on the surface from the drop locations, due to the interfacial forces that are inclined to drive the droplet outward. As multiple droplets have been cast onto the substrate surface, when the edges of the droplet come in contact with each other, they mix together, forming a noncircular drop with the concave contact line. The schematic of the drop casting method is shown in Figure 2.5. Deegan et al. (2000) have found that noncircular drops have uneven deposition rates, highly convex regions, stronger evaporating flux, and thus denser deposits. The capillary flow mechanism happens for contact-line deposition in between the droplets. This also shows that deposition can be predicted and controlled even if the chemical nature of solute, liquid, and substrate is not known. The films formed using this method are nonuniform because of the inconsistent drying conditions and uneasy control. These films are thicker at the center and thinner at the edges. The drop casting method is an alternative technique for polymer semiconductors, which are expensive and have poor solubility.

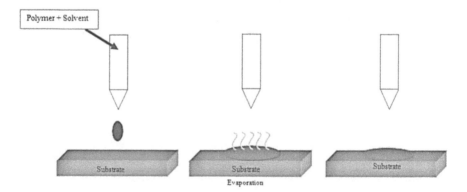

FIGURE 2.5 Drop casting technique.

2.7 DIP COATING TECHNIQUE

It is basically used as a corrosion prevention technique (Kaur et al., 2018). In this method, the object is dipped into the solution and polymer melt and taken out at a meticulous speed to get the uniform film thickness. The bottom of the object is usually of high thickness compared to the top, due to the drops of the excess solution formed at the bottom. It is used for tool handles, cleaning rubbers, balloons, and plastic caps.

2.8 FILM CASTING TECHNIQUE

The film casting method produces films of higher quality with more optical clarity and uniform thickness. This method has great industrial importance; hence, it has been extensively studied, both theoretically and experimentally. This method is used to manufacture thin films and coatings from a viscous polymer melt. In this method, the viscous polymer melt is ejected through a slit die to form a molten thin film, which is then taken up by a rotating chill roll. The ratio of the take-up speed V_D and the velocity of the film at the die exit V_E is referred to as the draw ratio D. When the film is taken up, the temperature of the film drops suddenly and the solidification of the film happens at that moment. The practical interest in the polymer film industry is to be capable to produce film-casted coatings that have reproducible properties in terms of width, thickness, and molecular orientation. The success of the film casting is limited by the draw resonance phenomenon. When the drawing ratio S exceeds a critical value D_c, in spite of the constant extrusion and constant drawing speed, a swinging behavior of the width and thickness of film is observed. This oscillatory behavior of the film was first observed by Ghijsels and Ente (1980) and Miller (1963). The destabilizing role of the elasticity can be understood in connection with the thinning rheological behavior of the material in both shear and extension (Burghelea et al., 2012).

In this chapter, we have discussed several techniques for the preparation of polymer coatings. Some of them have great potential for making polymer coatings at an industrial/commercial level. However, a few methods like solution casting and dip coating methods can be used in labs for experimenting with polymer coatings. Nowadays, more emphasis is given on the production of green coatings, i.e., pollution-free coatings, coatings without organic solvents, which are difficult to dispose of and are harmful to the environment.

REFERENCES

Acrivos, A., Shah, M.J., and Peterson E.E., 1960. On the flow of a non-Newtonian liquid on a rotating disk. *Journal of Applied. Physics*, 31(5): 963–968. https://doi.org/10.1063/1.1735785.

Alhulaifi, A.S., Buck, G.A. and Arbegast, W.J., 2012. Fundamentals of cold spraying. *Journal of Thermal Spray Technology*, 21(10): 852–862. https://doi.org/10.1063/1.4963926.

Alkhimov, A.P., Papyrin, A.N., Kosarev, V.F., Nesterovich, N.I. and Shushpanov, M.M., 1994. Gas-dynamic spraying method for applying a coating, US Patent 5302414.

Attinger, D., Zhao, Z. and Poulikakos, D., 2000. An experimental study of molten microdroplet surface deposition and solidification: Transient behavior and wetting angle dynamics. *Journal of Heat Transfer*, 122(12): 544–556. https://doi.org/10.1115/1.1287587.

Aziz, S.D. and Chandra, S., 2000. Impact, recoil and splashing of molten metal droplets. *International Journal of Heat and Mass Transfer*, 43(16): 2841–2857. https://doi.org/10.1016/S0017–9310(99)00350-6.

Aziz, F. and Ismail, A., 2015. Spray coating methods for polymer solar cells fabrication: A review. *Materials Science in Semiconductor Processing*, 39(9): 416–425. https://doi.org/10.1016/j.mssp.2015.05.019.

Bao, Y. and Gawne, D.T., 1991. Plasma spray deposition of Nylon 11 coatings. *IMF*, 69(5): 80–85. https://doi.org/10.1080/00202967.1991.11870898.

Bao, Y., Gawne, D.T. and Pearson, M., 1997, Plasma spray deposition and engineering applicants and magnesium hydroxide-polymide composite coatings, *IMF*, 75(4): 189–193. https://doi.org/10.1080/00202967.1997.11871170.

Bao, Y., Gawne, D.T., Vesely, D. and Bevis, M.J., 1994, Production of polymer matrix composite by thermal spraying. *IMF*, 72(3): 110–113. https://doi.org/10.1080/00202967.1994.11871033.

Bao, Y., Gawne, D.T. and Zhang, T., 1995. The effect of feedstock particle size n the heat transfer rates and properties of thermally sprayed polymer coatings, *IMF*, 73(5): 119–124. https://doi.org/10.1080/00202967.1995.11871072.

Bhola, R. and Chandra, S., 1999. Parameters controlling solidification of molten wax droplets falling on a solid surface. *Journal of Materials Science*, 34(9): 4883–4894. https://doi.org/10.1023/A:1004680315199.

Bornside, D.E., Macosko, C.W., Scriven, L.E., 1989. Spin coating: One-dimensional model. *Journal of Applied Physics*, 66(8): 5185–5193. https://doi.org/10.1063/1.343754.

Brogan, J., 2000. Thermal-spraying of polymers and polymer blends. *MRS Bulletin*, 25(5): 48–53. https://doi.org/10.1557/mrs2000.124.

Brogan, J. and Berndt, C., 1997. The coalescence of combustion-sprayed ethylene–methacrylic acid copolymer. *Journal of Materials Science*, 32(8): 2099–2106. https://doi.org/10.1023/A:1018526906452.

Burghelea, T.I., Grieß, H.J. and Münstedt, H.J., 2012. An in situ investigation of the draw resonance phenomenon in film casting of a polypropylene melt. *Journal of Non-Newtonian Fluid Mechanics*, 173(9): 87–96. https://doi.org/10.1016/j.jnnfm.2012.02.006.

Chang, C.C., Pai, C.L., Chen, W.C. and Jenekhe, S.A., 2005. Spin coating of conjugated polymers for electronic and optoelectronic applications. *Thin Solid Films*, 479(6): 254–260. https://doi.org/10.1016/j.tsf.2004.12.013.

Deegan, R.D., Bakajin, O., Dupont, T.F., Huber, G., Nagel, S.R. and Witten, T.A., 2000. Contact line deposits in an evaporating drop. *Physical Review. E, Statistical Physics, Plasmas, Fluids, and Related Interdisciplinary Topics*, 62(1 Pt B): 756–65. https://doi.org/10.1103/PhysRevE.62.756.

Dykhuizen, R.C., 1994. Review of impact and solidification of molten thermal spray droplets. *Journal of Thermal Spray Technology*, 3(10): 351–361. https://doi.org/10.1007/BF02658980.

Emslie, A.G., Bonner, F.T. and Peck, L.G., 1958. Flow of a viscous liquid on a rotating disk. *Journal of Applied Physics*, 29(4): 858–862. https://doi.org/10.1063/1.1723300.

Flack, W.W. et al., 1984. A mathematical model for spin coating of polymer resists. *Journal of Applied Physics*, 56(8): 1199–1206. https://doi.org/10.1063/1.334049.

Ghijsels, A. and Ente, J., 1980. *Draw Resonance Studies on Polypropylene Melts, Rheology*. Springer, pp. 15–24. https://doi.org/10.1007/978-1-4684-3746-1_3.

Hall, D.B., Underhill, P. and Torkelson, J.M., 1998. Spin coating of thin and ultrathin polymer films. *Polymer Engineering & Science*, 38(12): 2039–2045. https://doi.org/10.1002/pen.10373.

Hellstrom, S., 2007. Basic models of spin coating. Submitted as coursework for Physics, 210.

Ichiki, M., Zhang, L., Yang, Z., Ikehara, T. and Maeda, R., 2004. Thin film formation on non-planar surface with use of spray coating fabrication. *Microsystem Technologies*, 10(3): 360–363. https://doi.org/10.1007/BF02637104.

Kashirin, A.I., Klyuev, O.F. and Buzdygar, T.V., 2002, Apparatus for gas-dynamic coating, US Patent 6402050.

Kaur, H., Sharma, J., Jindal, D., Arya, R.K., Ahuja, S.K. and Arya, S.B., 2018. Crosslinked polymer doped binary coatings for corrosion protection. *Progress in Organic Coating*, 125(7): 32–39. https://doi.org/10.1016/j.porgcoat.2018.08.026.

Kausar, A., 2018. Polymer coating technology for high performance applications: Fundamentals and advances. *Journal of Macromolecular Science, Part A*, 55(8): 440–448. https://doi.org/10.1080/10601325.2018.1453266.

Lawrence, C.J., 1988. The mechanics of spin coating of polymer films. *The Physics of Fluids*, 31(9): 2786–2795. https://doi.org/10.1063/1.866986.

Liu, Y., Zhao, X., Cai, B., Pei, T., Tong, Y., Tang, Q. and Liu, Y., 2014. Controllable fabrication of oriented micro/nanowire arrays of dibenzo-tetrathiafulvalene by a multiple drop-casting method. Nanoscale, 6(5): 1323–1328. https://doi.org/10.1039/C3NR05680E.

Meyerhofer, D., 1978. Characteristics of resist films produced by spinning. *Journal of Applied Physics*, 49(6): 3993–3997. https://doi.org/10.1063/1.325357.

Miller, J.C., 1963. Swelling behavior in extrusion. *Journal of Polymer Engineering and Science*, 3(3): 134–137. https://doi.org/10.1002/pen.760030208.

Moridi, A., Hassani-Gangaraj, S.M., Guagliano, M. and Dao, M., 2014. Cold spray coating: Review of material systems and future perspectives. *Surface Engineering*, 30(26): 369–395. https://doi.org/10.1179/1743294414Y.0000000270.

Mostaghimi, J., Chandra, S., Ghafouri-Azar, R. and Dolatabadi, A., 2003. Modeling thermal spray coating processes: A powerful tool in design and optimization. *Surface and Coatings Technology*, 163(10): 1–11. https://doi.org/10.1016/S0257-8972(02)00686-2.

Norrman, K., Ghanbari-Siahkali, A. and Larsen, N.B., 2005. 6 Studies of spin-coated polymer films. *Annual Reports Section "C" (Physical Chemistry)*, 101(27): 174–201. https://doi.org/10.1039/B408857N.

Petrovicova, E. and Schadler, L.S., 2002. Thermal spraying of polymers. *International Materials Reviews*, 47(21): 169–190. https://doi.org/10.1179/095066002225006566.

Rickerby, D., 1996. Measurement of coating adhesion. In: K.H. Stern (Editor), *Metallurgical and Ceramic Protective Coatings*. Dordrecht: Springer, pp. 306–333. https://doi.org/10.1007/978-94-009-1501-5_12.

Riera-Galindo, S., Tamayo, A. and Mas-Torrent, M., 2018. Role of polymorphism and thin-film morphology in organic semiconductors processed by solution shearing. *ACS Omega*, 3(10): 2329–2339. https://doi.org/10.1021/acsomega.8b00043.

Saitoh, L., Babu, R.R., Kannappan, S., Kojima, K., Mizutani, T. and Ochiai, S., 2012. Performance of spray deposited poly [N-9″-hepta-decanyl-2,7-carbazole-alt-5,5-(4′, 7′-di-2-thienyl-2′, 1′, 3′-benzothiadiazole)]/[6,6]-phenyl-C61-butyric acid methyl ester blend active layer based bulk heterojunction organic solar cell devices. *Thin Solid Films*, 520(6): 3111–3117. https://doi.org/10.1016/j.tsf.2011.12.022.

Schwartz, L.W. and Roy, R.V., 2004. Theoretical and numerical results for spin coating of viscous liquids. *Physics of Fluids*, 16(3): 569–584. https://doi.org/10.1063/1.1637353.

Siemann, U., 2005. Solvent cast technology – a versatile tool for thin film production. In: N. Stribeck and B. Smarsly (Editors), *Scattering Methods and the Properties of Polymer Materials*. Berlin, Heidelberg: Springer Berlin Heidelberg, pp. 1–14. https://doi.org/10.1007/b107336.

Sukanek, P.C., 1985. Spin Coating. *Journal of Imaging Technology*, 11(6): 184–190. http://pascal-francis.inist.fr/vibad/index.php?action=getRecordDetail&idt=8518683.

Susanna, G., Salamandra, L., Brown, T.M., Di Carlo, A., Brunetti F. and Reale, A., 2011, Airbrush spray-coating of polymer bulk-heterojunction solar cells, *Solar Energy Materials Solar Cells*, 95(3): 1775–1778. https://doi.org/10.1016/j.solmat.2011.01.047.

Vak, D., Kim, S.S., Jo, J., Oh, S.H., Na, S.I., Kim, J. and Kim, D.Y., 2007. Fabrication of organic bulk heterojunction solar cells by a spray deposition method for low-cost power generation. *Applied Physics Letters*, 91(1–3): 081102. https://doi.org/10.1063/1.2772766.

3 Polymer Coatings Based on Nanocomposites

Liliana Licea-Jiménez, Ulises Antonio Méndez-Romero, Abraham Méndez-Reséndiz, Arturo Román Vázquez-Velázquez, Ricardo Antonio Mendoza-Jiménez, Diego Fernando Rodíguez-Díaz, and Sergio Alfonso Pérez-García
Centro de Investigación en Materiales Avanzados S.C., CIMAV

CONTENTS

3.1 INTRODUCTION

Based on the fact that coatings are present all around us, and due to the high demand for new materials and new technologies in our daily life that allow humankind to have comfort and a good quality of life, it is necessary to develop multifunctional materials, considering production processes that are environment-friendly. Such is the case for coatings, which are important for changing the characteristics of surfaces and providing new properties and functionalities. There is an ongoing trend of requests for coatings that cover several different functionalities at once; examples are combinations of extreme wetting behavior, special optical properties, and high mechanical robustness.

In this framework, nanotechnology has opened up new and promising possibilities for the development of coatings with functional features. In particular, the development of polymeric nanocomposites (consist of a reinforcement, mainly NPs and a polymer matrix) has become an opportunity to meet the demands of functional materials.

Polymer nanocomposites offer the possibility of new materials with a unique manifold of structure–property relationships. Current interest in nanocomposites has been generated due to the functionality that can be given by the appropriate combination of nanomaterials, resulting in properties that are not achieved with traditional materials.

Polymer coatings based on nanocomposites provide a real opportunity to make a difference across a wide variety of applications and markets, the application of nanocomposite technologies to further enhance current commercial products or add completely new properties to an existing technology. These coatings can improve product longevity, surface strength, and product performance, while enhancing energy efficiency performance. Thus, properties with incredible practical applications for mechanical, optical and electronic products are expected.

3.2 WHY ARE NANOCOMPOSITES ATTRACTIVE FOR POLYMER COATINGS?

There is a long list of characteristics that justify the use of nanocomposites for polymer coatings. Some of them are listed below:

- Because of the outstanding features of nanoparticles (NPs), advanced products and coatings can be produced by their incorporation into a polymer.
- The large specific surface area of NPs allows them to have enhanced reactivity, superior absorption, higher solubility, lower melting point, and enhanced electronic properties, such as quantum effects found on particles with particle size <10 nm (important for electronic and optoelectronic applications).
- NPs are invisible to the human eye; when they are incorporated in a polymer matrix, they do not affect visible light transmission, making them ideal materials for advanced clear coatings.
- No sacrifice in light transmission, which is critical in optical applications.
- In general, the improvement of the mechanical properties of a polymer coating (such as scratch resistance, wear, and adherence)
- Nanocomposites require very low loading of nanomaterial to yield significant improvement in mechanical and other desired properties.

3.3 GENERAL ASPECTS, CONSIDERATIONS, AND CHALLENGES

Coatings can be produced by the incorporation of NPs into polymer matrices, due to their characteristics such as particle size, surface area, and properties. NPs and nanomaterials are considered for use in selective coatings, self-cleaning coatings, coatings for energy applications, printed electronics, etc. To fabricate technologically relevant functional coatings, one needs to understand and control the interactions in different materials by manipulating interfaces at the nanoscale. As a consequence, it is very important to create an optimal NP–polymer interface through an appropriate nanometric scale design.

A proper incorporation of nanomaterials into the polymer is a key issue, since many of these properties are closely related to interfaces. Although there are significant advances in the manufacture of nanocomposite coatings, processing remains one of the challenges for the full utilization of the nanoscale properties of the particles. Also, one of the first difficulties is the dispersion of NPs. Consideration of the key issues enable a coating formulation with improved properties, as well as the optimization of the developed materials.

The development of an easy and innovative method to create and apply a functional coating is the main objective of various applications. The search for a high-quality and low-cost coating is a challenge for the application and transfer of technology, so the factors that directly influence the way to obtain the best properties must be understood. This type of coatings can be applied by different easy methods and economical materials such as sol–gel, immersion, spin coating, printing, and others. The selection of the appropriate method to create a functional coating depends on the mechanical and physical–chemical properties of the substrate used. Preparation of the substrate is one of the most important parameters in the procedure for coating application, and its results directly affect the quality of the final product. In this way, some threads represent additional costs to the general process; however, they represent improvements in the quality of the final product.

The performance evaluation of coatings is an important point to consider, together with its formulation. The characterization of materials and the analysis of the properties of the coating play a very important role.

The design of new functional coating materials, the characterization of novel and tailored physicochemical properties, and enhanced processing capability are among the most crucial challenges related to achieving coating products with smarter, stronger, and more durable characteristics.

The formulation strategy of polymer coatings based on nanocomposites depends on the application and different approaches can be followed. However, some general facts are as follows:

- Often NPs require functionalization (seldom used as prepared) and formulation before their integration into coatings.
- Both aqueous and nonaqueous solvents are used to disperse NPs.
- Often the type of coating dictates the chosen solvent.
- In most cases, nanocomposite coating formulation consist of dispersed particles, dispersion media, and other additives.

In Figure 3.1, a proposed approach to achieve a specific and functional polymer nanocomposite coating is presented.

As it has been mentioned, coatings can be described by their function. In the following sections, functional polymer coatings based on nanocomposites are described.

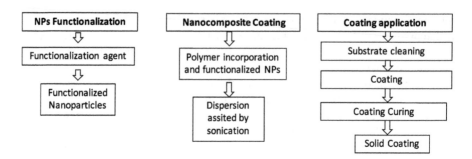

FIGURE 3.1   A schematic approach for a polymer coating based on nanocomposites.

3.4   SELF-CLEANING COATINGS BASED ON POLYMER NANOCOMPOSITES

Nowadays, the development of self-cleaning coatings has gained importance in the field of research, caused by the multiple commercial applications that can be achieved. Automotive, building, aerospace, and energy sectors are some of the fields where self-cleaning coatings have been used; these are mainly deposited on glass materials like windows, mirrors, lenses, and optical and outdoor applications (Cedillo-González et al. 2014).

Preparation of a polymer coating based on nanocomposites incorporating NPs (metal oxides, carbon-based, among others) allows improvement and confer new characteristics to the polymer coating. Such nanocomposites have the potential to be used in various applications, including interior and exterior of buildings, houses, textile, glass, ceramic, plastics, wood, as well as different kinds of transportation vehicles and structures (Ganbavle et al. 2011; Guldin et al. 2013; Wang et al. 2013).

These novel coatings are an alternative solution to specific requirements, for example, scratch resistance, self-cleaning, anti-icing, self-healing and antireflective properties, and anticorrosion (Davis et al. 2014; Niu and Wang 2009; Li, Du, and He 2010).

Wettability of coatings is a feature that has received interest for technological applications due to its unique superhydrophobic and superhydrophilic characteristics, such as self-cleaning, antifouling, and fluid drag reduction. Self-cleaning coatings have the ability to be cleaned out through natural processes (Viswanadam and Chase 2012; Decker et al. 1999). A hydrophilic surface possesses a large affinity to water, forming a layer of water that avoid the incrustation and deposition of dust, dirt, organic matter, and other different kinds of impurities. In a hydrophobic surface, there is water repellence forming water droplets on the surface. These waterdrops roll on the surface of the hydrophobic coating, taking dirt with it. The self-cleaning property is related to the contact angle of a water droplet to a solid surface (Figure 3.2). When the water contact angle is less than 90°, the surface can be classified as hydrophilic; when the water contact angle is more than 90°, the solid surface can be considered as a hydrophobic surface; and if the contact angle is above 150°, the solid surface is superhydrophobic. Accordingly, if the contact angle is near to 0°, the solid surface can be considered superhydrophilic (Bhushan and Jung 2011; Zhou et al. 2015; Lee et al. 2016).

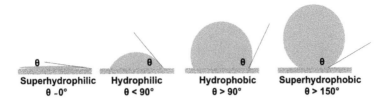

Superhydrophilic   Hydrophilic   Hydrophobic   Superhydrophobic
θ -0°   θ < 90°   θ > 90°   θ > 150°

FIGURE 3.2 Water contact angle classification.

The nanocomposite materials that can be used for self-cleaning coatings are those containing carbon-based nanomaterials and metal oxide NPs as reinforcements. Some aspects that should be considered are as follows: (1) the dispersion of the nanomaterials in the polymer matrix, that is, the nanoscale homogeneity of the materials, and (2) the functionalization of the nanomaterial to create an affinity between the reinforcement and the polymer matrix. The main carbon-based materials that can be used as reinforcement are graphene-derived materials, such as graphene oxide (GO) and reduced graphene oxide (rGO) (Velasco-Soto et al. 2016). On the other hand, metal oxide NPs such as SiO_2, TiO_2, CeO_2, ZnO, and Al_2O_3 are just some examples of NPs employed for the development of polymer coatings based on nanocomposites. Natural polymers and synthetic polymers like acrylic, polyurethane, rubber, polyethylene, polypropylene, and different kinds of resins are the most commonly used polymers.

3.4.1 SELF-CLEANING COATINGS BASED ON METAL OXIDE NPS/POLYMER NANOCOMPOSITES

Usually, superhydrophilic and hydrophilic coatings present features like high surface energy, which means that water is going to spread out along the surface coating, generating low water contact angle. These coatings have a similar surface energy than water, creating a phenomena that keep clean the surface when it is wet (Nakajima, Hashimoto, and Watanabe 2001).

Various metal oxide NPs have been used for self-cleaning coatings, such as TiO_2, SiO_2, ZnO, CeO_2, and Al_2O_3. This kind of NPs show catalytic and good optical transmission values, among other properties, that can be used and applied for this purpose. In addition, this type of coatings can be applied in various easy methods and economical materials such as sol–gel, immersion, spin coating, printing, and others.

Due to the strong tendency of NPs such as metal oxides to agglomerate, homogeneous dispersion of these materials in a polymer coating is extremely challenging. In order to overcome this problem and to enhance the reinforcement–polymer interaction, a functionalization or chemical modification of NPs is always required. The formulation of novel materials with modified surfaces by organic and inorganic functional groups has gained technological importance.

As previously mentioned, nanocomposite coatings allow designing new functionalities and tailoring physicochemical properties. Thus, it can be considered the combination of NPs. In a study of a binary nanocomposite system based on covalently functionalized TiO_2-SiO_2 (f-TiO_2-SiO_2) with trimethylolpropane triacrylate (TMPTA) and

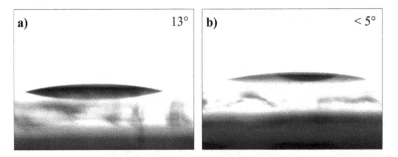

FIGURE 3.3 Water contact angles of (a) hydrophilic nanocomposite 0.1wt% f-TiO$_2$SiO$_2$/PAA and (b) superhydrophilic nanocomposite 0.5%wt% f-TiO$_2$-SiO$_2$/PAA. Reprinted with permission from Vázquez-Velázquez et al., "Functionalization effect on polymer nanocomposite coatings based on TiO$_2$–SiO$_2$ nanoparticles with superhydrophilic properties". *Nanomaterials* 8(6):369, 2018.

embedded in a poly acrylic acid (PAA), it was possible to produce a synergy between the constituents (Vázquez-Velázquez et al. 2018). This effect the wettability of the polymer nanocomposite achieving a hydrophilic (13°) and superhydrophilic (5°) coating. Figure 3.3 presents the water contact angles of 0.1wt% f-TiO$_2$-SiO$_2$/PAA and 0.5wt% f-TiO$_2$-SiO$_2$/PAA; here is shown the concentration effect of the modified NPs in the acrylic polymer: change from a hydrophilic (Figure 3.3a) to superhydrophilic (Figure 3.3b) behavior, through an increase in the concentration of NPs from 0.1wt% to 5wt%, respectively. Coatings with water contact angles near to 0° are useful for designing self-cleaning coatings.

In self-cleaning coatings, the counterpart of superhydrophilic coatings are the superhydrophobic ones. A superhydrophobic coating can be understood primarily because it is governed by a process similar to the lotus effect; this feature allows the waterdrop to remove and to carry particles of soiling from the surface of the material, keeping the surface clean (Valipour, Birjandi, and Sargolzaei 2014). The growing need to achieve a self-cleaning property through a superhydrophobic coating has generated the development of different strategies. It is known that to obtain these coatings, there are key issues: roughness at a nanometric scale and NPs with hydrophobic features and good polymeric affinity. The proper interaction permits surfaces with high water contact angle and the development of polymeric coatings–based nanocomposites. However, not all the materials present hydrophobic features, so chemical surface modification is required. These modifications can be carried on through a functionalization process, producing a chemical change on the material surface (Ma and Hill 2006).

As a case study, the development of a superhydrophobic coating through SiO$_2$ NPs chemically modified with a silane group (f-SiO$_2$) and its incorporation in a polyurethane matrix is described. The incorporation of the f-SiO$_2$ improves the wettability properties of a polyurethane coating. The water contact angle analysis showed that modified NPs increase the contact angle of the polyurethane coating (Figure 3.4a) from 79° to 98° for a concentration of 50wt% of NPs and 10wt% of polymeric matrix in the nanocomposite (Figure 3.4b); when the polymer concentration was 5wt%

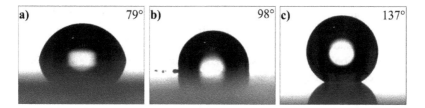

FIGURE 3.4 Water contact angles of (a) polyurethane coating, (b) 50wt% f-SiO$_2$/10wt%PU, and (c) 50wt% f-SiO$_2$/5wt%PU.

(Figure 3.4c), the water contact angle reached a value of 137°. The superhydrophobic features display important perspectives to be implemented in different industrial applications, for example, in outdoor weather-proof coatings, glass, textile, windows, solar panels, mirrors, and automotive (Paul and Robeson 2008).

3.4.2 Polymer Coatings Using Carbon-Based Nanomaterials

Recently, polymer coatings that incorporate graphene (or derivatives) have gained a commercial attention due to market demand for stronger, more durable, and environment-friendly solutions. Due to its unique properties, for example, chemically inert, impermeability, friction, and good antiwear properties, among others, the use of graphene is growing for application in protective and/or functionalized coatings.

Graphite can be oxidized to obtain graphite oxide, which can be easily exfoliated due to the oxygen atoms that aid to separate its layers, obtaining GO (Paredes et al. 2008). Assisted by reducing agents, rGO can be obtained, being a material with properties similar to graphene, but with remaining functional groups that were gained by oxidation, facilitating the functionalization of this material (Fernández-Merino et al. 2010) As mentioned before, to improve the dispersion, it is an important functionalization process. The chemical nature of the polymer must be known to find the appropriate functionalization agent.

For the formulation of a nanocomposite system using carbon-based nanomaterials in a polymeric coating, it is important to think about the final application. For instance, to develop a coating for outdoor use without affecting its optical properties, it is essential to have high resistance to abrasion and transparency to visible light. A suitable material for these functions is a polyurethane coating but with certain restrictions. Thus, a multifunctional polymer nanocomposite coating can be tailored. Recalling the critical aspects that have been already discussed, a possible approach could be as follows.

In order to achieve a good dispersion of carbon-based nanomaterials (GO and rGO), functionalization has to be done by choosing the proper coupling agent. Also, it is important to consider the correct solvent for both the functionalized nanomaterial and the polymer. In this case, both the GO and the rGO can be functionalized with a material containing long carbon chains giving a nonpolar character, which can be dispersed in organic solvents as well as in polyurethane.

In this coating, among other desired characteristics is its functionality, that is, being a self-cleaning or photocatalytic material. The use of additional nanomaterials can lead to obtain such characteristics. TiO$_2$ is a widely studied material with interesting properties such as high photocatalytic activity, strong oxidation power, low cost, chemical and thermal stability, and a band gap (3.2 eV) ideal to be photoactive with UV radiation (Momeni, Ahadzadeh, and Rahmati 2016). To add this NPs as a reinforcement to the polymer nanocomposite could render photocatalytic properties, enhancing the self-cleaning.

Based on the latter, a polyurethane-based nanocomposite coating can be formulated, where functionalized rGO and TiO$_2$ NPs act as a reinforcement. So, functional coatings can be obtained with either hydrophobic or hydrophilic properties. A polyurethane coating has a contact angle near to 70°, but with the addition of the reinforcement, the contact angle increases close to 90°, starting to become a hydrophobic material. As it can be seen in Figure 3.5, by adding the reinforcement at different concentrations, the contact angle changes. With the addition of just 0.1% of this reinforcement, the contact angle approaches a hydrophobic behavior. A proportional increment of the contact angle with respect to the concentration of the reinforcement could be expected. But at 0.5% of the reinforcement, a slight decrease of the contact angle is reflected. This can be attributed to several factors; for instance, by adding more reinforcement, it becomes difficult to achieve a stable dispersion of the nanomaterial, which can lead to a nonhomogeneous deposit of the coating in the substrate. This will result in differences in the contact angle across the coating. Hence, optimization in the design of the nanocomposite is essential to achieve the desired properties, in this case for self-cleaning.

A self-cleaning coating can also have hydrophilic behavior, but in order to achieve this kind of coating, different approach needs to be used. Utilizing GO as a reinforcement could be a better choice, since it is inherently hydrophilic in character and the only modification would be the addition of TiO$_2$ NPs for the photocatalytic

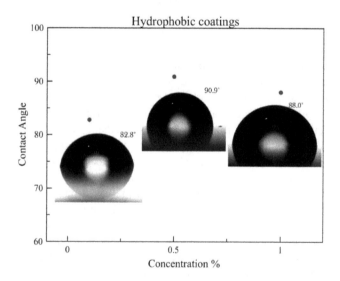

FIGURE 3.5   Contact angle of hydrophobic coatings of polyurethane with reduced graphene oxide (rGO) and TiO$_2$ as reinforcement at different concentrations.

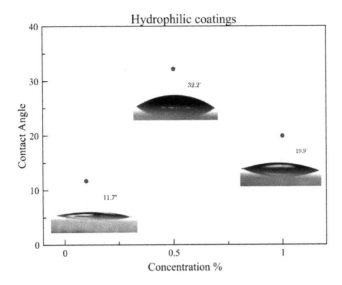

FIGURE 3.6 Contact angle of hydrophilic coatings of polyacrylic acid with GO and TiO_2 as reinforcement at different concentrations.

property (Oribayo et al. 2017). The nature of the polymer also needs to be hydrophilic, while giving optimal mechanical properties and being transparent. A polymer that can fit these parameters is polyacrylic acid. The hydrophilic nature of functional groups attached to the GO´s surface permit to be dispersed in solvents such as ethanol (Othman et al. 2019). As previously mentioned, dispersion of the reinforcement affect directly the properties of the composite; also the concentration of the reinforcement affects the dispersion.

As shown in Figure 3.6, different behaviors are distinguished. At a concentration of 0.1% of reinforcement, the contact angle dropped until 11.7°. But, when the concentration increased until 0.5%, the contact angle drastically increased above 30°, which is even higher than the contact angle of the polyacrylic coating (around 22°). Then, at a concentration of 1%, the contact angle drops. This erratic behavior can be attributed to various factors, such as a poor dispersion of the reinforcement, poor cleaning of the substrate, deposition method, and contamination of the formulation. Therefore, it is important to highlight that the design of a polymer coating based on nanocomposites demands to consider the proper nanomaterials, solvents, functionalization agent, dispersion, formulation, and even the application method.

3.5 POLYMER NANOCOMPOSITE COATINGS FOR ORGANIC SOLAR CELLS

For a sustainable society, the supply of cheap and green electricity is indispensable (Smalley 2005). For this purpose, one of the most explored technologies that eventually could overcome the energy problem is solar cells or photovoltaic cells (PVs). These devices convert the electromagnetic energy radiation from the sun into electrical energy, which, in principle, could meet the energy needs (Krebs 2008).

In general, conventional PVs are silicon-based p-type (electron-deficient) and n-type (electron-rich) semiconductor junctions, with maximum power conversion efficiency (PCE) around 25%, and the commercial ones around 16% (O'Regan and Gratzel 1991; Philipps and Bett 2016). However, drawbacks of these PVs are their high cost and fragility (Markavart 2005).

Based on the above, alternatives have been explored with organic compounds, i.e., organic photovoltaics (OPVs) (Thompson and Fréchet 2008). In OPVs, the most important component is the *active layer*, which basically is a conjugated polymer nanocomposite coating. The conjugated polymers are the matrix, while nanomaterials like fullerene derivatives $PC_{60}BM$, $PC_{71}BM$ and indene-C_{60} bisadduct (ICBA) (Wadsworth et al. 2019) are the reinforcements (Figure 3.7).

In order to obtain better PCE values in an OPV, there are some paths that can be addressed. One is the chemical modification and engineering of the polymers and nanomaterials, and the other is to find a way to produce better characteristics of the coating. These strategies can be used as a simple and effective modification strategy to optimize the active-layer nanocomposite morphology through the enhancement of crystallinity of the resulting molecules, improved planarity, better molecular orientation, etc. Dramatically improving the PCE enhances the morphological miscibility in the active-layer nanocomposite coating.

Here we present some of the strategies that can be taken in order to achieve a better and improved active layer performance.

3.5.1 D–A Strategy and Halogenation

D–A Strategy: The ultimate advantages of conjugated polymers are their tunable properties such as band gap, molecular energy levels, carrier mobility, morphology, among others. A general strategy to synthesize a conjugated polymer is by combining an

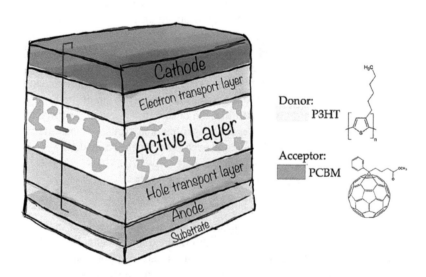

FIGURE 3.7   OPV general architecture.

electron-rich monomer (donor) and an electron-deficient monomer (acceptor), in order to polymerize a D–A alternating conjugated polymer. When the donor monomer is bonded to the acceptor monomer, it results in a D–A conjugated polymer that has a linear combination of molecular orbitals, i.e., the HOMO (highest occupied molecular orbital) of the polymer is similar to the HOMO of the donor monomer, while the LUMO (lowest unoccupied molecular orbital) of the polymer is similar to the LUMO of the acceptor monomer, resulting in a lower band gap and therefore capable of use in a wider solar spectrum, as seen in Figure 3.8.

Halogenation: This is another approach to improve and control the properties of the conjugated polymers; for example, fluorinated materials show enhanced extinction coefficient and higher charge carrier mobility compared to their nonfluorinated analogues. Also, fluorination as a simple and effective molecular modification strategy is used to optimize the active-layer nanocomposite morphology through the enhanced crystallinity of the resulting molecules, improved planarity, and better molecular orientation.

3.5.2 SMALL MOLECULES

In comparison with the traditional fullerene derivatives, small molecules (SM) possess higher reproducibility in synthesis and PV performance and more direct and reliable analysis in the structure–property relationships. Nevertheless, when compared to polymers, SM frequently exhibits lower extinction coefficient, lower hole mobility, and poorer coating-forming properties. To overcome these shortcomings of SM donors, researchers have made many successful attempts, especially the molecular fluorination (Fan et al. 2019). Some examples of conjugated polymers and SMs can be seen in Figure 3.9.

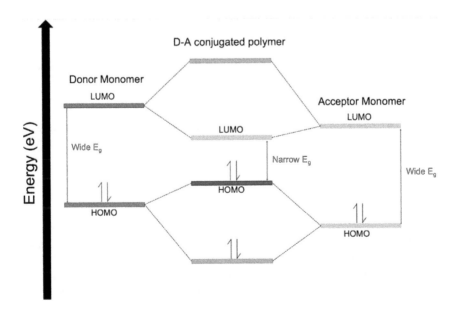

FIGURE 3.8   Band gap engineering by D-A strategy.

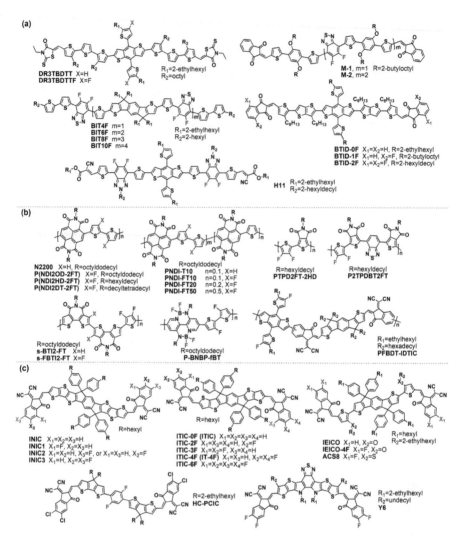

FIGURE 3.9 Chemical structures of representative (a) SM-donors, (b) polymer acceptors, and (c) SM-acceptors used in OPVs. Reprinted with permission from Fan et al., "Fluorinated photovoltaic materials for high-performance organic solar cells". *Chemistry – An Asian Journal*. Sep 16;14(18):3085–3095, 2019.

3.5.3 POLYMER NANOCOMPOSITE COATINGS WITH 2D MATERIALS

To obtain a polymer-based nanocomposite that can be applied as a coating and also provide the function of an active layer in a PV, one needs to consider what kind of acceptor materials can be used. In this regard, since the discovery of the one-atom thin graphene in 2004 (Novoselov et al. 2004), the so-called 2D materials have attracted much attention, because of their huge range of properties. Two-dimensional materials present strong in-plane covalent bonds, whereas they are stacked vertically by weak van der Waals (vdW) forces. The electronic, optical, mechanical, and thermal properties of these materials are strongly dependent on the number of layers and therefore the thickness, exhibiting a drastic change when they are in a monolayer or in few layers in comparison to its bulk counterpart (Sun et al. 2019). In order to achieve solution processing, 2D materials are often required to be chemically modified or functionalized. As a good example of this, one can consider graphene derivatives like GO and rGO, which have been already used in the OPVs as photoactive layer (Liu et al. 2008; Méndez-Romero et al. 2019). Figure 3.10 shows a noncovalent functionalization approach of GO with octadecylamine (ODA) from which the obtained GO-ODA can be easily transferred to organic solvents like oDCB or toluene (Méndez-Romero et al. 2019). With this method, apart from obtaining GO soluble in organic solvents, the introduction of the ODA groups prevents the restacking of the material due to steric hindrance.

Nevertheless, the introduction of oxygen functional groups induces a rise in GO's band gap up to ~3 eV (Méndez-Romero et al. 2019) and a high electrical resistance (~10^{11} Ω) (Méndez-Romero et al. 2019), which is detrimental for the carrier transport and final performance of the OPV devices. To overcome this issue, a reduction where some of the oxygen groups are removed can be conducted, thereby obtaining a material called rGO. Figure 3.11 shows an easy and straightforward method to conduct the chemical and thermal reduction of the GO-ODA to obtain rGO-ODA that maintains its solubility in organic solvents. In this way, the rGO-ODA can be mixed with some polymer, ensuring a good dispersion that later can be translated into obtaining a polymer nanocomposite coating without agglomeration or segregation, enabling it to be used as a photoactive layer.

Other 2D materials that are suitable to be used in OPV devices are the family of TMDCs because of their intrinsic semiconducting nature (Sun et al. 2019). Among TMDCs, one of the most studied materials, MoS_2 exhibits a band gap change from ~1.29 eV as a multilayer to ~1.9 eV when it is a monolayer (Singh et al. 2017). This band gap change also shifts from indirect to direct as the number of layers decreases, which is something desirable for PV applications (Splendiani et al. 2010).

As mentioned before, to make feasible the fabrication of the solar cells, liquid-phase processing is desired. Obtaining 2D TMDCs by liquid-phase exfoliation (LPE) has proved to be an interesting approach that would allow these materials to be combined with organic materials and form 2D organic nanocomposites, which can be coated and used in different parts of the OPVs like the photoactive layer.

3.6 NANOCOMPOSITE COATINGS FOR CONDUCTIVE INKS

In the last decades, printed electronics has become a low-cost, easy, quick, and reliable way to manufacture functional devices; e.g., sensors, photodetectors, solar cells, organic light-emitting diodes, and supercapacitors (Li, Lemme, and Östling 2014).

Therefore, the use of conductive inks is increasing in demand and need for better capabilities. Current printing techniques make their elaboration in a faster, cheaper, and easier way, becoming an option for the replacement of photolithography manufacturing systems that requires multiple manufacturing steps as well as the use of chemical reagents and expensive equipment (Kamyshny et al. 2005).

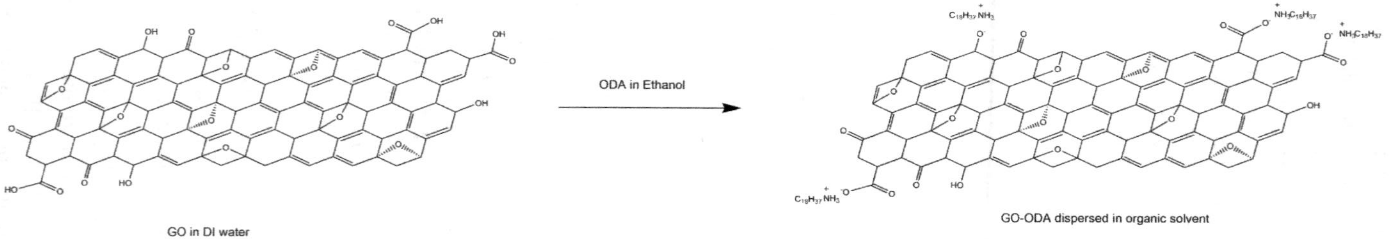

FIGURE 3.10 Electrostatic functionalization of GO with ODA, transferred to oDCB or toluene. Reprinted with permission from Méndez-Romero et al., "Functionalized reduced graphene oxide with tunable band gap and good solubility in organic solvents". *Carbon* 146:491–502, 2019.

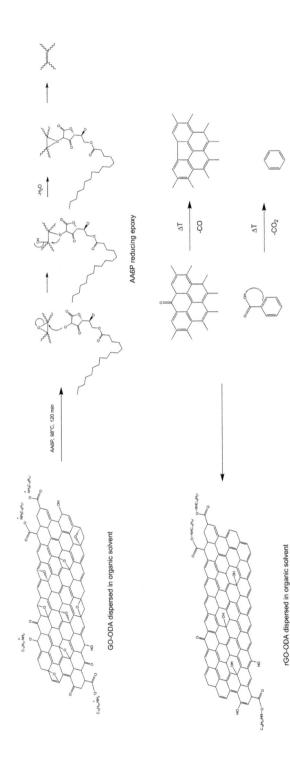

FIGURE 3.11 Chemical and thermal reduction of functionalized graphene oxide to generate functionalized reduced graphene oxide (rGO-ODA). Reprinted with permission from Méndez-Romero et al., "Functionalized reduced graphene oxide with tunable band gap and good solubility in organic solvents". *Carbon* 146:491–502, 2019.

The conductive inks as conductive coatings are manufactured with a polymer matrix and the incorporation of metallic particles such as Ag, Cu, Au, and Al, whose conductivities are on the order of 10^7 S/m (Kamyshny and Magdassi 2014). But, these kinds of inks require further steps and curing at high temperatures, making them not optimal for many applications.

Therefore, the development of new nanocomposite materials with the addition of nanostructures belonging to the 2D materials family such as graphene, carbon nanotubes (CNT), transition metal dichalcogenides (TMDs), black phosphorous (BP), and hexagonal boron nitride (h-BN) have become relevant for their novel electrical, thermal, mechanical, and optical properties, which are attributed to the quantum confinement of its nature due to their dimensions (McManus et al. 2017).

Additionally, 2D materials can be dispersed in a wide variety of organic solvents without chemically reacting with the environment, avoiding changes in their intrinsic structure (restacking of the layers). This allows formulation of an ink organic solvents as stable and uniform coatings, and additives can be used without modifying the properties of these materials, generating inks with rheological properties adjustable to different types of printing techniques or substrates.

Actually, there are many printing techniques such as *inkjet printing, screen printing, rotary screen, flexographic printing*, and *gravure printing*; each of these techniques gives us different characteristics in the resolution and performance of printed patterns (Hu et al. 2018), as we can see in Figure 3.12.

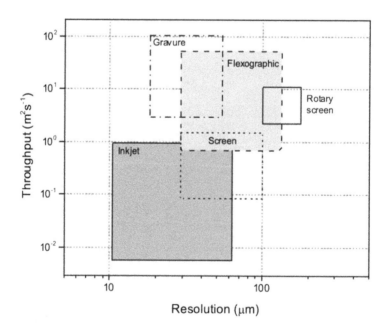

FIGURE 3.12 Comparison between the resolution and the production speed of the different printing techniques. Reprinted with permission from G. Hu et al., "Functional inks and printing of two-dimensional materials," *Chem. Soc. Rev.* 47(2015):3265–3300, 2018.

In order to use any of these printing techniques, we must have certain considerations, such as the size and nature of the substrate, ink concentration, as well as the volume of ink needed for each of the different printing techniques.

An ink system based on polymer nanocomposites possess three main components:

- Functional materials that give the main characteristics – in this particular case, a polymer nanocomposite of metallic NPs or 2D materials.
- A solvent, sometimes called a vehicle, where the material is dissolved or dispersed depending on the nature of the first.
- An additive that modifies the rheological properties of viscosity and surface tension.

Functional material: The amount of material present in an ink is defined by its solubility, dispersibility, and stability (e.g., graphene, rGO, carbon nanotubes). For inks that do not contain dissolved materials, the main problem is agglomeration and subsequent coagulation and sedimentation of the materials. The stability of a system is related to the balance between attractive forces and van der Waals forces, which must be counteracted by repulsive forces. There are two main mechanisms to overcome this issue: (1) stabilization by adsorbed charges on the surface that forms a layer, avoiding interaction between particles and (2) stabilization by steric effects, where interactions are prevented by forming a physical barrier through adsorption of a molecule on the surface of the particle (Arao and Kubouchi 2015).

Solvents: The solvent is the vehicle that dissolves or suspends the functional material and other components of the ink. Some inks are water based, they are widely used on substrates like paper and textiles, due to ink–substrate compatibility. However, they have disadvantages in their formulation, due to their need of stabilizers, additives, and buffers to adjust the rheological parameters (Hutchings and Martin 2012).

Solvent-based inkjet inks are the most widespread in industrial graphic applications because of their fast-drying time, print quality, image durability, and compatibility with a wide range of substrates (e.g., metal, glasses, ceramics, plastics). In many formulations, mixtures of various solvents are used, which enable tailoring ink properties such as viscosity, evaporation rate, and surface tension; these properties facilitate obtaining homogenous coatings free of defects (Kamyshny and Magdassi 2012).

Additives: It modifies the physicochemical properties of the medium to make it compatible with some of the processing methods; in some inks, it maintains the stability of the dispersed material. Some additives are as follows:

- Humidifiers mainly are used in water-based inks to prevent drying and clogging of the ink in the nozzle at the air–ink interface. The main chemical characteristic that a humectant must satisfy is the presence of hydrophilic groups in its structure, such as hydroxy, carboxy, or amine, in order to adsorb water and delay the evaporation of the solvent during the expulsion of the ink.
- Surfactants, sometimes called wetting agents, serve two major purposes within ink formulation. First, they are used to stabilize the dispersion of

particles in the medium avoiding material agglomeration. Second, it is the control of surface tension, a very important feature for using an ink in a printing system.

- Adhesion promoters are used to generate compatibility between the substrate and the ink. They may simply dry and solidify by solvent evaporation or can require some curing in order to produce cross-linking. The most common polymers used are acrylics, alkyds, cellulose, and resins such as vinyl chloride/vinyl acetate, acrylic resins, rubber resins, and polyketone resins (Hutchings and Martin 2012; Zhan et al. 2017).

Rheological performance: The precise formulation of the ink components defines the physical properties of the inks. Rheology is the science of flow and deformation. Flow properties of coatings are critical for the proper application and appearance of films. Depending on how stress is applied to a fluid, there are several types of flow. Of major importance in coatings is flow under shear stress (Jones, Nichols, and Pappas 2017). The two physical properties that dominate the behavior of an ink in an injection system are viscosity and surface tension (Kamyshny and Magdassi 2012).

Viscosity: It is defined as the resistance that a fluid opposes to the movement when a tangential force (shear or shear force) is applied to the surface. The fluid forms a gradient of velocities (Figure 3.13), because the external zones of the liquid perceive a greater deformation compared to the internal zones of the fluid (Coussot 2012).

Viscosity is a parameter for classifying inks and their use in any printing process. Inks can be separated into two groups of low viscosity and high viscosity. Each group needs a different printing process, as shown in Table 3.1 (Żołek-Tryznowska 2016).

Surface tension: It is a manifestation of the intermolecular forces within the liquids, which make them stay together and is defined as the amount of energy needed to increase the surface of a liquid. In the matter of conductive inks, this rheological property is related to the ability of the ink to coat the substrate once the ink–substrate interaction occurs. The ink must be in a range between 25 and 50 m/Nm, this range is adjusted to the characteristics of inks used in injection systems (Hutchings and Martin 2012; Cummins and Desmulliez 2012; Lyklema 1999).

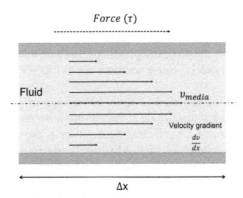

FIGURE 3.13   Scheme of a fluid velocity profile.

TABLE 3.1

Typical Ink Composition for Common Printing Technologies

Technique	Pigment (wt%)	Binder (wt%)	Solvent (wt%)	Additive (wt%)	Viscosity (mPa s)
Inkjet	5–10	5–20	65–95	1–5	4–30
Screen	12–20	45–65	20–30	1–5	1k–10k
Gravure	12–17	20–35	60–65	1–2	100–1k
Flexo	12–17	40–45	25–45	1–5	1k–2k

Source: Data from Hu, G., J. Kang, L.W. Ng, et al. 2018. Functional inks and printing of two-dimensional materials. *Chemical Society Reviews* 47 (9): 3265–3300.

3.7 CONCLUSIONS

Over the last decades, the role of functional coatings has considerably changed. Polymer coating technology has evolved by the increase in scientific and technological understanding of important principles. The development and design of polymer nanocomposites–enabled coatings has gained research importance as they offer the promise of incremental and disruptive improvements to products and processes. Polymer nanocomposite coatings represent properties with incredible practical applications for mechanical, optical, and electronic products. Polymer coatings based on nanocomposites offer significant product performance and cost-saving advantages with functional features.

REFERENCES

Arao, Y., and M. Kubouchi. 2015. High-rate production of few-layer graphene by high-power probe sonication. *Carbon* 95: 802–808.

Bhushan, B., and Y. C. Jung. 2011. Natural and biomimetic artificial surfaces for superhydrophobicity, self-cleaning, low adhesion, and drag reduction. *Progress in Materials Science* 56 (1): 1–108.

Cedillo-González, E.I.R., M. Montorsi, M. Montorsi, P. Falcaro, and C. Siligardi. 2014. Self-cleaning glass prepared from a commercial TiO$_2$ nano-dispersion and its photocatalytic performance under common anthropogenic and atmospheric factors. *Building and Environment* 71: 7–14.

Coussot, P. 2012. Introduction to the rheology of complex fluids. In *Understanding the Rheology of Concrete* (pp. 3–22). Woodhead Publishing.

Cummins, G., and P.Y. Marc Desmulliez. 2012. Inkjet printing of conductive materials: A review. *Circuit World $V 38* (4): 193–213.

Davis, A., Y.H. Yeong, A. Steele, I.S. Bayer, and E. Loth. 2014. Superhydrophobic nanocomposite surface topography and ice adhesion. *ACS Applied Materials & Interfaces* 6 (12): 9272–9279.

Decker, E.L., B. Frank, Y. Suo, and S. Garoff. 1999. Physics of contact angle measurement. *Colloids and Surfaces A: Physicochemical and Engineering Aspects* 156 (1): 177–189.

Fan, Q., U.A. Méndez-Romero, X. Guo, E. Wang, M. Zhang, and Y. Li. 2019. Fluorinated photovoltaic materials for high-performance organic solar cells. *Chemistry – An Asian Journal* 14, 3085.

Fernández-Merino, M.J., L. Guardia, J.I. Paredes, et al. 2010. Vitamin C is an ideal substitute for hydrazine in the reduction of graphene oxide suspensions. *The Journal of Physical Chemistry C* 114 (14): 6426–6432.

Ganbavle, V.V., Uzma K.H. Bangi, S.S. Latthe, S.A. Mahadik, and A.V. Rao. 2011. Self-cleaning silica coatings on glass by single step sol–gel route. *Surface and Coatings Technology* 205 (23): 5338–5344.

Guldin, S., P. Kohn, M. Stefik, et al. 2013. Self-cleaning antireflective optical coatings. *Nano Letters* 13 (11): 5329–5335.

Hu, G., J. Kang, L.W. Ng, et al. 2018. Functional inks and printing of two-dimensional materials. *Chemical Society Reviews* 47 (9): 3265–3300.

Hutchings, I.M., and G.D. Martin. 2012. *Inkjet Technology for Digital Fabrication*. John Wiley & Sons, United Kingdom.

Jones, F.N., M.E. Nichols, and S.P. Pappas. 2017. *Organic Coatings: Science and Technology*. John Wiley & Sons.

Kamyshny, A., and S. Magdassi. 2012. *Chapter 12:* Inkjet Ink Formulations, In *Inkjet-based Micromanufacturing*, edited by J.G. Korvink, P.J. Smith, and D.-Y. Shin (388 pp.). John Wiley & Sons, Germany.

Kamyshny, A., and S. Magdassi. 2014. Conductive nanomaterials for printed electronics. *Small* 10 (17): 3515–3535.

Kamyshny, A., M.B. Moshe, S. Aviezer, and S. Magdassi. 2005. Ink-jet printing of metallic nanoparticles and microemulsions. *Macromolecular Rapid Communications* 26 (4): 281–288.

Krebs, F.C. 2008. *Polymer Photovoltaics: A Practical Approach*. Vol. 175, Bellingham, WA: SPIE Press.

Lee, J.H., E.J. Park, D.H. Kim, M.-G. Jeong, and Y.D. Kim. 2016. Superhydrophobic surfaces with photocatalytic activity under UV and visible light irradiation. *Catalysis Today* 260:32–38.

Li, X., X. Du, and J. He. 2010. Self-cleaning antireflective coatings assembled from peculiar mesoporous silica nanoparticles. *Langmuir* 26 (16): 13528–13534.

Li, J., M.C. Lemme, and M. Östling. 2014. Inkjet printing of 2D layered materials. *ChemPhysChem* 15 (16): 3427–3434.

Liu, Z., Q. Liu, and Y. Huang, et al. 2008. Organic photovoltaic devices based on a novel acceptor material: Graphene. *Advanced Materials* 20 (20): 3924–3930.

Lyklema, J. 1999. The surface tension of pure liquids: Thermodynamic components and corresponding states. *Colloids and Surfaces A: Physicochemical and Engineering Aspects* 156 (1): 413–421.

Ma, M., and R.M. Hill. 2006. Superhydrophobic surfaces. *Current Opinion in Colloid & Interface Science* 11 (4): 193–202.

Markavart, T., and Castañer, L. 2005. *Solar Cells: Materials, Manufacture and Operation*. Oxford: Elsevier.

McManus, D., S. Vranic, and F. Withers, et al. 2017. Water-based and biocompatible 2D crystal inks for all-inkjet-printed heterostructures. *Nature Nanotechnology* 12 (4): 343.

Méndez-Romero, U.A., S.A. Pérez-García, X. Xu, E. Wang, and L. Licea-Jiménez. 2019. Functionalized reduced graphene oxide with tunable band gap and good solubility in organic solvents. *Carbon* 146: 491–502.

Momeni, M. M., I. Ahadzadeh, and A. Rahmati. 2016. Nitrogen, carbon and iron multiple-co doped titanium dioxide nanotubes as a new high-performance photo catalyst. *Journal of Materials Science: Materials in Electronics* 27 (8): 8646–8653.

Nakajima, A., K. Hashimoto, and T. Watanabe. 2001. Recent studies on super-hydrophobic films. *Monatshefte für Chemie / Chemical Monthly* 132 (1): 31–41.

Niu, J.J., and J.N. Wang. 2009. A novel self-cleaning coating with silicon carbide nanowires. *The Journal of Physical Chemistry B* 113 (9): 2909–2912.

Novoselov, K.S., A.K. Geim, and S.V. Morozov, et al. 2004. Electric field effect in atomically thin carbon films. *Science* 306 (5696): 666–669.

O'Regan, B., and M. Gratzel. 1991. A low-cost, high-efficiency solar cell based on dye-sensitized colloidal TiO$_2$ films. *Nature* 353 (6346): 737–740.

Oribayo, O., X. Feng, G.L. Rempel, and Q. Pan. 2017. Synthesis of lignin-based polyurethane/graphene oxide foam and its application as an absorbent for oil spill clean-ups and recovery. *Chemical Engineering Journal* 323: 191–202.

Othman, N.H., M.C. Ismail, M. Mustapha, N. Sallih, K.E. Kee, and R.A. Jaal. 2019. Graphene-based polymer nanocomposites as barrier coatings for corrosion protection. *Progress in Organic Coatings* 135: 82–99.

Paredes, J.I., S. Villar-Rodil, A. Martínez-Alonso, and J.M.D. Tascón. 2008. Graphene oxide dispersions in organic solvents. *Langmuir* 24 (19): 10560–10564.

Paul, D.R., and L.M. Robeson. 2008. Polymer nanotechnology: Nanocomposites. *Polymer* 49 (15): 3187–3204.

Philipps, S. P., and A. W. Bett. 2016. Current status of concentrator photovoltaic (cpv) technology 2016. [cited February 02 2016].

Singh, E., K.S. Kim, G.Y. Yeom, and H.S. Nalwa. 2017. Atomically thin-layered molybdenum disulfide (MoS2) for bulk-heterojunction solar cells. *ACS Applied Materials & Interfaces* 9 (4): 3223–3245.

Smalley, R.E. 2005. Future global energy prosperity: The terawatt challenge. *Mrs Bulletin* 30 (06): 412–417.

Splendiani, A., L. Sun, and Y. Zhang, et al. 2010. Emerging photoluminescence in monolayer MoS2. *Nano Letters* 10 (4): 1271–1275.

Sun, J., Y. Choi, Y.J. Choi, et al. 2019. 2D–Organic hybrid heterostructures for optoelectronic applications. *Advanced Materials.* 31, 1803831.

Thompson, B.C., and J.M.J. Fréchet. 2008. Polymer–fullerene composite solar cells. *Angewandte Chemie International Edition* 47 (1): 58–77.

Valipour M.N., F.C. Birjandi, and J. Sargolzaei. 2014. Super-non-wettable surfaces: A review. *Colloids and Surfaces A: Physicochemical and Engineering Aspects* 448: 93–106.

Vázquez-Velázquez, A.R., M.A. Velasco-Soto, S.A. Pérez-García, and L. Licea-Jiménez. 2018. Functionalization effect on polymer nanocomposite coatings based on TiO$_2$–SiO$_2$ nanoparticles with superhydrophilic properties. *Nanomaterials* 8 (6): 369.

Velasco-Soto, M. A., J. A. León-Gil, J. Alvarez-Quintana, et al. 2016. Chapter 8- carbon polymer nanocomposites. In *Nanocolloids*, edited by M.S. Domínguez, and C.R. Abreu. Amsterdam: Elsevier.

Viswanadam, G., and G.G. Chase. 2012. Contact angles of drops on curved superhydrophobic surfaces. *Journal of Colloid and Interface Science* 367 (1): 472–477.

Wadsworth, A., M. Moser, A. Marks, et al. 2019. Critical review of the molecular design progress in non-fullerene electron acceptors towards commercially viable organic solar cells. *Chemical Society Reviews* 48: 1596–1625.

Wang, F.J., S. Lei, J.F. Ou, M.S. Xue, and W. Li. 2013. Superhydrophobic surfaces with excellent mechanical durability and easy repairability. *Applied Surface Science* 276: 397–400.

Zhan, Z., J. An, Y. Wei, and H. Du. 2017. Inkjet-printed optoelectronics. *Nanoscale* 9 (3): 965–993.

Zhou, Y., M. Li, X. Zhong, X. Zhu, P. Deng, and H. Liu. 2015. Hydrophobic composite coatings with photocatalytic self-cleaning properties by micro/nanoparticles mixed with fluorocarbon resin. *Ceramics International* 41 (4): 5341–5347.

Żołek-Tryznowska, Z. 2016. 6- rheology of printing inks. In *Printing on Polymers*, edited by J. Izdebska, and S. Thomas. William Andrew Publishing, United Kingdom.

4 Mechanism, Anti-Corrosion Protection and Components of Anti-Corrosion Polymer Coatings

Akarsh Verma
Indian Institute of Technology

Naman Jain
Meerut Institute of Engineering and Technology

Shweta Rastogi
G. B. Pant University of Agriculture and Technology

Vaishally Dogra
Graphic Era Hill University

Sanjay Mavinkere Rangappa and Suchart Siengchin
King Mongkut's University of Technology North Bangkok

Rokbi Mansour
M'sila University

CONTENTS

4.1   INTRODUCTION

In today's world, researchers are striving to increase the efficiency, durability, reliability, etc. of materials which have application in different fields. For certain application, surface modification is required by coating to enhance the performance and efficiency and product life of materials. Coating is a film forming process in which organic or inorganic materials may form a hard or soft film, depending upon the process. Sometimes pigments are also used, and a combination of resin and pigment is called binder. To form liquid solution, binder is dissolved into solvent and a combination of solvent, resin and pigment is called vehicle. After employed on the surface, the solvent gets evaporated and the remaining resin and pigment stick on the material surface. Resistance against the environmental condition such as corrosion can be achieved by coating. We can also use coating as designing factor view only. Different types of coating materials can be employed, such as polymers, metals and ceramics, each having its own advantages and disadvantages. Metallic coating can be done to increase electrical and thermal conductivity with high hardness. However, metallic coating suffers from environment contamination and holds higher processing cost. On the other hand, polymer coating has low processing cost and is corrosion protective, electric insulated, etc., but suffers from low hardness and thermal stability.

4.1.1   Film Forming from Solvent Evaporation

In this mechanism, polymer is dissolved or emulsified into solvent which may be organic or water. The polymer with solvent sprayed onto the substrate surface changes into droplet form (refer to Figure 4.1). Then solvent is evaporated by drying the surface, followed by interpenetration of polymer chain. Solution (polymer plus solvent) viscosity plays an important role in the interpenetration of polymer chain [1]. Solvation is also considered as important parameter which measure coating [2]. To improve the interaction of polymer with substrate surface, random coil structure with higher end-to-end distance is preferred that enhance adhesion [3]. Another critical factor is the rate of solvent evaporation. Low rate of evaporation results in overwetted substrate surface, while higher evaporation rate dried the polymer droplet before impinging. To maintain proper evaporation rate, many actors such as temperature, air amount, pressure and relative humidity (in case water solvent) must be controlled.

The atomized solution droplet first makes contact with the substrate surface. Then it starts spreading over the substrate followed by interpenetration/film forming, as shown in Figure 4.1. During film forming process, different surface tension is involved such as solid–air, solid–liquid and liquid–air interfaces [4–6]. Finally, wettability of the coating solution which is also an important measure or the adhesion of polymer with the substrate surface occurs [7–9].

4.1.2   Cross-Linked Film Forming Process

Another film forming mechanism is through cross-linking which is used to fabricate the high-performance coatings. The coating films obtained by cross-linking are thermosetting in nature. In this mechanism, cross-linking agent reacts with the polymer

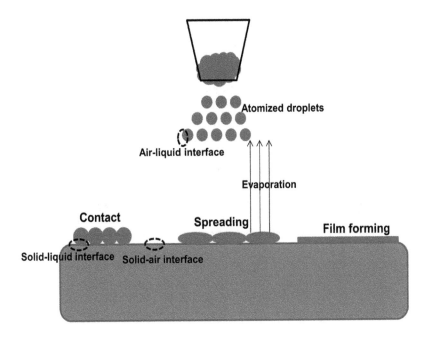

FIGURE 4.1 Film forming mechanism by solvent evaporation.

to form three-dimensional cross-linked networks. The method was not recently employed as drying oils (such as tung oil or linseed oil) were used without any solvent while developing the paint. Film formation takes place by oxidative cross-linked mechanism. Natural oils were used which contain almost 50% fatty acid mixed with resin. The curing reaction is facilitated by the metallic salts or driers. Resistance against moisture and chemical is attained by the film, when oil reacts by the oxygen present in the atmosphere to form cross-linked networks. Chemical cross-linking is also another technique which is employed by many researchers in recent years [10,11]. In this mechanism, two or more monomers are cross-linked with each other to form a three-dimensional network.

4.2 COATING FOR MEDICAL INSTRUMENTS

Coating in biomedical instruments has been done to improve mechanical strength, wear resistance, corrosion protection and electrical conductivity and to modify the surface chemistry. In this section, a review of application of polymeric coating on medical instrument such as orthopaedic materials, cardiovascular stents, drugs delivery, etc. has been done. Promising material for the fabrication of orthopaedic material was magnesium (Mg) and its alloys because of its similar elastic modulus and compressive yield strength as compared to bone. Moreover, they have high fracture toughness, low density and non-toxic corrosion product [10,11]. However, the reduction of corrosion of pure Mg is one of the major areas in which many researchers have worked. Luo et al. [11] applied the conducting polymer poly(3,4-ethylenedioxy-thiophene) (PEDOT) coating on Mg rod to avoid corrosion which also release drug.

Moreover, anti-inflammatory drug dexamethasone can also be loaded on PEDOT coating. Xu and Yamamoto [12] studied the effect of coating of poly (ε-caprolactone) (PCL), poly(L-lactic acid) (PLLA) and poly(glycolic acid) (PGA) on cytocompatibility of Mg and its alloy. PLLA coating films provide better adhesive strength as compared to other polymers to Mg substrate. In addition, for seven-day incubation, cytocompatibility of Mg and its alloy was improved by all polymers. Abdal-hay et al. [13] overcome the galvanic corrosion of Mg by coating it with poly lactic acid (PLA) through electrospinning and dip coating. Dip coating resulted in coating thickness of around 11–13.06 μm, whereas electrospinning resulted in around 30.4 μm. No delamination of coating films occurred after two days of immersion, but after seven days intermediate layer was clearly seen.

In 1994, bare metal stents were used to treat coronary artery disease, but they have major issue of stent thrombosis and restenosis [14]. Siroliums and paclitaxel drugs were used with the coating materials for first-generation drug-eluting stents made up of stainless steel [15,16]. Later, Bege et al. [15] reduced stent thrombosis by applying coating of poly(ethylene carbonate) (PEC) and paclitaxel with the help of spray coating. The amount of paclitaxel varies with the increase in coating layer, and within two days drug is released 100% via diffusion. Everolimus and zotarolimusin drugs were used with polymer coating in the second-generation drug-eluting stent.

4.3 ANTI-CORROSION METAL COATING

Another most important issue is corrosion of metals which affect their applications [17–19]. Corrosion environment such as water immersion for long hours, burial in soil, continuous contact with the pollution or ultraviolet radiation may result in metal degradation. Therefore, requirement of anti-corrosion coating becomes an important issue for the researchers. According to ISO 12944 [20], corrosion environment is divided into three types: atmospheric (industrial, marine and rural) [21], splash zone (region near the sea waterline where water waves make contact periodically) [22] and immersion (freshwater, seawater and soil). In corrosion, the metal and environment physicochemical reaction takes place and results in electrochemical reaction. In rusting, mechanism of iron is presented in Figure 4.2.

Polymeric coating on metal surface is considered as one of the important measures taken by many scientists in past years. Debarry in 1985 [23] applied electroactive polyaniline coating on ferrite stainless steels. Coating results in protection of anode in mineral acids by developing the immobilized polyaniline film. Due to low wear resistance and hardness of polymer, plain polymer coating is not long-lasting. To overcome this effect, polymers were blended or reinforced with nanoparticles which improve the durability and hardness of coating films. Conradi et al. [24] in 2014 applied epoxy-nanoparticles silica coating on austenite stainless steel. Silica nanoparticles improve the hardness of film by reining the microstructure of coated matrix. The above coating film shows good barrier to chloride ion-rich environment.

Coating films used for anti-corrosive purpose consist of multilayers and each layer has a specific purpose. Top coat, intermediate coats (one or many depends upon the situation) and primer are main constituents of the coating films. The major functions of the top coat which remain exposed to external environment are as follows: it

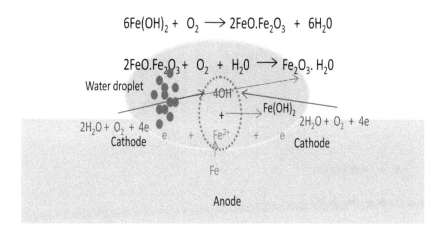

$$6Fe(OH)_2 + O_2 \longrightarrow 2FeO.Fe_2O_3 + 6H_2O$$

$$2FeO.Fe_2O_3 + O_2 + H_2O \longrightarrow Fe_2O_3.H_2O$$

FIGURE 4.2 Corrosion mechanism of iron.

should have high resistance to weathering condition and high wear resistance, resist UV radiation and provide glow and colour to surface. On the other hand, primer is one which interacts with substrate. The major function of primer is to provide protective coating on substrate and form good adhesion with substrate. Intermediate coat in general is used to increase the thickness and transport the aggressive species to metal surface. Anti-corrosive coatings are basically classified into three categories:

a. *Barrier coating protection*: In this protection, impermeability of ions results in very low electric conductance at the substrate–coating interface [25–27]. This results in transfer of minimum amount of corrosion current between cathode and anode [28]. Thus, action of oxygen and water from the environment inhibit the barrier coating [29]. Ritter and Rodriguez in 1982 [30] proposed that accumulation of chloride ion, metal surface roughness and surface oxide thickness are the major factors which affect the phenomenon of metal corrosion. The corrosion phenomenon can be initiated due to defect in coating, rupture of coating films or by solubility of impurity present on surface. Due to any of the above reasons, the development of anodes and cathodes result in iron dispersion into solution which combine with hydroxide and get precipitated. This hydroxide corrosion products form the hollow cone. Where the amount of oxygen is higher, it results in new cathodic site. Some ions penetrate into the coating films. At anode, the hydrolysis of $Fe(H_2O)_6^{2+}$ produces H_3O^+. To maintain neutrality at the surface, chlorine ions migrate into the hollow cone, as shown in Figure 4.3. The migration of chloride ions results in Na^+ in bulk, which penetrates the cathodic site. As deterioration of coating film increases, further breakthrough between anodic and alkaline will occur.

b. *Sacrificial coating protection*: It works on galvanic corrosion principle. In this protection, substrate metal is coated with metal or alloy which enriches the film making it electrochemically more active than the base material [31,32], as shown in Figure 4.4. Vilche et al. [31] evaluated the effect of

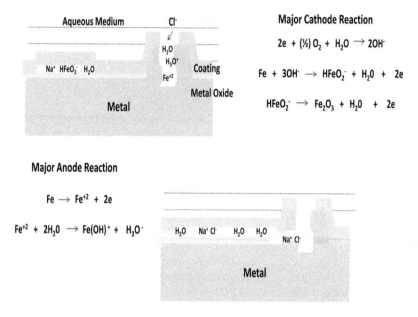

FIGURE 4.3 Accumulation of chloride ion, metal surface roughness and surface oxide thickness are the major factors which affect phenomenon of metal corrosion.

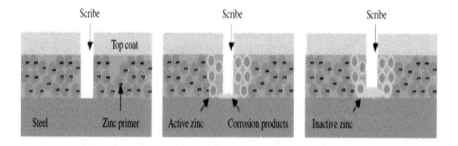

FIGURE 4.4 The working mechanisms of a zinc-rich coating system.

zinc-rich pants (ZRP) for protection of naval steel from corrosion when exposed to artificial seawater for 70 days. ZRP epoxypolyamine-amide coating acts as porous electrode due to electrical contact between reactive pigment particles and steel surface. Moreover, spherical zinc provides more effective cathodic protection.

c. *Inhibitive coating protection*: In this protection, a metallic complex [33–37] insoluble layer was built at the surface of a substrate which acts as the barrier. These metallic complexes prevent the transportation of aggressive species to substrate surface. Cohen in 1976 [33] studied the corrosion behaviour of specimens when exposed to aerated chloride having phosphate and nitride solutions. Result shows that specimen dipped into phosphate medium corroded over large area. Moreover, specimens can be repaired by three different mechanisms: (1) re-oxidizing by dissolving oxygen into the deoxidized metal; (2) forming insoluble compounds of ferrous by oxidizing the ionic ferrous product and (3) producing ionic ferrous components at the anode through anodic oxidation.

4.4 COMPONENTS OF ANTI-CORROSION COATING

A wide variety of materials and chemicals with each component having specific purposes are mixed together to form a coating layer. Coating mainly consists of five components: solvents, binders, additives, pigments and extenders. These five components are discussed in detail below.

4.4.1 SOLVENTS

A major constituent of coating is solvent which has special function to dissolve or disperse the other component such as binder and pigment. Moreover, to facilitate the dipping or spraying technique, solvent reduces the viscosity of coating liquid [38]. Another important factor is wettability of specimen which also play an important role in the selection of solvent. Organic solvents such as aliphatic and aromatic hydrocarbons, alcohols and glycol esters were used. The major disadvantage of organic solvent is that at room temperature and pressure, it gets easily volatile. There are many factors such as being environment-friendly, curing, substrate nature, pigmentation, curing, application, etc. which play a role in the selection process of solvent. Waterborne coating is also used where water is taken as solvent which make it eco-friendly. Polymers containing alcohols or glycol ethers are completely soluble in water to form water-soluble paints, which have high wetting, good pigment and high level of corrosion protection. On the other hand, some insoluble resin results in the formation of small clusters or emulsion/latex such as resin includes polyvinyl acetate, alkyds, acrylics etc. But in some cases, water is also counted as a disadvantage for waterborne coating. Development of foam, brush mars due to insufficient time to make corrections to the freshly applied paint, development of fungal or bacterial growth, etc. are also other major disadvantages [39,40]. In another type of coating, no solvent is used; this is called solvent-free coating, also known as powder coating. It is further classified into thermoplastic in which resin (thermoplastic) at elevated temperatures melt and flow over substrate while keeping same chemical composition. Another type is thermosetting in which upon curing process, resin (thermosetting) flow and melt over specimen. Polyesters, acrylic-epoxies, polyester-epoxies, acrylics, and polyurethane [41] are the major examples. High efficiency, durable finish, environment-friendly and recycling are the major advantages of solvent-free coating. On the other hand, high cost for oven process and special spray equipment are the disadvantages.

4.4.2 BINDERS

Component which ensures the high mechanical strength, adhesion of polymer to the metal and cohesion with the coating films is known as binder. The basic structure and permeability of the corrosion-resistant coating films depend upon the binders. Binders form the cross-linking structure during the curing process to ensure the properties of coating films. Sometimes anti-corrosion coating is also named on binder. The process of transformation of coating product from liquid to solid is known as film formation mechanism and it is governed in three ways:

1. By physical evaporation of the solvent through drying, i.e. transformation of liquid coating product in solid state by physically drying process. Physically drying process is further divided into two groups on the basis of agent applied to liquefy the binder. A film forming mechanism through physical drying process occurs in three steps: first, water is evaporated; followed by deformation and merging of polymer particles; and in the end gradual development of cohesive strength through joining of the nearby polymer particles [42].
2. By chemical reaction, evaporation of solvent in which binder chemically reacts with resin reactant and curing agent (also termed as cross-linker). On the basis of chemical curing process, binders are used in corrosion resistance coating which is further divided into three groups: coating in which oxygen is absorbed from the air and reacts in the presence of catalyst known as oxidative curing mechanism; In the second type, reaction occurs between particles like zinc silicates and moisture from air during curing reaction, known as moisture curing mechanism; in the third type, curing agent and binder react with each other in the presence of catalyst, which is known as two-component mechanism.
3. By combination of both, i.e. physical drying and chemical reaction evaporation of solvent.

4.5   DIFFERENT BINDERS IN COATING TECHNOLOGY

Epoxy: Due to high resistance to chemical, moisture and heat, the epoxy is most widely used in the coating industry [43–53]. On the other hand, chalking and yellowish appearance of epoxy coating films when exposed to ultraviolet radiation is a problem which limits its application to primer or intermediate coating. Molecular weight plays an important role is deciding the properties of epoxy coating. Reactive groups are higher in case of low-molecular-weight uncreative epoxy as compared to higher-molecular-weight uncreative epoxy. Therefore, curing reaction results in higher cross-linking density which results in good hardness and resistance to solvent is also high, whereas wettability (due to the presence of more hydroxyl groups), flexibility and fracture toughness are low as compared to high-molecular-weight epoxy [54]. Figure 4.5 represents the curing reaction of epoxy with amine. Most commonly

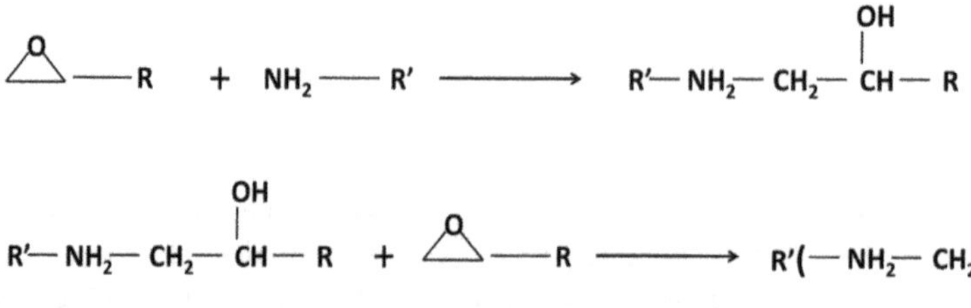

FIGURE 4.5   Curing reaction of epoxy with amine group.

used cross-linking agents are aliphatic amines because of their high corrosion resistance, whereas amides and polyamines are the other replacements. Curing process is more controllable and film obtained is more flexible than the amines agents (having moderate resistance to corrosion, acid and solvent) [55]. In some cases, modified polyamides or polyamidoamines is also used as curing agents to render water solubility by salting with acetic acid [56]. In another waterborne epoxy technology, predispersion of solid epoxy in water or cosolvents is used. In these types, water is added to the mixture of resin and curing agent which results in migration of curing agent into dispersed epoxy particles from aqueous phase. Under wetting condition, electrodeposition through cation provides highest corrosion resistance due to excellent adhesion in waterborne epoxy [57,58].

Acrylic: Good photostability, high adhesion and film forming properties of acrylic copolymer and polymer increase their application in coating industry. The main reason of growing interest of acrylic is high resistance to ultraviolet radiation, hydrolysis, environment stability and chemical inertness [59]. But they have poor performance in acid rain due to hydrolysis and is replaced by modified acrylics such as silane-modified acrylics, epoxy acid acrylics and carbamates. Higher durability, adhesion and quick drying are the major advantages of acrylic solvent-borne coating as compared to waterborne coating. Solvent evaporation mechanism takes place when hardening of coating film occurs. The most commonly used cross-linking agents are isocyanates and amino resin, whereas identical curing process occurs for water- and solvent-borne coating which may be accelerated by catalyst (tertiary amines and organometallic compounds). The advantage of waterborne coating over solvent-borne coating is low VOC emission.

Siloxane: Siloxanes are the polymers made up of silicone–oxygen (Si-O) backbone in which each silicone atom has phenyl, methyl or ethyl group and occurs in both linear and cyclic forms, as shown in Figure 4.6. Siloxane coating films have superior colour and gloss retention, ultraviolet resistance and low VOCs emission, but on the other hand it suffers from low mechanical properties due to their brittle nature. As seen from its chemical structure, siloxanes are already in oxide state; therefore, it prevents further oxidation. Polysiloxane may also react with organic compounds or metals to form hybrid coating to overcome the limitation and increase industry application. In 2005, Ahmed et al. [60] fabricated the siliconized epoxy matrix by blending epoxy resin (bisphenol A) with polydimethyl siloxane (hydroxyl-terminated) by

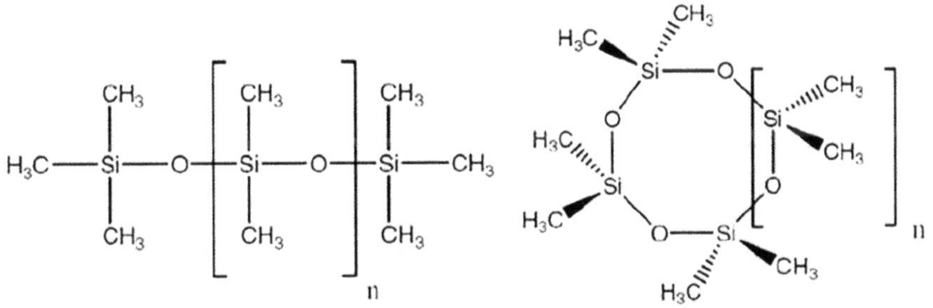

FIGURE 4.6   Linear and cyclic form of siloxane.

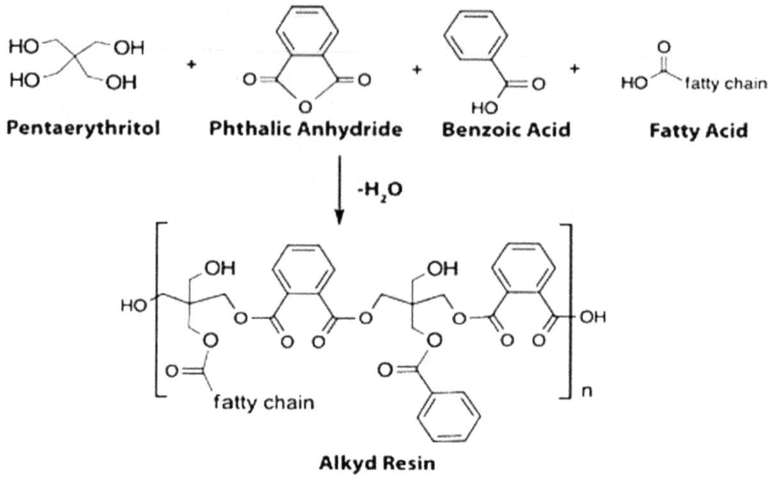

FIGURE 4.7 Synthesis of alkyd resin.

employing open-ring polymerization technique. Results show that coating film had good high gloss value and colour retention and also resists to acidic, organic solvent and saline environment. Many researchers also studied inorganic zinc silicate for coating of metallic substrates as anti-corrosion coating. Corrosion resistance is obtained through cathodic protection of zinc silicate, and curing mechanism takes help of atmospheric moisture [61]. Minimum RH of about 50%–60% is required to perform the curing process [62]. Reaction between steel substrate and metallic zinc will occur to some extent.

Alkyd: They are oil-modified polyester synthesized by condensation of dibasic acid and polyhydric alcohol, as shown in Figure 4.7. Alkyd resin consists of unreactive acid and hydroxyl group and by interaction with oxygen or other molecules, they form solid binder. During curing phenomena, oxidative reaction occurs in alkyd coating in which oxygen from air with certain catalyst is involved. Oxygen from air reacts with hydrogen atom of the diallylic methylene which promotes addition of series in hydrocarbon chain [63]. The disadvantage of alkyd coating is that ester lineage reacts with alkaline materials which result in the formation of carbolic acid salt and alcohol; this process is known as saponification [64]. To make alkyd compatible with water medium, certain modifications are made. Typically, prepolymers are used which are rich in hydroxyl group or malinization of the fatty acids (with polyol and dibasic acid).

Polyurethane: Most common process or synthesis of polyurethane through reaction is between polyol and diisocyanate, as shown in Figure 4.8. High resistance to weathering is one of the major advantages of polyurethane coating as to epoxy coating. Moreover, polyurethane coating exposed to ultraviolet radiation has less change in colour and gloss of coating [65]. Therefore, it is often used as top coat where materials have higher exposure to sunlight but may also be used as intermediate and primer coating. Another major advantage is that it has high resistance to scratch due to hydrogen bonding between urethane linkages which may be rebuilt after deformation and the phenomenon is known as self-healing. On the other hand, polyurethane

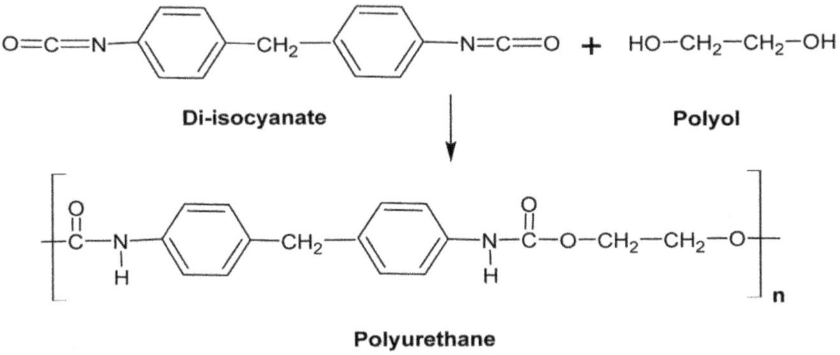

FIGURE 4.8 Reaction for synthesis of polyurethane.

coating suffer from poor resistance to mechanical strain and high-temperature degradation [60]. Both processes may be employed, i.e. moisture-cured and two-component film forming. To overcome the above problem, cross-linking may be used which can enhance abrasion, strength and resistance to solvent and acidic medium.

REFERENCES

1. Bauer, K.H., Lehmann, K., Osterwald, H.P., Rothgang, G., 1998. *Coated Pharmaceutical Dosage Forms.* CRC Press, Boca Raton, FL.
2. Kramer, P.A., Banker, G.S., Nadkarni, P.D., Kildsig, D.O., 1975. Effect of surface roughness and coating solvent on film adhesion to tablets. *Journal of Pharmaceutical Sciences, 64*(9), pp. 1554–1557.
3. Felton, L.A., McGinity, J.W., 1999. Adhesion of polymeric films to pharmaceutical solids. *European Journal of Pharmaceutics and Biopharmaceutics, 47*(1), pp. 3–14.
4. Porter, S.C., 2012. Coating of pharmaceutical dosage forms. In: Allen, L.V.E.A. (Ed.), *Remington the Science and Practice of Pharmacy.* Pharmaceutical Press, London, pp. 977–987.
5. Bruce, H.F., Sheskey, P.J., Garcia-Todd, P., Felton, L.A., 2011. Novel low-molecular-weight hypromellose polymeric films for aqueous film coating applications. *Drug Development and Industrial Pharmacy, 37*, pp. 1439–1445.
6. Pearnchob, N., Siepmann, J., Bodmeier, R., 2003. Pharmaceutical applications of shellac: Moisture-protective and taste-masking coatings and extended-release matrix tablets. *Drug Development and Industrial Pharmacy, 29*(8), pp. 925–938.
7. Khan, H., Fell, J.T., Macleod, G.S., 2001. The influence of additives on the spreading coefficient and adhesion of a film coating formulation to a model tablet surface. *International Journal of Pharmaceutics, 227*(1–2), pp. 113–119.
8. Felton, L.A., Austin-Forbes, T., Moore, T.A., 2000. Influence of surfactants in aqueous-based polymeric dispersions on the thermo-mechanical and adhesive properties of acrylic films. *Drug Development and Industrial Pharmacy, 26*(2), pp. 205–210.
9. Zografi, G., Johnson, B.A., 1984. Effects of surface roughness on advancing and receding contact angles. *International Journal of Pharmaceutics, 22*(2–3), pp. 159–176.
10. Kirkland, N.T., 2012. Magnesium biomaterials: past, present and future. *Corrosion. Engineering, Science and Technology, 47*(5), pp. 322–328.
11. Luo, X., Cui, X.T., 2011. Electrochemical deposition of conducting polymer coatings on magnesium surfaces in ionic liquid. *ActaBiomater, 7*(1), pp. 441–446.

12. Xu, L., Yamamoto, A., 2012. Characteristics and cytocompatibility of biodegradable polymer film on magnesium by spin coating. *Colloids Surfaces B, 93*, pp. 67–74.

13. Abdal-hay, A., Barakat, N.A.M., Lim, J.K., 2013. Influence of electrospinning and dip-coating techniques on the degradation and cytocompatibility of Mg-based alloy. *Colloids Surfaces A, 420*, pp. 37–45.

14. Martin, D.M., Boyle, F.J., 2011. Drug-eluting stents for coronary artery disease: a review. Medical engineering & physics. *Medical Engineering & Physics, 33*(2), pp. 148–163.

15. Bege, N., Steinmüller, S.O., Kalinowski, M., Reul, R., Klaus, S., Petersen, H., Curdy, C., Janek, J., Kissel, T., 2012. Drug eluting stents based on Poly (ethylene carbonate): optimization of the stent coating process. *European Journal of Pharmaceutics and Biopharmaceutics, 80*(3), pp. 562–570.

16. Byrne, R.A., Kastrati, A., Massberg, S., Wieczorek, A., Laugwitz, K.-L., Hadamitzky, M., Schulz, S., Pache, J., Fusaro, M., Hausleiter, J., Schömig, A., Mehilli, J., 2011. Biodegradable polymer versus permanent polymer drug-eluting stents and everolimus-versus sirolimus-eluting stents in patients with coronary artery disease: 3-year outcomes from a randomized clinical trial. *Journal of the American College of Cardiology, 58*(13), pp. 1325–1331.

17. Koch, G.H., Brongers, M.P.H., Thomson, N.G., Virmani, Y.P., Payer, J.H., 2002. Corrosion cost and preventive strategies in the United States. *Materials Performance, 65*, p. 1.

18. Fragata, F., Salai, R.P., Amorin, C., Almeida, E., 2006. Compatibility and incompatibility in anticorrosive painting—the particular case of maintenance painting. *Progress in Organic Coatings, 56*, p. 257.

19. Pandey, M.D., Nessim, M.A., 1996. Reliability-based inspection of post-tensioned concrete slabs. *Canadian Journal of Civil Engineering, 23*, p. 242.

20. ISO 12944. 1998. International Standards Organization, Geneve.

21. Bardal, E., 2005. *Corrosion and Protection.* Springer-Verlag, London.

22. Husain, A., Al-Shamali, O., Abduljaleel, A., 2004. Investigation of marine environmental related deterioration of coal tar epoxy paint on tubular steel pilings. *Desalination, 166*, p. 295. doi: 10.1016/j.desal.2004.06.084.

23. DeBerry D.W., 1985. Modification of the electrochemical and corrosion behavior of stainless steels with an electroactive coating. *Journal of the Electrochemical Society, 132*, pp. 1022–1026.

24. Conradi M., Kocijan A., Kek-Merl D., Zorko, M. Verpoest, I., 2014. Mechanical and anticorrosion properties of nanosilica-filled epoxy-resin composite coatings. *Applied Surface Science, 292*, pp. 432–437.

25. Dickie, R.A., Smith, A.G., 1980. How paint arrests rust. *Chemtech, 10*, p. 31.

26. Bacon, C.R., Smith, J.J., Rugg, F.G., 1948. Electrolytic resistance in evaluating protective merit of coatings on metals. *Industrial & Engineering Chemistry, 40*, p. 161. doi: 10.1021/ie50457a041.

27. Kittelberger, W.W., Elm, A.C., 1952. Diffusion of chloride through various paint systems. *Industrial & Engineering Chemistry Research, 44*, p. 326.

28. Hare, C., 1989. Barrier coatings. *Journal of Protective Coatings and Linings, 6* ,p. 59.

29. Thomas, N.L., 1991. The barrier properties of paint coatings. *Progress in Organic Coatings, 19*, p. 101.

30. Ritter, J.J., Rodriguez, M.J., 1982. Corrosion phenomena for iron covered with a cellulose nitrate coating. *Corrosion, 38*, p. 223.

31. Vilche, J.R., Bucharsky, E.C., Guidice, C., 2002. Application of EIS and SEM to evaluate the influence of pigment shape and content in ZRP formulation on the corrosion prevention of naval steel. *Corrosion Science, 44*, p. 1287. doi: 10.1016/S0010-938X(01)00144-5.

32. Hare, C., Steele, M., Collins, S.P., 2001. Zinc loadings, cathodic protection, and post-cathodic protective mechanisms in organic zinc-rich metal primers. *Journal of Protective Coatings and Linings, 18*, p. 54.

33. Cohen, M., 1976. The breakdown and repair of inhibitive films in neutral solution. *Corrosion*, *32*, p. 12.
34. Romagnoli, R., Vetere, V.F., 1995. Heterogeneous reaction between steel and zinc phosphate. *Corrosion*, *51*, p. 116.
35. Meng, Q., Ramgopal, T., Frankel, G.S., 2002. The influence of inhibitor ions on dissolution kinetics of Al and Mg using the artificial crevice technique. *Electrochemical and Solid-State Letters*, *5*, p. B1. doi: 10.1149/1.1429542.
36. Rafey, S.A.M., Abd El Rehim, S.S., 1996. Inhibition of chloride pitting corrosion of tin in alkaline and near neutral medium by some inorganic anions. *Electrochimica Acta*, *42*, p. 667.
37. Schmucki, P., Virtanen, S., Isaacs, H.S., Ryan, M.P., Davenport, A.J., Bö hni, H., Stenberg, T., 1998. Electrochemical behavior of Cr2O3/Fe2O3 artificial passive films studied by in situ XANES. *Journal of the Electrochemical Society*, *145*, p. 791. doi: 10.1149/1.1838347.
38. Lambourne, R., Strivnes, T.A., 1999. *Paint and Surface Coatings—Theory and Practice*. Woodhead, Cambridge.
39. Kiil, S., 2006. Drying of latex films and coatings: Reconsidering the fundamental mechanisms. *Progress in Organic Coatings*, *57*, p. 236.
40. Schwartz, J., 1992. The importance of low dynamic surface tension in water-borne coatings. *Journal of Coatings Technology*, *64*, p. 65.
41. Gaschke, M., Dreher, B., 1976. Review of solvent-free liquid epoxy coating technology. *Journal of Coatings Technology*, *48*, p. 46.
42. Daniels, E.S., Klein, A., 1991. Development of cohesive strength in polymer films from latices: Effect of polymer chain interdiffusion and crosslinking. *Progress in Organic Coatings*, *19*, p. 359.
43. Oichi, M., Takamiy, K., Kiyohara, O., Nakanishi, T., 1998. Effect of the addition of aramid-silicone block copolymer on phase structure and toughness of cured epoxy resins modified with silicone. *Polymer*, *39*, p. 725.
44. Verma, A., Singh, V.K., 2019. Mechanical, microstructural and thermal characterization of epoxy-based human hair–reinforced composites. *Journal of Testing and Evaluation*, *47*(2), pp. 1193–1215.
45. Verma, A., Negi, P., Singh, V.K., 2019. Experimental analysis on carbon residuum transformed epoxy resin: Chicken feather fiber hybrid composite. *Polymer Composites*, *40*(7), pp. 2690–2699.
46. Verma, A., Negi, P., Singh, V.K., 2018. Experimental investigation of chicken feather fiber and crumb rubber reformed epoxy resin hybrid composite: Mechanical and microstructural characterization. *Journal of the Mechanical Behavior of Materials*, *27*(3–4), 1–24.
47. Verma, A., Negi, P., Singh, V.K., 2018. Physical and thermal characterization of chicken feather fiber and crumb rubber reformed epoxy resin hybrid composite. *Advances in Civil Engineering Materials*, *7*(1), pp. 538–557.
48. Verma, A., Joshi, K., Gaur, A., Singh, V.K., 2018. Starch-jute fiber hybrid biocomposite modified with an epoxy resin coating: Fabrication and experimental characterization. *Journal of the Mechanical Behavior of Materials*, *27*(5–6), pp. 1–16.
49. Verma, A., Budiyal, L., Sanjay, M.R., Siengchin, S., 2019. Processing and characterization analysis of pyrolyzed oil rubber (from waste tires)-epoxy polymer blend composite for lightweight structures and coatings applications. *Polymer Engineering & Science*, *59*(10), pp. 2041–2051.
50. Verma, A., Baurai, K., Sanjay, M.R., Siengchin, S., 2020. Mechanical, microstructural, and thermal characterization insights of pyrolyzed carbon black from waste tires reinforced epoxy nanocomposites for coating application. *Polymer Composites*, *41*(1), pp. 338–349.

51. Verma, A., Singh, V.K., Experimental characterization of modified epoxy resin assorted with almond shell particles. *ESSENCE-International Journal for Environmental Rehabilitation and Conservation*, p. 36.
52. Verma, A., Parashar, A., Packirisamy, M., 2018. Atomistic modeling of graphene/hexagonal boron nitride polymer nanocomposites: A review. *Wiley Interdisciplinary Reviews: Computational Molecular Science*, 8(3), p. e1346.
53. Verma, A., Gaur, A., Singh, V.K., 2017. Mechanical properties and microstructure of starch and sisal fiber biocomposite modified with epoxy resin. *Materials Performance and Characterization*, 6(1), pp. 500–520.
54. Levita, G., De Petris, S., Marchetti, A., Lazzeri, A., 1991. Crosslink density and fracture toughness of epoxy resins. *Journal of Materials Science*, 6, p. 2348.
55. Salem, L.S., 1996. Epoxies for steel. *Journal of Protective Coatings and Linings*, 77, 77–98.
56. Wegmann, A., 1993. Novel waterborne epoxy resin emulsion. *Journal of Coatings Technology*, 65, p. 27.
57. Miskovic-Stankovic, V.B., Drazic, D.M., Teodorovic, M.J., 1995. Electrolyte penetration through epoxy coatings electro- deposited on steel. *Corrosion Science*, 37, p. 241.
58. Mišković-Stanković, V.B., Zotović, J.B., Kačarević-Popović, Z., Maksimović, M.D., 1999. Corrosion behaviour of epoxy coatings electrodeposited on steel electrochemically modified by Zn–Ni alloy. *Electrochimica Acta*, 44(24), pp. 4269–4277.
59. Carretti, E., Dei, L., 2004. Physicochemical characterization of acrylic polymeric resins coating porous materials of artistic interest. *Progress in Organic Coatings*, 49(3), pp. 282–289.
60. Ahmad, S., Gupta, A.P., Sharmin, E., Alam, M., Pandey, S.K., 2005. Synthesis, characterization and development of high performance siloxane-modified epoxy paints. *Progress in Organic Coatings*, 54(3), pp. 248–255.
61. Munger, C.G., 1994. The chemistry of zinc silicate coatings. *Corrosion Prevention and Control*, 41(6), pp. 140–142.
62. Socha, R.P., Pommier, N., Fransaer, J., 2007. Effect of deposition conditions on the formation of silica-silicate thin films. *Surface and Coatings Technology*, 201(12), pp. 5960–5966.
63. Van Gorkum, R., Bouwman, E., 2005. The oxidative drying of alkyd paint catalysed by metal complexes. *Coordination Chemistry Reviews*, 249(17–18), pp. 1709–1728.
64. Aigbodion, A.I., Okieimen, F.E., Obazee, E.O., Bakare, I.O., 2003. Utilisation of maleinized rubber seed oil and its alkyd resin as binders in water-borne coatings. *Progress in Organic Coatings*, 46(1), pp. 28–31.
65. Howarth, G.A., 2003. Polyurethanes, polyurethane dispersions and polyureas: Past, present and future. *Surface Coatings International Part B: Coatings Transactions*, 86(2), pp. 111–118.

5 Morphology of Polymer Coatings

Jyoti Sharma, Manju Rawat, Daisy Sharma, Sanjeev Kumar Ahuja, Avinash Chandra, and Sanghamitra Barman
Thapar Institute of Engineering and Technology

Raj Kumar Arya
Dr. B. R. Ambedkar National Institute of Technology

CONTENTS

5.1 INTRODUCTION

Coatings are a part of daily life in the form of architectural, automotive, or protection, functional or decorative. The combinations of conductive and biodegradable polymers help in enhancing the properties of the polymeric coatings that have potential applications in electrodes, antistatic coatings, biosensors, membranes, light-emitting diodes, etc. (Srinivasarao et al., 2001; Yabu et al., 2005). The main purpose of the coatings is to provide the desired appearance and to protect the substance from corrosion, scratches, abrasion, chemical attack, and temperature (Sharma et al., 2017a).

The coatings must be compatible with the substrate to avoid failures such as wetting or solvent incompatibility in future. The failures such as coating defects (blessing, cissing, and flaking), slow chemical reactions (concrete and galvanized steels), and mismatches in mechanical properties are due to the improper selection of the system, improper drying, over-coating of the coats, difference between the osmotic gradients between the coatings and environment, etc. These failures can be corrected by repainting of the coatings, keeping a check on the parameters that affect the properties of the coatings (Pathania et al., 2018). The selection of coatings should be done by keeping in mind their desired properties, work requirements, safety, compatibility, environmental restrictions, and cost. After the application of the coating to a substrate, it must become thin, dense, and adherent, resulting in the desired properties

(Sharma et al., 2019b). The desired or ideal coating properties include resistance to abrasion, heat, external weather, water, fuel, and chemical reactions. The drying time and the ease of application are also the desired properties of the coatings.

The polymers are dissolved in organic solvents and deposited on the substrate via different techniques such as dip casting, spin coating, solution casting, ink-jet printing, and drop casting to make the desired structures and films. The processing of coatings depends on the solvent and the concentration of the polymeric solutions (Sharma et al., 2017a).

Polymers provide the coatings the capacity to stick to the substrate, most of their resistance to chemicals, and flexibility. These materials are also capable of controlling porosity ratio, morphology, large-area processing, and adherence to the flexible substrate. It also makes the polymer films fulfill the requirements of a wide range of applications (Sharma et al., 2019c). The amorphous and crystalline nature of the polymers predicts the morphology of the coatings to a large extent.

The phase separations of coatings/blends/films are due to their immiscibility and affect the domains of morphology to a great extent. This results in the strong effect on the electrical and mechanical properties of the coatings (Hirtz et al., 2013). Thermodynamically and kinetically, the species involved in making the coatings affects their stability, the nature of the polymeric chains, the thickness of the film, preparation, homogeneity, processing, the substrate on which the coating is to be applied, and the environmental properties such as temperature or solvent vapor (Sharma et al., 2019a).

Polymer coatings are incomplete without their implicational use and structural analysis. The surface morphology of the composites has attracted much interest due to its different nature from the bulk materials (Richards and Peace, 1999). Polymers such as linear homopolymers (Pitois and Francois, 1999), copolymers (Francois et al., 1999), and coil–coil block polymers (Widawski et al., 1994) have been known to form porous structures.

Polymer morphology provides a platform to understand the polymer structure that includes branching, molecular weight, crystallinity, cross-linking, and so on. The crystalline or amorphous nature of the polymers used in making coatings can be easily understood. Thermoplastic polymers used for making coatings generally are semicrystalline, a combination of both crystalline and amorphous regions. Thermoplastics are highly influenced by their morphology. The morphology of the coatings determine their porosity, symmetric or asymmetric nature, and dense or phase separation behavior.

Studying the surface chemistry of the polymeric films helps in controlling the surface morphology of the coatings. Micro-contact printing is used to explain the surface templates and to provide the formation of micron-scale patterns in the polymer coatings. This technique can be used in the restructuring of semiconductor wafers, X-ray photolithography, and adsorbed nanoparticles and block copolymers. During annealing, the changes in domain structures give insights into the factors and phenomena that control the thin film blend morphology.

Each and every type of coatings such as symmetric and asymmetric, porous and non-porous, and dense and phase separated has grabbed the attention of the researchers because of the emerging field of coatings (Fan et al., 1999). For example, asymmetric membranes are mainly used in the liquid and gas and liquid because the top layer gives the barrier to the external particles and the small pores also gives the good mechanical strength (Jung et al., 2004). These types of membranes include a variety of structures with large to fine pores, forming the skin on the top of the membrane (Ladewig and Al-Shaeli, 2017). The factors such as surface composition can be controlled or changed by changing the molecular weight, concentration, surface free energy, and architecture (Liu and Messmer, 2003). They have many industrial applications such as ultra-filtration, reverse osmosis, and gas separation (Vaessen et al., 2002). The phenomena such as desorption and adsorption, and the applicable use of coatings as modifiers, adhesives, and so on, can be predicted by the morphology of the polymeric coatings (Adiga et al., 2009; Qian and Zhang, 2011).

5.1.1 FACTORS AFFECTING THE MORPHOLOGY OF COATINGS

Factors that affect the morphology of polymeric coatings are as follows:

a. *Concentration*:
   Concentration is the most important factor/parameter that affects not only the morphology of the coatings, but also the physical and chemical properties of the coatings, too. Polymers tend to aggregate when their concentration reaches a critical point. The critical concentration depends upon the polymer's molecular weight, chemical structure, solvent, and temperature. The intermolecular forces or van der Waals force leads to the aggregation between the polymer chains. A higher concentration of polymer results in higher aggregation, leads to gelling of the polymer, and changes the morphology of the polymer completely. Simha and coworkers (Simha and Utracki, 1967) discussed that the concentration regions would be characterized by the product of the concentration (c) and the intrinsic viscosity (η). If c. $\eta < 1$, then the solution would show an ideal behavior; if c. $\eta = 1 \sim 4$, then the solution would show deviation from the ideal behavior; and if c. $\eta \sim 4$–10, then interpolymer interactions would become possible.

b. *Type of blends*:
   Another factor that affects the morphology is the type of blends that is used in making coatings. Polymeric coatings can be made up of mixture of blends, being either binary or ternary mixtures. In ternary blends, the effect of two bounding interfaces provides extra complexities to morphology of the blends. The phase change behavior in binary (Binder, 1994; Brézin et al., 1976; Krausch, 1995) and ternary blends (Gersappe et al., 1994; Israels et al., 1995; Nauman and He, 1994) has been studied in the past. The third component (copolymer, surfactants, additives, or adhesives) in the binary mixtures changes the properties of the coatings (Nauman and He, 1994). Phase-separated films, dewetting in polymer films, and pattern formation are attracting much interest due to their vast importance from biological adhesion to thin-film optics and electronics devices. In electronic devices, a change in photoluminescence and electroluminescence occurs due to the change in the morphological behavior of films or the aggregation pattern

of the polymer chains in the thin films (Bradley et al., 1992). These effects can be altered by changing the solvent or concentration, or the coating techniques.

c. *Processing*:

The processing conditions also affect the morphology of the coatings. The phase size, shape, and dimension of the morphology of the coatings can be altered by controlling the processing conditions. For immiscible polymer blends, the non-equilibrium behavior of the processing operations employed for the fabrication determined the morphological behavior. In *melt processing*, the non-equilibrium morphologies prepared in the melt are frozen are either done by subsequent cooling, or by reactive processing to the point of vitrification. In another *solvent-based processing*, solvent evaporation leads to phase structures, and their improvement is noted at the point at which solvent removal leads to vitrification. During processing, polymer mixtures tend to phase separation either by the temperature change or by the solvent removal. The possibilities or the resultant morphology obtained from various processing technologies is present in abundance (Decker et al., 2003; Krebs, 2009; Torrey and Bordia, 2008; Wengeler et al., 2011).

d. *Phase separation in thin films*:

Porous polymeric membranes are mostly prepared by the controlled phase separation of polymer solutions. The phase separation of the polymeric solutions can be done in two ways: non-solvent-induced phase separation and thermally induced phase separation (TIPS). Immersion precipitation, precipitation from the vapor phase, and dry casting of polymer solutions are the processes responsible for the non-solvent-induced phase separation. Another process of the non-solvent-induced phase separation contains three-component systems, and the components of the system are a polymer, a solvent and a non-solvent (Matsuyama et al., 2000).

It has been reported that the properties of the phase-separated films were different from those of the bulk polymer films. The complications occur due to the presence of the interfaces of the film/air and substrate/film (Arya, 2011; Arya, 2013; Hsu and Yao, 2014; Sharma et al., 2017a; Sharma et al., 2017b). The bulk miscible blend of polymers was found to undergo phase separation when the blend was in the form of thin film with thickness being less than twice the radius of the gyration of an unperturbed chain, $2R_g$ (Li et al., 2004; Tanaka et al., 1995). The temperature for phase separation, the Flory–Huggins parameter χ (Nesterov et al., 1991), and the breakup mechanism (Li et al., 1994) were also responsible for the different properties compared to the bulk. The phase-separated films are very sensitive from the various parameters such as film thickness, molecular weight of polymer (Sharma et al., 2019a), substrate (polymer substrate interactions) (Affrossman et al., 1998; Slep et al., 1998), the relevant solvent parameter (polymer solubility (Prashar et al., 2019a), evaporation rate (Prashar et al., 2019b), presence of additives (Salata et al., 2019), and many more.

Some According to some mechanisms, a few components of the system having lower surface energies will prefer to migrate to free surface; on the other hand, the component with higher affinity to the substrate could segregate to the substrate interface. Such preferable segregation is known as vertical phase separation that has gradient distribution in the thickness direction and creates layered structures (Liu et al., 2016). However, the neutral substrate causes a lateral phase separation, in which the phase boundaries are perpendicular to the surface. A substrate having ordered pattern will provide the component with a higher affinity to the corresponding area copying the pattern on the substrate, creating an ordered, instead of random, pattern of phase separation (Sharma et al., 2019c). The other mechanism that has a better compatibility between compositions results in uniform morphologies with a smaller size scale (Gutmann et al., 1999). Changing the molecular weight affects the compatibility; smaller molecular weight enhances the miscibility of polymer blend, causing a decrease in the size of micro-domains (Jung et al., 2004). For a low-molecular-weight polymer, the difference in the surface energy between main chain groups and chain end groups particularly influenced the phase-separated structure. The segregation occurs at the surface when the main chain groups are greater than the chain end groups (Ade et al., 1999). This effect is noticeable because of the decrease in the Mn; there is an increase in the number of chain end groups. The molecular weight affects the movement of polymer chains and hence the morphology of phase separation. Mostly, thin polymer films were prepared using solvent techniques such as spin coating and solution casting in which the solvent plays a crucial role in determining the phase-separated morphology. In preparation of coatings, phase separation generally occurs in immiscible blends by the loss of solvent during evaporation. The different solubility parameters of the polymers result in demixing or immiscibility of the polymers as the solvent evaporates. In the non-equilibrium state, the increase in the viscosity of the film captures the phase-separated structure in place. A component having higher solubility improves at the film surface and results in a vertical phase separation. This effect further results in a vertical phase separation with the higher surface tension component on top of the film. As for selective solvent, the phase domain with better solubility is still swollen when the phase domains of the component with less solubility is solidified. Further evaporation of the solvent collapses the swollen phase to a level below that of the vitrified phase and results in relief structures. For the same kind of solvent, the solvent with lower vapor pressure will take a longer time to reach polymer solidification; therefore, the phase separation process is closer to the thermodynamic equilibrium state (Cui et al., 2006; Walheim et al., 1997).

The importance of the study of the morphology of the coatings helps in understanding the mechanism of separation, protection, decoration, filtration, and function. The membrane used for filtration has its importance in solid–liquid separations for colloidal particles, gas mixtures, multi-component solutions, and macromolecules (Cheryan, 1998). Techniques such as stretching, sintering, template leaching, and dip coating are constructed to design the desired morphology. Phase inversion processes such as solvent evaporation, thermal and immersion precipitation, and precipitation from the vapor phase are used to prepare asymmetric membranes.

The membranes of flat sheet type can be formed by coating a porous mechanical support with a thin film of polymeric solution.

5.1.2 Background on Morphology of Coatings

The evaporation of the solvent results in 3D or 2D arrangements of hexagonally ordered pores (Pitois and Francois, 1999). This principle that leads to these types of highly ordered hexagonal pores is Marangoni convection (de Boer et al., 2000; Karthaus et al., 2000; Yabu et al., 2003). The phase separation of a polymer–solvent–non-solvent system by vapor phase is discussed in the literature (Matsuyama et al., 2000).

The morphology of the mixture of polymers was discussed by Burns and Kim (1988) who examined the blends of poly(methyl methacrylate), polystyrene, and methylene chloride (MeCl) prepared using solution casting method. The morphology of the blends was characterized using scanning electron microscopy (SEM). Keeping the molecular weight of polystyrene constant, that is 24,500, the effects of change in the molecular weight of poly(methyl methacrylate) from 39,300 to 56,900 were studied. The SEM images showed a highly ordered structure in the case of higher molecular weight of the poly(methyl methacrylate) with polystyrene than in the case of lower molecular weight of poly(methyl methacrylate).

The complexities created in the morphology of the blend with the addition of one more polymer were examined by Walheim et al. (1999). Their study discussed the phase morphology of the ternary blends (PS–PMMA–PVP) prepared using spin casting onto the nonpolar substrates. The morphological behavior studied via atomic force microscopy (AFM) indicated that the third species (PMMA) in the polymer intercalates between the PS–PVP interfaces and formed a wetting layer. The compatibility of the solvent with the ternary blends played the key role in forming the phase-separated morphology. A selective solvent, that is THF, was used to explain the behavior of the phase-separated morphology. The addition of PMMA formed the quasi-two-dimensional network that separated the PVP and PS domains and led to no wetness and enriched phases at the air or substrate surface.

The coating morphology plays a key role at a nanoscale level, too. It plays an important role in transducers based on surface acoustic wave (SAW) or surface plasmon resonance (SPR). Rapid expansion of supercritical solution (RESS) is a method used to deposit nanoscale polymeric coatings on miniature devices. The coating morphology can be changed by controlling the RESS processing conditions. Levit et al. (2004) deposited the nanoscale polymeric coatings of PDMS by RESS and air brush. The RESS-deposited films exhibited no visible macroscopic imperfections in contrast to air brush in which a series of islands were present with an average size of 10–100 μm. Therefore, the control of the morphology of the coatings facilitates enhancement of the properties of both optical- and acoustic-based chemical sensors.

To obtain coatings with honeycomb structures, star, comb, and block copolymers were used. Cheng et al. (2004) and Cheng et al. (2005a) reported the synthesis of dendronized polymethacrylate-*b*-poly(ethylene oxide) and their block copolymers via atom transfer radical polymerization. The SEM and TEM showed the formation of rods due to the steric hindrance caused by the dendrons. The amphiphilic hybrid linear dendritic rod di-block copolymers were assembled into complex micelles in solution, forming Langmuir–Blodgett films on air–water interfaces and monolayers at surfaces. Further, the authors prepared the honeycomb macroporous films from amphiphilic dendronized block copolymer (Cheng et al., 2005b). The factors such as the concentration of copolymers and the relative humidity in the atmosphere and substrates affected the morphology of the films. Mica, metal-coated glass slide, and glass were used to prepare the porous polymer films. The honeycomb structures are formed on the mica and glass substrates, but not on the silica substrates because the hydrophilic silica made the mold weaker and the formation of pores in the films was difficult to make. The highly ordered structures were formed in the case of high relative humidity, that is 95%, and they decreased with lower relative humidity.

Further study on the morphology of immiscible polymer blend thin films was carried out by Wang and Koberstein (2004). They investigated the morphologies of the thin films prepared from the immiscible mixtures of polystyrene with poly(*tert*-butyl acrylate) (PtBA) using spin coating technique. The morphological behavior of PS/PtBA thin films was analyzed using atomic force microscopy, and the results showed that island-like structures were formed when the ratio of PS to PtBA was less than 1 and holes were formed when that ratio was larger than or equal to 1. François et al. (1995) prepared the first honeycomb structures of solid films of polystyrene solutions on glass plates. Hexagonal 2D arrays of holes with 1–5 μm diameter were formed.

Coffey and Ginger (2005) studied the nanoscale structure in thin polymer films using dip-pen nanolithography (DPN). DPN technique was known to control nanoscale phase separation in films that are 20–60nm thick with sub-150-mm lateral positions. They used the DPN to study the heterogeneous nucleation on surfaces and to screen size-dependent interactions between films and surface features. The spin-coated film of poly-3-hexylthiophene–polystyrene was deposited on the specially prepared substrate of 16-mercaptohexadecanoic acid that was deposited on gold via DPN. The unpatterned gold was passivated using benzenethiol. Atomic force microscopy (AFM) showed domains of rich poly-3-hexylthiophene on the top of the 16-mercaptohexadecanoic acid. Fluorescence microscopy and conductive probe AFM gave the composition of the domains. The results from optical micrographs, AFM, and X-ray photoelectron spectroscopy also showed that on benzenethiol, the films of poly-3-hexylthiophene–polystyrene were laterally homogeneous and, on 16-mercaptohexadecanoic acid, phase-separated films were formed.

The probing inside any material or the surface of the sample is also of keen interest and challenge to the researchers/scientists. The probing is done to investigate the compositional changes with depth or to find out the interfaces that are buried over the overlying material. A new technique ultra-low-angle microtomy (ULAM), developed by Hinder et al. (2005), imparted a cross-sectional, ultra-low-angle taper via polymeric materials such as paints and coatings. ULAM produced tapers that could be used in conjunction with X-ray photoelectron spectroscopy or time-of-flight secondary ion mass spectrometry for compositional depth profiling or buried interface analysis. The authors compared the results of SEM analysis of polyurethane (PU) primer bulk and polyvinylidene fluoride (PVdF) topcoat and their blended multilayered coatings prepared from ULAM processing and a simple microtome knife. The bulk PVdF and PU prepared from ULAM processing showed the two morphological regions each through SEM that were flatter surface contained with smaller granules (~1 μm in diameter). These granules were pigments in PVdF and PU coating formulations. No smearing of the coatings at the PVdF/PU interface clarifies that the ULAM cleaved the coatings in a very precise manner. ULAM also revealed the well-defined boundary in the interfacial region between the coatings.

The characterization of the morphology of polymeric coatings helps in understanding the strong correlation between charge carrier mobility that are measured with thin-film transistors and the number average molecular weight of the conjugated polymers. Kline et al. (2005) studied the morphology of the high-molecular-weight and low-molecular-weight polymeric films of poly-3-hexylthiophene with atomic force microscopy. P3HT was prepared by dissolving it in chloroform, and films were prepared using spin coating method. Poly-3-hexylthiophene with low molecular weight and low mobility showed a highly ordered nanorod structure, while that with high molecular weight and high mobility showed a less ordered isotropic nodule structure.

Wong et al. (2006) prepared polystyrene-block-poly(dimethylacrylamide) (PS-b-PDMA) films using airflow, casting on water, cold stage, and emulsion methods using water droplets as templates. The SEM images revealed the formation of irregular low-quality pores for linear PS. Comb polystyrene formed ordered hexagonal pore arrays in all the techniques and gave the best result with the water casting technique. The amphiphilic di-block PS-b-PDMA did not form films with water casting technique due to the interaction between the hydrophilic block and water droplets. The regular honeycomb structures of the PS-b-PDMA were formed between the flow rates of 0.04 and 0.2 L/min.

Tanaka et al. (2008) discussed the morphology of the polystyrene–poly(methyl methacrylate)–toluene composite and how the morphology changed with the change in the molecular weight. They used gel permeation chromatography to check the molecular weight distributions and observed the blends using the optical micrographs, scanning electron microscopy (SEM), and transmission electron microscopy (TEM) and also measured the interfacial tension by using pendant drop method. The SEM images showed that the degree of protrusion increased with increasing molecular weight and had the snowman-like structure. The TEM micrographs showed the Janus-like morphology with an increase in the molecular weight, which results in the decrease in the interfacial tension. The phase separation occurred at a higher molecular weight of the polymer with lower weight percentage.

The hierarchical ordered porous structures that appeared on the polymeric films have potential applications in environmental sensors, photonic bandgap materials, and patterned light-emitting diodes (Park and Kim, 2004). Breath-figures patterns are the most common structures observed in the morphology of the polymeric coatings. These figures are highly ordered and patterned arrays of micrometer-sized defects in the film and are formed when droplets of solvents condensed onto the polymer solution during the drying of film (Peng et al., 2004). Zander et al. (2007) observed breath-figures patterns in polystyrene films. PS with varying molecular weights (Mn = 10, 19.8, 51, 97.1, 160, 411, and 670 kDa) was dissolved in dichloromethane and was casted in the controlled humidity chamber. At 62% relative humidity, the morphology of the surface was significantly different; that is, at low molecular weight, the films had a large number of defected pores all over the surface with uninterrupted ordering of less than 0.7 μm variation between the pores. However, at high molecular weight, that is 411 and 650 kDa, there was an increase in the pore size. This was due to the high viscosity of the solution because high MW of polymer resulted in the ability of the droplets to properly organize on the polymer surface, creating larger pores and variation in the pores of up to 4 μm.

Ghannam et al. (2007) discussed the self-assembled honeycomb structures of polystyrene with cationic end group that formed star-like structures with carbon disulfide. The films were spread on organic supports (PVC sheets) and inorganic supports (mica and glass). The SEM images showed that the hexagonal structures had a homogeneous pore size of 4 µm on mica and glass. Less ordered structures with much greater pore polydispersity and homogeneous pores of size 1.5 µm were observed on PVC sheets.

Sharifian (2011) studied the conductive polymer system of polyaniline (PANI)–starch–polystyrene–THF composites. FTIR, UV, DSC, and SEM analyses were performed, and their biodegradability tests were performed. The size of PANI played a key role in the granular structure of PANI/starch blends. Nanoscale PANI also showed a high conductivity in PANI/starch blends. All the characterizations showed that the PANI/starch blend provided a higher conductivity as compared to pure starch.

Other than the formational changes in the morphology of the coatings, the colorimetric properties of the coatings are also affected by the morphology. The change in or bleaching of color due to the presence of sufficient electric potential or the process of electron transfer is known as electrochromism. The coatings of conductive polymers are best suited for this type of study due to their high coloration efficiency, and have the light transmissive electrochromic displays used for storage and optical information, and high contrast ratios. Mortimer et al. (2009) studied the colorimetric properties of the electrochromic conjugated di-substituted polymer films of poly(3,4-propylenedioxythiophene) (PProDOT-(Hx)$_2$ and (PProDOT-(2-MeBu)$_2$). The coatings were prepared by spray coating deposition on the indium tin oxide (ITO)-coated glass, and their morphological studies were done by atomic force microscopy. The morphology of (PProDOT-(Hx)$_2$ was smoother with a lower surface area in contrast to (PProDOT-(2-MeBu)$_2$). For (PProDOT-(Hx)$_2$ coatings, smoother and lower surface area was observed. This resulted in the formation of hysteresis from reduction to oxidation, and the perceived color depended on the film thickness and the direction in which the potential was changed. The minimal hysteresis graph was observed in the (PProDOT-(2-MeBu)$_2$) films, which suggests that the anions move in and out of the films more easily.

Nicho et al. (2010) studied the morphology of polystyrene (PS)–poly(3-octylthiophene) (P3OT)–toluene thin films using atomic force microscope (AFM). The results of the investigation declared that PS–P3OT had a lamellar structure having pit formation. The pit depth varied from 46 to 70 nm for 20%–40% of P3OT in PS and reached up to 74 nm at nearly 60% of P3OT in PS.

Ferrari et al. (2011) studied the effect of solvents on the breath-figures prepared from linear polystyrene solutions. The solvents used for making coatings were carbon disulfide, acetone, chloroform, dichloromethane, tetrahydrofuran, methyl ethyl ketone, ethyl acetate, and toluene. The solvents were selected by keeping in mind that they must have partial miscibility or immiscibility in water and a greater or lower density than water. Poorly ordered and irregular breath-figures were obtained from acetone, tetrahydrofuran, ethyl acetate, and acetone; on the other hand, ordered breath-figures were obtained from the solvent using dichloromethane, chloroform, and carbon disulfide.

The morphology of the coatings was also affected by the drying rate and drying conditions of the films. For example, drying rate faster than the phase separation rate results in no formation of membranes. High drying rate with small coating thickness leads to dense polymeric film from the top due to skinning of the film. After the skinning formation, the non-solvent will penetrate into the skin from weak spot and create voids (Shojaie et al., 1994a, 1994b).

The simulation and modeling study were done for the formation of symmetric and asymmetric membranes by Arya (2013) via dry-cast process. He selected the cellulose acetate–water–acetone system, in which water acts as the non-solvent because of the low diffusion out of the other two. He examined the effect of polymer concentration, coating thickness, non-solvent concentration, and air velocity on the formation of membranes. With the increase in the non-solvent (water) concentration, the decrease in the thickness of the coatings results in the increase in the void volume. The thermodynamic instability between the acetone and water created phase separation; hence, asymmetric membranes were formed. Increasing the polymer concentration also showed that the solvent–substrate interaction entered the two-phase region and the solution–air would be out of the phase region forming asymmetric membranes. The thinner coatings formed the thicker skin layer and caused the sublayer structure with a high porosity at the bottom of the membrane. High flow rate favored the phase separation.

Borah et al. (2013) reported the morphology of the block copolymer thin films of polystyrene-b-polydimethylsiloxane (PS-b-PDMS) and polystyrene-b-poly(methyl methacrylate) (PS-b-PMMA) in toluene prepared via spin coating method. The disordered patterns having more defects with the change in the molecular weight of block polymers lower (5000) to higher (104,000) was seen. The annealing procedure also affected the morphology of the phase separation of the high-molecular-weight PS-b-PDMS and PS-b-PMMA.

Veerabhadraiah et al. (2017) fabricated polyvinyl acetate/dimethylformamide (PVAc/DMF) fibers using the electrospinning method. The morphological characterization from SEM of the thin films of PVAc/DMF showed that with the increase in the flow rate from 0.5 to 0.7 ml/h, the number of beads increases with uniform and relatively smoother surface and the fiber density increases. The diameter of the beads formed also increased from 25 and 44 nm to 58 and 67 nm.

Recently, Sharma et al. (2019c) have prepared the coatings of a ternary system, that is polystyrene–polyethylene glycol–chlorobenzene, using solution casting technique. They compared the morphology of the coatings prepared using single thick layer and layer-by-layer techniques. The top side had the porous structure, while the bottom was of the dense nature. By controlling the polymer content, the size of the pore could also be controlled. With an increase in the polymer content, there was a decrease in the pore diameter of the holes in the coatings. The doubling of either of the polymer favored the decrease in the pore diameter of the holes. The prepared coatings were porous in nature in the case of layer-by-layer assembly with uniform pore distribution. The single thick layer was relatively dense as compared to the layer-by-layer coatings. In their another study, the effect of molecular weight of another ternary system, polystyrene–poly(methyl methacrylate)–ethylbenzene, was examined. The effects of molecular weight on the morphological behavior and the glass transition temperature of multi-polymer solvent PS–PMMA–EB coatings were investigated. The average molecular weight of both the polymers resulted in the formation of phase separation (Sharma et al., 2019a). Table 5.1 summarizes the work done and the images of the morphology of the coatings.

TABLE 5.1
Summarized Morphology of the Coatings in the Literature

Authors' Name	Observations	Technique's Name	Results
Walheim et al. (1999)	A ternary system of PS–PMMA–PVP prepared using spin casting method on nonpolar substrates.	Atomic force microscopy (AFM)	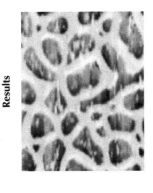

FIGURE 5.1 Atomic force microscopy images of a PS/PMMA/PVP film (1:1:1) on the SAM-covered surface. (Reprinted with permission from Walheim et al. (1999).)

Levit et al. (2004)	Nanoscale coatings of PDMS by air brush and RESS. The results showed no visibility of macroscopic imperfections via RESS method, but in air brush method, a series of islands were present with an average size of about 10–100 μm.	Scanning electron microscopy (SEM)	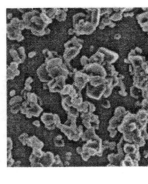

FIGURE 5.2 SEM image of RESS particles. (Reprinted with permission from Levit et al. (2004).)

(Continued)

TABLE 5.1 (*Continued*)
Summarized Morphology of the Coatings in the Literature

Authors' Name	Observations	Technique's Name	Results
Cheng et al. (2005b)	Morphology of the synthesized dendronized polymethacrylate-*b*-poly(ethylene oxide) and their block copolymers using atom transfer radical polymerization.	Atomic force microscopy (AFM) and scanning electron microscopy (SEM)	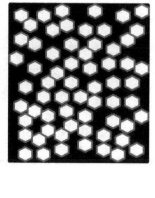
Wang and Koberstein, (2004)	They investigated the morphologies of the thin films prepared from immiscible mixtures of polystyrene with poly(*tert*-butyl acrylate) (PtBA).	Atomic force microscopy (AFM)	

FIGURE 5.3 The honeycomb structures observed in the films on the mica surface. (Redrawn from Cheng et al. (2005b).)

FIGURE 5.4 AFM topographic images of the PS/PtBA films. (Reprinted with permission from Wang and Koberstein (2004).)

(*Continued*)

TABLE 5.1 (Continued)
Summarized Morphology of the Coatings in the Literature

Authors' Name	Observations	Technique's Name	Results
Coffey and Ginger (2005)	The morphology of the spin-coated film poly-3-hexylthiophene–polystyrene deposited on the specially prepared substrate of 16-mercaptohexadecanoic acid was studied.	Atomic force microscopy	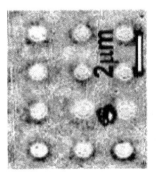
Kline et al. (2005)	They investigated the morphology of the high-molecular-weight and low-molecular-weight polymeric films of poly-3-hexylthiophene.	Atomic force microscopy (AFM)	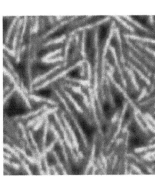

FIGURE 5.5 AFM image of P3HT/PS blend. (Reprinted with permission from Coffey and Ginger (2005).)

FIGURE 5.6 Topography of atomic force microscopy images of low-molecular-weight films. (Reprinted with permission from Kline et al. (2005).)

(Continued)

TABLE 5.1 (*Continued*)
Summarized Morphology of the Coatings in the Literature

Authors' Name	Observations	Technique's Name	Results
			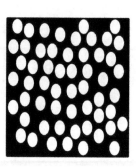 FIGURE 5.7 Topography of atomic force microscopy images of high-molecular-weight films. (Reprinted with permission from Kline et al. (2005).)
Wong et al. (2006)	They investigated the films of polystyrene-block-poly(dimethylacrylamide) (PS–b–PDMA) using airflow, casting on water, cold stage, and emulsion methods using water droplets as templates.	Scanning electron microscopy (SEM)	FIGURE 5.8 SEM images of honeycomb-structured porous films obtained through solvent evaporation of an emulsion/solution of comb polystyrene. (Redrawn from Wong et al. (2006).)

(*Continued*)

TABLE 5.1 (*Continued*)
Summarized Morphology of the Coatings in the Literature

Authors' Name	Observations	Technique's Name	Results
Zander et al. (2007)	They observed the breath-figures in the polystyrene films.	Optical micrographs	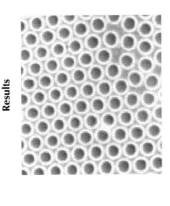 **FIGURE 5.9** Series of optical micrographs of 1 wt.% dicarboxy-terminated PS breath-figures. (Reprinted with permission from Zander et al. (2007).)
Ghannam et al. (2007)	The highly ordered self-assembled honeycomb structures of polystyrene with cationic end group that formed a star-like arrangement with carbon disulfide were observed.	Scanning electron microscopy (SEM)	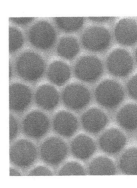 **FIGURE 5.10** SEM images of the honeycomb structure of polystyrene on glass. (Reprinted with permission from Ghannam et al. (2007).)

(Continued)

TABLE 5.1 (Continued)
Summarized Morphology of the Coatings in the Literature

Authors' Name	Observations	Technique's Name	Results
Sharifian (2011)	The ternary system of polyaniline (PANI)–starch–polystyrene–THF composites was studied.	Scanning electron microscopy	

FIGURE 5.11 SEM micrographs of PANI/starch blend. (Reprinted with permission from Sharifian (2011).)

(Continued)

TABLE 5.1 (*Continued*)
Summarized Morphology of the Coatings in the Literature

Authors' Name	Observations	Technique's Name
Mortimer et al. (2009)	The colorimetric properties of the electrochromic conjugated di-substituted polymer films (poly (3,4-propylenedioxythiophene) (PProDOT-(Hx)$_2$ and (PProDOT-(2-MeBu)$_2$) were studied.	Atomic force microscopy (AFM)

Results

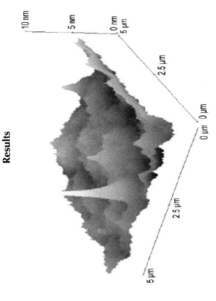

FIGURE 5.12 Three-dimensional AFM height images of a thinner PProDOT-(Hx)$_2$ film. (Reprinted with permission from Mortimer et al. (2009).)

(Continued)

TABLE 5.1 (*Continued*)
Summarized Morphology of the Coatings in the Literature

Authors' Name	Observations	Technique's Name	Results
Nicho et al. (2010)	The morphology of polystyrene (PS)–poly(3-octylthiophene) (P3OT)–toluene thin films was studied.	Atomic force microscopy (AFM)	

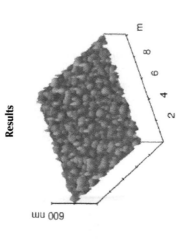

FIGURE 5.13 AFM micrographs of PS/P3OT films spin-cast from toluene. (Reprinted with permission from Nicho et al. (2010).)

(Continued)

TABLE 5.1 (*Continued*)
Summarized Morphology of the Coatings in the Literature

Authors' Name	Observations	Technique's Name
Ferrari et al. (2011)	The effect of solvents in preparing breath-figures from linear polystyrene solutions in solvents acetone, ethyl acetate, chloroform, carbon disulfide, methyl ethyl ketone, dichloromethane, toluene, and tetrahydrofuran was investigated.	Scanning electron microscopy (SEM)

Results

FIGURE 5.14 SEM images of PS solutions in chloroform. (Reprinted with permission from Ferrari et al. (2011).)

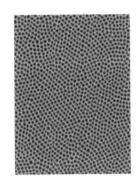

FIGURE 5.15 SEM images of PS solutions in dichloromethane. (Reprinted with permission from Ferrari et al. (2011).)

(*Continued*)

TABLE 5.1 (*Continued*)
Summarized Morphology of the Coatings in the Literature

Authors' Name	Observations	Technique's Name

Results

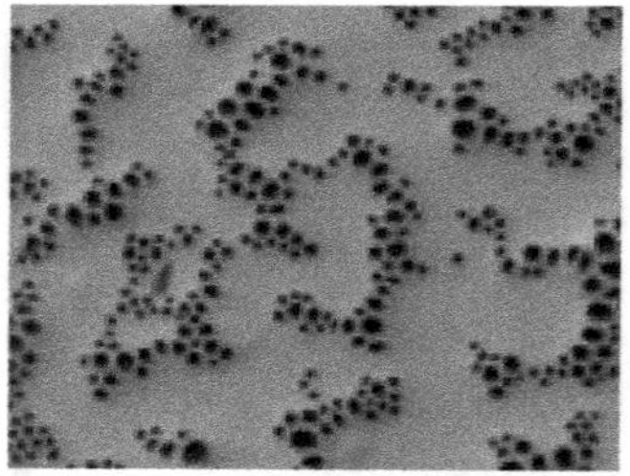

FIGURE 5.16 SEM images of PS solutions in carbon disulfide. (Reprinted with permission from Ferrari et al. (2011).)

FIGURE 5.17 SEM images of PS solutions in methyl ethyl ketone. (Reprinted with permission from Ferrari et al. 2011).)

(*Continued*)

TABLE 5.1 (*Continued*)
Summarized Morphology of the Coatings in the Literature

Authors' Name	Observations	Technique's Name
Arya (2013)	He studied the multi-component system cellulose acetate–water–acetone in which water acts as the non-solvent because of the low diffusion out of the other two. He examined the effect of non-solvent concentration, polymer concentration, coating thickness, and air velocity on the formation of membranes.	Modeling and simulation (theoretical study)

Results

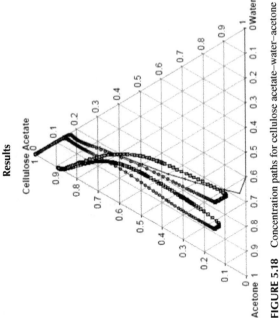

FIGURE 5.18 Concentration paths for cellulose acetate–water–acetone systems. (Reprinted with permission from Arya (2013).)

(*Continued*)

TABLE 5.1 (*Continued*)

Summarized Morphology of the Coatings in the Literature

Authors' Name	Observations	Technique's Name
Borah et al. (2013)	The morphology of the block copolymer thin films of polystyrene-b-polydimethylsiloxane (PS-b-PDMS) and polystyrene-b-poly(methyl methacrylate) (PS-b-PMMA) in toluene was investigated.	Scanning electron microscopy (SEM)
Veerabhadraiah et al. (2017)	They fabricated polyvinyl acetate/dimethylformamide (PVAc/DMF) fibers using the electrospinning method.	Scanning electron microscopy (SEM)

Results

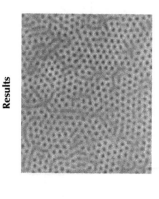

FIGURE 5.19 SEM images of the PS-b-PMMA. (Reprinted with permission from Borah et al. (2013).)

FIGURE 5.20 SEM images of electrospun fibers of PVAc/DMF. (Reprinted with permission from Veerabhadraiah et al. (2017).)

(*Continued*)

TABLE 5.1 (*Continued*)
Summarized Morphology of the Coatings in the Literature

Authors' Name	Observations	Technique's Name	Results
Sharma et al. (2019c)	They prepared the coatings of the ternary system, that is polystyrene–poly(ethylene glycol)–chlorobenzene, using solution casting technique. They compared the morphology of the coatings prepared from layer-by-layer and single thick layer techniques. Films of medium thickness give symmetric films.	Scanning electron microscopy (SEM)	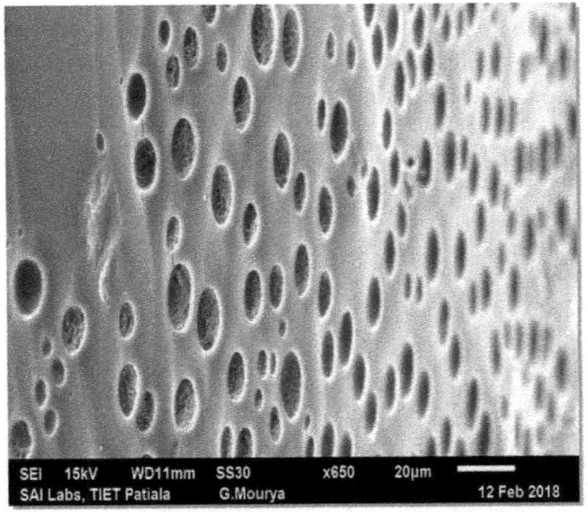

FIGURE 5.21 SEM image showing top surface image, respectively, of percentage having polystyrene-polyethylene glycol-6000-chlorobenzene. (Reprinted with permission from Sharma et al. (2019).)

(Continued)

TABLE 5.1 (Continued)

Summarized Morphology of the Coatings in the Literature

Authors' Name	Observations	Technique's Name	Results
Sharma et al. (2019a)	The effect of molecular weight of another ternary system, polystyrene–poly(methyl methacrylate)–ethylbenzene system, was examined. The effect of molecular weight on the morphological behavior of multi-polymer solvent PS–PMMA–EB coatings was investigated.	Scanning electron microscopy (SEM)	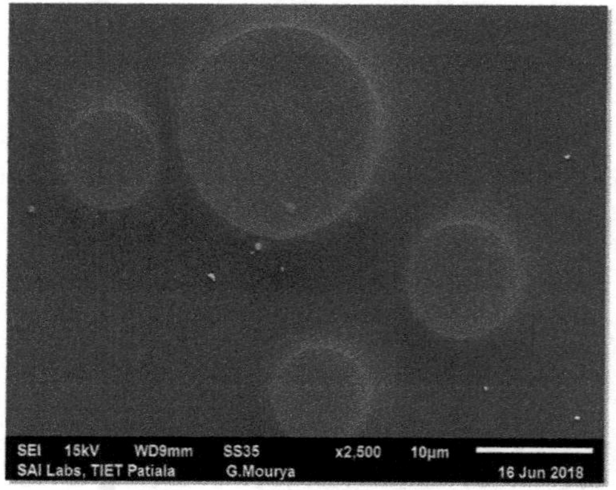

FIGURE 5.22 Top surface of the PSPM-2 films having percentage of PS (192000)-PM (350000)-EB. (Reprinted with permission from Sharma et al. (2019a).)

5.2 CONCLUSIONS

This chapter summarized the various types of polymeric coatings prepared using different coating techniques and different types of polymers and hence the characterization of them using different techniques. This also enables us to understand how a small change in either the environmental or processing conditions can change the overall morphology of the coatings to a great extent. These coatings need to be analyzed for their better applicability in the modern era.

REFERENCES

Ade, H., Smith, A.P., Qu, S., Ge, S., Sokolov, J. and Rafailovich, M., 1999. Phase segregation in polymer thin films: Elucidations by X-ray and scanning force microscopy. *EPL (Europhysics Letters)*, 45(4): 526, https://doi.org/10.1209/epl/i1999-00198-7.

Adiga, S.P., Jin, C., Curtiss, L.A., Monteiro-Riviere, N.A. and Narayan, R.J., 2009. Nanoporous membranes for medical and biological applications. *Wiley Interdisciplinary Reviews: Nanomedicine and Nanobiotechnology*, 1(5): 568–581, https://doi.org/10.1002/wnan.50.

Affrossman, S., AO'Neill, S. and Stamm, M., 1998. Topography and surface composition of thin films of blends of polystyrene with brominated polystyrenes: Effects of varying the degree of bromination and annealing. *Macromolecules*, 31(18): 6280–6288, https://doi.org/10.1021/ma971676q.

Arya, R.K., 2011. Calibration curves to measure concentrations in multicomponent polymeric coatings using confocal Raman spectroscopy. *International Journal of Chemical Engineering and Applications*, 2(6): 421, http://www.ijcea.org/papers/145-A657.pdf.

Arya, R.K., 2013. Phase separation in multicomponent polymer-solvent-nonsolvent coatings. *South African Journal of Chemical Engineering*, 18(1): 30–40, https://www.ingentaconnect.com/content/sabinet/chemeng/2013/00000018/00000001/art00003.

Binder, K., 1994. *Phase Transitions in Polymer Blends and Block Copolymer Melts: Some Recent Developments, Theories and Mechanism of Phase Transitions, Heterophase Polymerizations, Homopolymerization, Addition Polymerization*. Springer, pp. 181–299, https://doi.org/10.1007/BFb0017984.

Borah, D., Senthamaraikannan, R., Rasappa, S., Kosmala, B., Holmes, J.D. and Morris, M.A., 2013. Swift nanopattern formation of PS-b-PMMA and PS-b-PDMS block copolymer films using a microwave assisted technique. *Acs Nano*, 7(8): 6583–6596, https://doi.org/10.1021/nn4035519

Bradley, D., Brown, A.R., Burn, P.L., Friend, R.H., Holmes, A.B. and Kraft, A., 1992. *Electro-Optic Properties of Precursor Route Poly (Arylene Vinylene) Polymers, Electronic Properties of Polymers*. Springer, pp. 304–309, https://doi.org/10.1007/978-3-642-84705-9_56.

Brézin, E., Le Guillou, J. and Zinn-Justin, J., 1976. *Phase Transitions and Critical Phenomena*. Domb and hi-S-Green Eds. (Acadeiuic, New York, 197G) Vol, 5, https://doi.org/10.1016/0021-9614(75)90197-4.

Burns, C.M. and Kim, W.N., 1988. Solution blending of polystyrene and poly (methyl methacrylate). *Polymer Engineering & Science*, 28(21): 1362–1372, https://doi.org/10.1002/pen.760282105.

Cheng, C., Tang, R., Zhao, Y. and Xi, F., 2004. Synthesis of dendronized poly (methacrylates) and their diblock copolymers by atom transfer radical polymerization. *Journal of Applied Polymer Science*, 91(4): 2733–2737, https://doi.org/10.1002/app.13483.

Cheng, C.X., Tang, R.P. and Xi, F., 2005a. Multiple morphologies from a novel diblock copolymer containing dendronized polymethacrylate and linear poly (ethylene oxide). *Journal of Polymer Science Part A: Polymer Chemistry*, 43(11): 2291–2297, https://doi.org/10.1002/pola.20703.

Cheng, C.X., Tian, Y., Shi, Y.Q., Tang, R.P. and Xi, F., 2005b. Porous polymer films and honeycomb structures based on amphiphilic dendronized block copolymers. *Langmuir*, 21(14): 6576–658, https://doi.org/10.1021/la050187d.

Cheryan, M., 1998. *Ultrafiltration and Microfiltration Handbook.* CRC Press, https://books.google.co.in/books?id=LpiuJVxJS_AC&printsec=frontcover&redir_esc=y#v=onepage&q&f=false.

Coffey, D.C. and Ginger, D.S., 2005. Patterning phase separation in polymer films with dip-pen nanolithography. *Journal of the American Chemical Society*, 127(13): 4564–4565, https://doi.org/10.1021/ja0428917.

Cui, L., Ding, Y., Li, X., Wang, Z. and Han, Y., 2006. Solvent and polymer concentration effects on the surface morphology evolution of immiscible polystyrene/poly (methyl methacrylate) blends. *Thin Solid Films*, 515(4): 2038–2048, https://doi.org/10.1016/j.tsf.2006.04.045.

de Boer, B., Stalmach, U., Nijland, H. and Hadziioannou, G., 2000. Microporous honeycomb-structured films of semiconducting block copolymers and their use as patterned templates. *Advanced Materials*, 12(21): 1581–1583, https://doi.org/10.1002/1521–4095.

Decker, C., Masson, F. and Schwalm, R., 2003. Dual-curing of waterborne urethane-acrylate coatings by UV and thermal processing. *Macromolecular materials and Engineering*, 288(1): 17–28, https://doi.org/10.1002/mame.200290029.

Fan, S., Chapline, M.G., Franklin, N.R., Tombler, T.W., Cassell, A.M. and Dai, H., 1999. Self-oriented regular arrays of carbon nanotubes and their field emission properties. *Science*, 283(5401): 512–514, https://doi.org/10.1126/science.283.5401.512.

Ferrari, E., Fabbri, P. and Pilati, F., 2011. Solvent and substrate contributions to the formation of breath figure patterns in polystyrene films. *Langmuir*, 27(5): 1874–1881, https://doi.org/10.1021/la104500j.

Francois, B., Ederle, Y. and Mathis, C., 1999. Honeycomb membranes made from C60 (PS) 6. *Synthetic Metals*, 103(1–3): 2362–2363, https://doi.org/10.1016/S0379-6779(98)00697-3.

François, B., Pitois, O. and François, J., 1995. Polymer films with a self-organized honeycomb morphology. *Advanced Materials*, 7(12): 1041–1044, https://doi.org/10.1002/adma.19950071217.

Gersappe, D., Irvine, D., Balazs, A.C., Liu, Y., Sokolov, J., Rafailovich, M., Schwarz, S. and Peiffer, D.G., 1994. The use of graft copolymers to bind immiscible blends. *Science*, 265(5175): 1072–1074, https://doi.org/10.1126/science.265.5175.1072.

Ghannam, L., Manguian, M., François, J. and Billon, L., 2007. A versatile route to functional biomimetic coatings: Ionomers for honeycomb-like structures. *Soft Matter*, 3(12): 1492–1499, https://doi.org/10.1039/B710282H.

Gutmann, J.S., Müller-Buschbaum, P., Schubert, D., Stribeck, N. and Stamm, M., 1999. Influence of the blend compatibility on the morphology of thin polymer blend films. *Journal of Macromolecular Science—Physics*, 38(5–6): 563–576, https://doi.org/10.1080/00222349908248121.

Hinder, S.J., Lowe, C., Maxted, J.T. and Watts, J.F., 2005. The morphology and topography of polymer surfaces and interfaces exposed by ultra-low-angle microtomy. *Journal of Materials Science*, 40(2): 285–293, https://doi.org/10.1007/s10853-005-6081-7.

Hirtz, M., Lyon, M., Feng, W., Holmes, A.E., Fuchs, H. and Levkin, P.A., 2013. Porous polymer coatings as substrates for the formation of high-fidelity micropatterns by quill-like pens. *Beilstein Journal of Nanotechnology*, 4(1): 377–384, https://doi.org/10.3762/bjnano.4.44.

Hsu, S.-T. and Yao, Y.L., 2014. Effect of film formation method and annealing on morphology and crystal structure of poly (L-lactic acid) films. *Journal of Manufacturing Science and Engineering*, 136(2): 021006, https://doi.org/10.1115/1.4025909.

Israels, R., Jasnow, D., Balazs, A.C., Guo, L., Krausch, G., Sokolov, J. and Rafailovich, M., 1995. Compatibilizing A/B blends with AB diblock copolymers: Effect of copolymer molecular weight. *The Journal of Chemical Physics*, 102(20): 8149–8157, https://doi.org/10.1063/1.469226.

Jung, B., Yoon, J.K., Kim, B. and Rhee, H.-W., 2004. Effect of molecular weight of polymeric additives on formation, permeation properties and hypochlorite treatment of asymmetric polyacrylonitrile membranes. *Journal of Membrane Science*, 243(1–2): 45–57, https://doi.org/10.1016/j.memsci.2004.06.011.

Karthaus, O., Maruyama, N., Cieren, X., Shimomura, M., Hasegawa, H. and Hashimoto, T., 2000. Water-assisted formation of micrometer-size honeycomb patterns of polymers. *Langmuir*, 16(15): 6071–6076, https://doi.org/10.1021/la0001732, https://doi.org/10.1021/ma047415f.

Kline, R.J., McGehee, M.D., Kadnikova, E.N., Liu, J., Fréchet, J.M. and Toney, M.F., 2005. Dependence of regioregular poly (3-hexylthiophene) film morphology and field-effect mobility on molecular weight. *Macromolecules*, 38(8): 3312–3319.

Krausch, G., 1995. Surface induced self assembly in thin polymer films. *Materials Science and Engineering: R: Reports*, 14(1–2): v-94, https://doi.org/10.1016/0927-796X (94)00173-1.

Krebs, F.C., 2009. Fabrication and processing of polymer solar cells: A review of printing and coating techniques. *Solar Energy Materials and Solar Cells*, 93(4): 394–412, https://doi.org/10.1016/j.solmat.2008.10.004.

Ladewig, B. and Al-Shaeli, M.N.Z., 2017. *Fundamentals of Membrane Processes, Fundamentals of Membrane Bioreactors*. Springer, pp. 13–37, https://doi.org/10.1007/978-981-10-2014-8_2.

Levit, N., Pestov, D. and Tepper, G.C., 2004. Influence of polymer coating morphology on microsensor response, chemical and biological point sensors for homeland defense. *International Society for Optics and Photonics*, pp. 203–211, https://doi.org/10.1117/12.516195.

Li, L., Sosnowski, S., Chaffey, C.E., Balke, S.T. and Winnik, M.A., 1994. Surface morphology of a polymer blend examined by laser confocal fluorescence microscopy. *Langmuir*, 10(8): 2495–2497, https://doi.org/10.1021/la00020a001.

Li, X., Wang, Z., Cui, L., Xing, R., Han, Y. and An, L., 2004. Phase separation of PS/PVME blend films induced by capillary force. *Surface Science*, 571(1–3): 12–20, https://doi.org/10.1016/j.susc.2004.07.052.

Liu, J., Yu, X., Xue, L. and Han, Y., 2016. Morphology control of polymer thin Films. *Polymer Morphology*: 299–316, https://doi.org/10.1002/9781118892756

Liu, Y. and Messmer, M.C., 2003. Surface structures and segregation of polystyrene/poly (methyl methacrylate) blends studied by sum-frequency (SF) spectroscopy. *The Journal of Physical Chemistry B*, 107(36): 9774–9779, https://doi.org/10.1021/jp0277513.

Matsuyama, H., Nishiguchi, M. and Kitamura, Y., 2000. Phase separation mechanism during membrane formation by dry-cast process. *Journal of Applied Polymer Science*, 77(4): 776–783, https://doi.org/10.1002.

Mortimer, R.J., Graham, K.R., Grenier, C.R. and Reynolds, J.R., 2009. Influence of the film thickness and morphology on the colorimetric properties of spray-coated electrochromic disubstituted 3, 4-propylenedioxythiophene polymers. *ACS Applied Materials & Interfaces*, 1(10): 2269–2276, https://doi.org/10.1021/am900431z.

Nauman, E.B. and He, D.Q., 1994. Morphology predictions for ternary polymer blends undergoing spinodal decomposition. *Polymer*, 35(11): 2243–2255, https://doi.org/10.1016/0032–3861(94)90757-9.

Nesterov, A., Horichko, V. and Lipatov, Y., 1991. Phase separation of poly (vinyl acetate)-poly (methyl methacrylate) mixtures in thin films. *Die Makromolekulare Chemie, Rapid Communications*, 12(10): 571–574, https://doi.org/10.1002/marc.1991.030121001.

Nicho, M., Peña-Salgado, D. and Altuzar-Coello, P., 2010. Morphological and physicochemical properties of spin-coated poly (3-octylthiophene)/polystyrene composite thin films. *Thin Solid Films*, 518(7): 1799–1803, https://doi.org/10.1016/j.tsf.2009.09.036.

Park, M.S. and Kim, J.K., 2004. Breath figure patterns prepared by spin coating in a dry environment. *Langmuir*, 20(13): 5347–5352, https://doi.org/10.1021/la035915g.

Pathania, A., Sharma, J., Arya, R.K. and Ahuja, S., 2018. Effect of crosslinked polymer content on drying of binary polymer—solvent coatings. *Progress in Organic Coatings*, 114: 78–89, https://doi.org/10.1016/j.porgcoat.2017.10.002.

Peng, J., Han, Y., Yang, Y. and Li, B., 2004. The influencing factors on the macroporous formation in polymer films by water droplet templating. *Polymer*, 45(2): 447–452, https://doi.org/10.1016/j.polymer.2003.11.019.

Pitois, O. and Francois, B., 1999. Formation of ordered micro-porous membranes. *The European Physical Journal B-Condensed Matter and Complex Systems*, 8(2): 225–231, https://doi.org/10.1007/s100510050685.

Prashar, A., Sharma, J., Ahuja, S., Chandra, A. and Arya, R.K., 2019a. Quaternary polymeric coating as an alternative to minimize the use of costly or toxic solvents: Poly (styrene)-ethylbenzene-toluene-*p*-xylene system. *Progress in Organic Coatings*, 127: 319–329, https://doi.org/10.1016/j.porgcoat.2018.11.023.

Prashar, A., Sharma, J., Ahuja, S., Chandra, A. and Arya, R.K., 2019b. Quaternary polymeric coatings as alternative to minimize the use of single solvent: Poly (methyl methacrylate)–ethylbenzene–toluene–acetone system. *Progress in Organic Coatings*, 132: 343–352, https://doi.org/10.1016/j.porgcoat.2019.03.028.

Qian, L. and Zhang, H., 2011. Controlled freezing and freeze drying: A versatile route for porous and micro-/nano-structured materials. *Journal of Chemical Technology & Biotechnology*, 86(2): 172–184, https://doi.org/10.1002/jctb.2495.

Richards, R.W. and Peace, S.K., 1999. *Polymer Surfaces and Interfaces III*. John Wiley, https://doi.org/10.1002/1097-0126(200007)49:7<806::AID-PI419>3.0.CO;2-W.

Salata, R., Pellegrene, B. and Soucek, M., 2019. Migration of fluorinated alkyd and fluorinated tung oil additives for partially self-stratifying coatings. *Progress in Organic Coatings*, 133: 406–417, https://doi.org/10.1016/j.porgcoat.2019.04.017.

Sharifian, I., 2011. Conductive and biodegradable polyaniline/starch blends and their composites with polystyrene. *Iranian Polymer Journal*, 20(4): 319–328, https://www.researchgate.net/publication/267996883.

Sharma, J., Ahuja, S. and Arya, R.K., 2019a. Effect of molecular weight on morphology and thermal properties of poly (styrene)-poly (methyl methacrylate)-ethylbenzene coatings. *Progress in Organic Coatings*, 132: 468–474, https://doi.org/10.1016/j.porgcoat.2019.04.001

Sharma, J., Ahuja, S. and Arya, R.K., 2019b. Experimental designing of polymer-polymer-solvent coatings: Poly (styrene)-poly (ethylene glycol)-chlorobenzene coating. *Progress in Organic Coatings*, 128: 181–195, https://doi.org/10.1016/j.porgcoat.2018.11.036.

Sharma, J., Arya, R.K. and Ahuja, S., 2019c. Drying induced phase separation in poly (styrene)–poly (ethylene glycol)–chlorobenzene system. *Journal of Porous Materials*, 26(4): 1043–1057, https://doi.org/10.1007/s10934-018-0704-2.

Sharma, J., Arya, R.K., Ahuja, S. and Bhargava, C.K., 2017a. Residual solvent study in polymer– polymer—solvent coatings: Poly (styrene)—Poly (methyl methacrylate)—tetrahydrofuran coatings. *Progress in Organic Coatings*, 113: 200–206 https://doi.org/10.1016/j.porgcoat.2017.09.010.

Sharma, J., Tewari, K. and Arya, R.K., 2017b. Diffusion in polymeric systems–A review on free volume theory. *Progress in Organic Coatings*, 111: 83–92, https://doi.org/10.1016/j.porgcoat.2017.05.004.

Shojaie, S.S., Krantz, W.B. and Greenberg, A.R., 1994a. Dense polymer film and membrane formation via the dry-cast process Part I. Model development. *Journal of Membrane Science*, 94(1): 255–280, https://doi.org/10.1016/0376-7388(93)E0228-C.

Shojaie, S.S., Krantz, W.B. and Greenberg, A.R., 1994b. Dense polymer film and membrane formation via the dry-cast process part II. Model validation and morphological studies. *Journal of Membrane Science*, 94(1): 281–298, https://doi.org/10.1016/0376-7388(93) E0229-D.

Simha, R. and Utracki, L., 1967. Corresponding state relations for the Newtonian viscosity of polymer solutions. II. Further systems and concentrated solutions. *Journal of Polymer Science Part A-2: Polymer Physics*, 5(5): 853–874, https://doi.org/10.1002/pol.1967.160050505.

Slep, D., Asselta, J., Rafailovich, M.H., Sokolov, J., Winesett, D.A., Smith, A.P., Ade, H., Strzhemechny, Y., Schwarz, S.A. and Sauer, B.B., 1998. Phase separation of polystyrene and bromo– polystyrene mixtures in equilibrium structures in thin films. *Langmuir*, 14(17): 4860–4864, https://doi.org/10.1021/la9804132.

Srinivasarao, M., Collings, D., Philips, A. and Patel, S., 2001. Three-dimensionally ordered array of air bubbles in a polymer film. *Science*, 292(5514): 79–83, https://doi.org/10.1126/science.1057887.

Tanaka, K., Yoon, J.-S., Takahara, A. and Kajiyama, T., 1995. Ultrathinning-induced surface phase separation of polystyrene/poly (vinyl methyl ether) blend film. *Macromolecules*, 28(4): 934–938, https://doi.org/10.1021/ma00108a021.

Tanaka, T., Nakatsuru, R., Kagari, Y., Saito, N. and Okubo, M., 2008. Effect of molecular weight on the morphology of polystyrene/poly (methyl methacrylate) composite particles prepared by the solvent evaporation method. *Langmuir*, 24(21): 12267–12271, https://doi.org/10.1021/la802287s.

Torrey, J.D. and Bordia, R.K., 2008. Processing of polymer-derived ceramic composite coatings on steel. *Journal of the American Ceramic Society*, 91(1): 41–45, https://doi.org/10.1111/j.1551-2916.2007.02019.x.

Vaessen, D., McCormick, A. and Francis, L., 2002. Effects of phase separation on stress development in polymeric coatings. *Polymer*, 43(8): 2267–2277, https://doi.org/10.1016/S0032–3861(02)00042–3.

Veerabhadraiah, A., Ramakrishna, S., Angadi, G., Venkatram, M., Ananthapadmanabha, V.K., NarayanaRao, N.M.H. and Munishamaiah, K., 2017. Development of polyvinyl acetate thin films by electrospinning for sensor applications. *Applied Nanoscience*, 7(7): 355–363, https://doi.org/10.1007/s13204-017-0576–9.

Walheim, S., Böltau, M., Mlynek, J., Krausch, G. and Steiner, U., 1997. Structure formation via polymer demixing in spin-cast films. *Macromolecules*, 30(17): 4995–5003, https://doi.org/10.1021/ma9619288.

Walheim, S., Ramstein, M. and Steiner, U., 1999. Morphologies in ternary polymer blends after spin-coating. *Langmuir*, 15(14): 4828–4836, https://doi.org/10.1021/la981467e.

Wang, P. and Koberstein, J.T., 2004. Morphology of immiscible polymer blend thin films prepared by spin-coating. *Macromolecules*, 37(15): 5671–5681, https://doi.org/10.1021/ma049664+.

Wengeler, L., Schmidt-Hansberg, B., Peters, K., Scharfer, P. and Schabel, W., 2011. Investigations on knife and slot die coating and processing of polymer nanoparticle films for hybrid polymer solar cells. *Chemical Engineering and Processing: Process Intensification*, 50(5–6): 478–482, https://doi.org/10.1016/j.cep.2010.11.002.

Widawski, G., Rawiso, M. and François, B., 1994. Self-organized honeycomb morphology of star-polymer polystyrene films. *Nature*, 369(6479): 387, https://doi.org/10.1038/369387a0.

Wong, K.H., Hernández-Guerrero, M., Granville, A.M., Davis, T.P., Barner-Kowollik, C. and Stenzel, M.H., 2006. Water-assisted formation of honeycomb structured porous films. *Journal of Porous Materials*, 13(3–4): 213–223, https://doi.org/10.1007/s10934-006-8007-4.

Yabu, H., Takebayashi, M., Tanaka, M. and Shimomura, M., 2005. Superhydrophobic and lipo-phobic properties of self-organized honeycomb and pincushion structures. *Langmuir*, 21(8): 3235–3237, https://doi.org/10.1021/la050013w.

Yabu, H., Tanaka, M., Ijiro, K. and Shimomura, M., 2003. Preparation of honeycomb-pat-terned polyimide films by self-organization. *Langmuir*, 19(15): 6297–6300, https://doi.org/10.1021/la034454w.

Zander, N.E., Orlicki, J.A., Karikari, A.S., Long, T.E. and Rawlett, A.M., 2007. Super-hydrophobic surfaces via micrometer-scale templated pillars. *Chemistry of Materials*, 19(25): 6145–6149, https://doi.org/10.1021/cm0715895.

6 Spectroscopic Analysis of Polymer Coatings

Jyoti Sharma, Manju Rawat, Daisy Sharma, Sanjeev Kumar Ahuja, Avinash Chandra, and Sanghamitra Barman
Thapar Institute of Engineering and Technology

Raj Kumar Arya
Dr. B.R Ambedkar National Institute of Technology

CONTENTS

6.1 INTRODUCTION

Polymeric coatings are made up of polymers (macromolecules) dissolved in solvents and are prepared using various techniques such as solution casting, spin coating, dip coating, drop casting, and film casting. Polymeric thin films are used in a wide range of devices from microelectronics, light-emitting diodes (LEDs), and high-density information storage media to medical implants (Sharma et al., 2017). To have the proper application of any material, one should be very much sure about its compositional and functional properties. The materials at nanoscale and microscale levels are very difficult to analyze at times, and the analysis should be done very carefully and accurately.

Finding the appropriate equipment or technique to have a better understanding of any natural or synthetic materials is always demanding. The spectroscopic techniques bring information at a molecular level that is unavoidable when characterizing any materials, for example nanomaterials, fillers, polymers, and composites. Spectroscopic techniques such as solid-state nuclear magnetic resonance (NMR) (Separovic et al., 2001), infrared spectroscopy (Tang et al., 2007), Raman spectroscopy (Sharma et al., 2019), X-ray photoelectron spectroscopy (Casaletto et al., 2002), and electrochemical impedance spectroscopy (Kaur et al., 2018) have the potential to analyze any material at molecular level (Nagaraja et al., 2019). The evaluation of any material, its interaction with the host, and the compositional and functional properties can also be done using these spectroscopic techniques.

6.1.1 BRIEF INTRODUCTION TO THE SPECTROSCOPY

Spectroscopy is basically defined as the study of the interaction between the electromagnetic radiation and the matter. Electromagnetic radiations are produced from the oscillation of magnetic field and electric charge present on the atom. These can be characterized with their wavelengths or frequencies or wavenumber. Generally, molecular spectroscopy originated from the study of dispersion of visible light by its wavelength, through a prism. The spectral transitions have the important property of position, which is measured in terms of their wavenumber, frequency, width, and intensity.

Spectroscopy is not restricted to the laboratory only, and it has many applications in every field of chemical research and analysis. The wide application of spectroscopy varies from single data point measurements to the structural analysis of complex macromolecules at conformational and configurational level of polymers. The sensitivity of a spectroscopic method can be easily detected by computer-assisted data handling. Through a spectroscopic technique, one can easily analyze the structural differences in polymers at various degrees of stress and is able to analyze the chemical structures at the interfaces and surfaces. Any change in the subcomponent level, either at atomic or at molecular level, gives rise to a change in physical phenomena, which are analyzed and characterized by spectroscopic techniques. The interaction of matter with electromagnetic radiation from microwaves to X-rays gives powerful tools to the researchers. Techniques such as UV–visible absorption of fluorescence help in characterizing an atom's energy levels at bonding conditions. Another spectroscopy, that is mass spectroscopy, allows only the empirical formula and the precise measurements of the mass of a molecule and its fragments. NMR measures magnetic moment that occurs due to the spins of the nuclei of atoms. Infrared and Raman spectroscopy techniques are, respectively, based on the rotational and vibrational energies of the atoms.

Spectroscopy is a very sensitive technique. It works on the molecular level. The collision between atoms and molecules leads to the deformation of any particle and, hence, perturbation, to some extent. This is the reason why molecular interactions are more severe in liquids than in gases and exhibit sharper lines for gases than for liquids. In solids, the randomness of atoms and molecules in any direction is very less and so the solid-phase spectra are often sharp. In liquids and gases, the molecular motion causes the absorption and emission frequencies to show Doppler shift. The randomness of molecules shifts to both high and low frequencies, resulting in the broadening of the spectral lines.

Spectroscopy is a very sensitive technique because it deals with the vibration, rotation, transition change of atoms or molecules in a compound. By changing a very small atom, we can change the entire molecule completely. It is clear now that a subtle structural change can have a great impact and, with the advantages of the spectroscopic techniques, we can have a unique fingerprint of the molecule.

Spectroscopic techniques are bringing information at nanoscale and molecular levels that is totally unavoidable when focusing on polymers, composites, and fillers. The spectroscopic techniques such as solid-state NMR, fluorescence spectroscopy, Raman spectroscopy, and infrared spectroscopy (Bokobza, 2017) have the complete potential for the analysis and evaluation of the filler surface, the interaction between the polymer and the filler particles, and the dynamics of polymer chains at the polymer–filler interface. Nanometer-scale particles formed in the shape of rods, spheres, and plates in a host polymeric medium have gained intense research attention in the development of rubber nanocomposites because they enhance the mechanical, thermal, and electrical properties of the polymer. Polymer coatings can be characterized using these spectroscopic techniques, and each technique can give valuable data and information that can be used further from the application point of view. Every technique has its unique advantages. For example, Raman spectroscopy of polymeric coatings can be used to easily obtain a good spectrum without damaging the sample and with easy handling (Arya et al., 2016).

6.1.2 CLASSIFICATION OF THE SPECTROSCOPIC TECHNIQUES

The classification of the spectroscopic techniques is shown in Figure 6.1.

6.1.2.1 Infrared Spectroscopy

Infrared spectroscopy originated from the transitions produced between the vibrational energy levels of a molecule by the absorption of radiations that belong to the infrared regions. IR spectra are generally shown by those molecules in which a change in the dipole moment was observed. The range of IR region is between 500 and 4000 cm^{-1}.

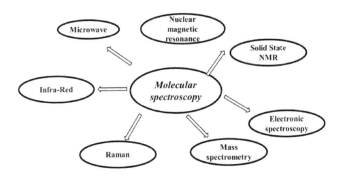

FIGURE 6.1 Classification of molecular spectroscopy.

6.1.2.1.1 Instrumentation

FTIR relies on the frequencies forming the spectrum. It consists of a sample, a laser reference, two mirrors, a beam splitter, a detector and two strategic mirrors as an interferometer, and a detector. A schematic of the FTIR spectrometer is shown in Figure 6.2 (Kumar et al., 2015).

The energy source strikes the beam splitter and then forms two types of beams of almost the same intensity. Out of the two beams, one beam strikes the fixed mirror and then comes back to beam splitter, and the other strikes the moving mirror. A change in the moving mirror makes the total path length variable with respect to the path length taken by the fixed mirror beam. When both the beams are combined at the beam splitter, their path lengths produce destructive and constructive interference, an interferogram.

Then, the combined beam after the beam splitter passes through the sample. The sample absorbs the characteristic wavelengths and eliminates the specific wavelengths with the help of an interferogram. The detector examines the variation of energy versus time for all wavelengths.

Polymers are generally composed of compounds of carbon and hydrogen, that is hydrocarbons. In polymers, carbon atoms are bonded with each other to form long chains, which are thus known as the backbone of the polymer. Polymers that contain only carbon and hydrogen atoms are polystyrene, polyethylene, polybutylene, polypropylene, and polymethylpentene. Sometimes, these hydrocarbons are attached to halogens also; for example, Teflon has fluorine and polyvinyl chloride (PVC) has chlorine attached to the all-carbon backbone. Some of the characteristic frequencies of the functional groups that are present in polymers are represented in Table 6.1. Figure 6.3 shows the FTIR of styrene and polystyrene (Hermán et al., 2015).

FTIR helps in predicting the compositional behavior of the polymers used to prepare polymeric coatings. In the field of polymeric coatings, IR helps in showing the significant changes in the chemical structure of the homopolymeric and

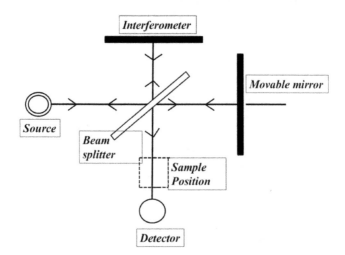

FIGURE 6.2 Instrumentation of infrared spectroscopy (redrawn from Kumar et al., 2015).

TABLE 6.1

Characteristic Wavenumbers of Some Functional Groups

Groups	Approximate Frequencies (cm⁻¹)
–OH	3600
–CH₂	2970 (asymmetric stretch)
	2870 (symmetric stretch)
	1460 (asymmetric deformation)
	1375 (symmetric deformation)
–CH	3000–2850 (stretch)
	1470–1450 (bend)
–C=C	1680–1640
	1100–1050
–C–Cl	725
	1750–1600
	3060

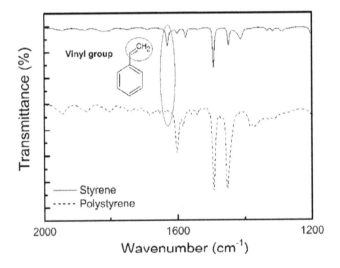

FIGURE 6.3 FTIR spectra of styrene and polystyrene (reprinted with permission from Hermán et al., 2015).

polyaniline–polypyrrole composite coatings (Rajagopalan and Iroh, 2003). FTIR is mostly used as a tool for the study of polymer surfaces (Culler et al., 1984). IR spectra can also give information about the formation of hydrogen bond between α-cyanoacrylate polymer and porous oxide layer on anodically oxidized aluminum surfaces (Suetaka, 1983). Overall, IR spectra give information about the functional groups present in any molecules or composites and also tell about the interfaces and surfaces of a polymer. Saure et al. (1998b) used the FTIR technique to investigate the drying of the polymeric coatings. They prepared coatings of polyvinyl acetate in

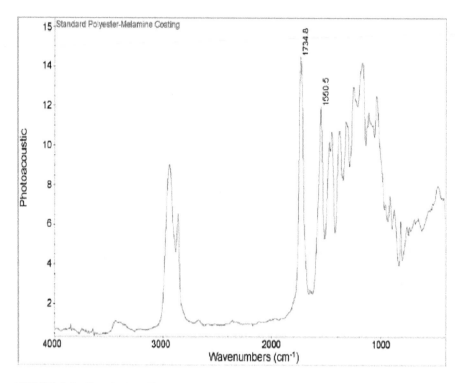

FIGURE 6.4 Standard PA-FTIR of polyester-melanine coatings (reprinted with permission from Zhang et al., 2009).

methanol or benzene on an aluminum substrate at 25°C and 40°C with an air velocity of 5–70 cm/s parallel to the coating surface. Zhang et al. (2009) used a modified version of the PA-FTIR technique, that is step-scan PA-FTIR, for the depth profiling study of two multilayered coatings, which is shown in Figure 6.4. One of the top coats was of polyvinylidene fluoride (PVdF) on a primer polyester melamine coating, and the other was of polyethylene terephthalate (PET) on a polyester melamine coating.

6.1.2.2 Raman Spectroscopy

Raman spectroscopy is also applied for the characterization of polymers. The revolutionary developments in Raman instrumentation made it possible to collect the spectra easily, and it is much more affordable to use than in the past. Polymerization can be characterized using Raman spectra. It is also possible to see how much more sensitive the Raman spectrum is over the IR to the presence of the unpolymerized >C=C< band. Also, orienting the sample and analyzing the polarization, the orientation of the polymers can be deduced. Raman spectra help in monitoring the crystallinity of the polymers. For example, polyethylene terephthalate has a >C=O band that tends to sharpen significantly in the crystalline form.

For a molecule to be Raman active, a rotation or vibration in the molecule should cause a change in polarizability. When an electromagnetic radiation of frequency υ consisting of photons having energy, hυ, strikes the molecule of a sample, then two possibilities are there: Both the collision is perfectly elastic and photon will deflect without any energy change, which is known as Rayleigh scattering. Another possibility is that the collision is inelastic and the molecule gains or loses energy related to the rotational and vibrational energy levels, and this is known as Raman scattering. When a photon is incident on a molecule and then interacts with the electric dipole of that molecule, then this effect is called Raman effect.

6.1.2.2.1 Instrumentation

Figure 6.5 illustrates the Raman spectrometer (Barkur et al., 2015). The laser beam is passed through the cell, generally a quartz or a narrow glass filled with sample. The scattered light from the sample is collected by the lens and passed into a grating monochromator. The signal is measured by a sensitive photomultiplier and is processed by a computer that plots the Raman spectrum.

The arrangement in the Raman spectrometer is also helpful in another type of findings in polymeric coatings. The pinhole arrangement in the Raman spectrometers is used to scatter only the specific types of signals. Figure 6.6 shows the schematic of the confocal Raman spectrometer. The laser beam is focused through the objective lens on a point within the sample at z-plane marked as red. In addition, a pinhole diaphragm is located at the focal length of the objective to make a confocal arrangement. The solid-lined z-plane represents the depth at which the laser is focused, while the dashed-lined planes represent the out-of-focus plane. The beams that are reflected from the focal plane travel through the pinhole aperture and are collected by the detector. The confocal arrangement cuts off the signals from the out-of-focus planes and provides the Raman signals purely from the point of focus.

The focal plane can be moved up and down manually or with the help of a z-motor with a resolution of 1 μm. The base of the coating is located by looking at the picture of the substrate by adjusting the position of platform up and down. The Raman spectra of the substrate must be noisy or give the signal of the substrate material

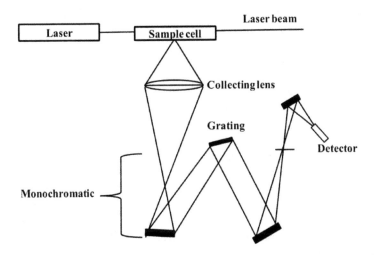

FIGURE 6.5 Schematic diagram of a Raman spectrometer (redrawn from Barkur et al., 2015).

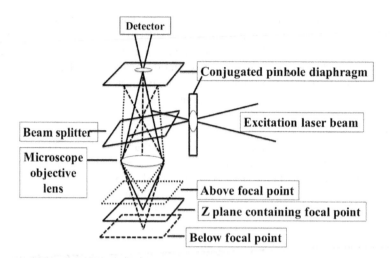

FIGURE 6.6 Schematic representation of the confocal Raman spectrometer (redrawn from Kaiser Optical Systems, Inc., Confocal Raman Microscopy, Raman Products Technical Note, No. 1350).

only, to make sure that the actual focus of the laser beam is on the substrate. The platform is further moved up by 1μm to recheck the spectra if they show any Raman signal, which confirms that the previous position was at the substrate and the laser is focused within the sample. Similar procedure is followed for the coating–air inter-face, too. Detailed depth profiling procedure is given elsewhere (Sharma et al., 2019). A Raman spectrum of pure polystyrene is shown in Figure 6.7 (Sharma et al., 2019), in which a 20X dry lens having a numerical aperture of 0.4 was used. During the measurement, the laser beam passes the space between objective lens and the sample. So, there will be a mismatch between the refractive indexes of both the air and the sample. This mismatch gives rise to some difference between the actual depth of penetration and the position at which the laser is focused. The difference is known as depth resolution and is a function of depth of focus, numerical aperture of the objec-tive lens, and refractive indexes.

In polymer–solvent solution, the diffusion plays an important role and its study is of key interest in the ongoing research. The study of the depth-dependent concentra-tion helps us to understand the processes such as diffusion, corrosion, oxidation, ion implantation, and interface segregation (Hofmann, 1976). The factors such as the penetrant, concentration, temperature, and nature of the polymer and solvent affect the dynamics of the diffusion as well as the overall drying behavior of the coatings (Siebel et al., 2015). The diffusion coefficient is known to decrease with decreasing solvent content by several orders of magnitude.

Earlier, the depth profiling of the coating was studied with the help of ATR-Fourier transform infrared spectroscopy (Saure and Gnielinski, 1994; Saure et al., 1998a), and this technique was limited to few microns. Nowadays, confocal Raman micros-copy is gaining much interest since it analyzes the sample without any deformation. Confocal Raman microscopy is widely used to image, map, and analyze the physi-cal and chemical properties in three dimensions. Confocal Raman spectroscopy is

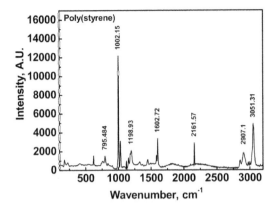

FIGURE 6.7 Raman spectra of pure poly(styrene) of molecular weight 192,000 (reprinted with permission from Sharma et al., 2019).

efficient in the depth profiling of laminates, coatings, membranes, and composites (Williams et al., 1994). Confocal Raman spectroscopy has been successfully used to measure the concentration profiles in dry and wet coatings of rubbery (Arya, 2014; Arya and Vinjamur, 2013; Schabel et al., 2004; Schabel et al., 2003a, 2003b; Siebel et al., 2017; Siebel et al., 2015) as well as glassy nature (Arya et al., 2016).

Everall (2000) derived the expression for predicting the spatial depth, shape, and position of the Raman response along the z-axis of poly(ethylene terephthalate) coated with UV-cured acrylate-based coating. He concluded that the depth of focus changed with the point of focus, and, hence, linearly increased with the depth.

Siebel et al. (2015) measured the concentration-dependent diffusion coefficients directly from the spectroscopic data of single film drying experiments using inverse micro-Raman spectroscopy. Arya et al. (2016) measured the concentration profiles of dried binary coatings of the poly(methyl methacrylate)–ethylbenzene and polystyrene–p-xylene systems using confocal Raman spectroscopy. Arya (2014) also studied the concentration profiles of the wet polymeric coatings of polystyrene–tetrahydrofuran, poly(methyl methacrylate)–tetrahydrofuran, and polystyrene–p-xylene systems using confocal Raman spectroscopy.

Arya and Vinjamur (2013) measured the concentration profiles in two ternary polymer–solvent systems using confocal Raman spectroscopy, that is polystyrene–tetrahydrofuran–p-xylene and poly(methyl methacrylate)–ethylbenzene–tetrahydrofuran systems. Siebel et al. (2017) studied the drying of quaternary coatings of polyvinyl acetate–toluene–methanol–dichloromethane. Their results showed that the initial coating composition affected the drying behavior, and the thickness was more in the case of equal loading as compared to that for the varying solvent loadings. They measured the concentration with respect to depth using confocal Raman spectroscopy.

Schlotter and Rabolt (1984) described in detail the surface of polymer films by waveguide Raman spectroscopy (WRS). The WRS showed the poly(vinyl alcohol)/poly(methyl methacrylate) (PVA/PMMA) laminate for $m=0$, 1, and 2 modes. (Here, m is the mode of vibration.) Yeo et al. (2009) did the probing of mixtures of the polyisoprene (PI) and polystyrene (PS) films using Raman spectroscopy.

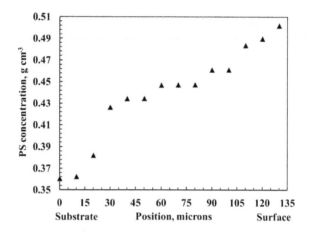

FIGURE 6.8 Depth profiling concentration of poly(styrene) in the poly(styrene) – tetrahydrofuran and poly(methyl methacrylate) – tetrahydrofuran system having initial percentages 5.2%, 6.25%, and 88.55%, respectively. Initial coating thickness was 935 μm. Drying temperature was 25°C (reprinted with permission from Sharma et al., 2019).

Recently, Sharma et al. (2019) have used the confocal Raman spectroscopy technique to measure the depth profiles of the multipolymer–solvent system. They successfully used the Raman binary calibration curves of polystyrene–tetrahydrofuran and poly(methyl methacrylate)–tetrahydrofuran for the depth profiling of the ternary system polystyrene–poly(methyl methacrylate)–tetrahydrofuran. Figure 6.8 shows the depth profiling concentration of polystyrene in the polystyrene–tetrahydrofuran and poly(methyl methacrylate)–tetrahydrofuran systems.

6.1.2.3 Electronic Spectra

These spectra originated from the electronic transitions in a molecule by the absorption of radiations falling in visible and UV regions. The electronic spectra in the visible region and ultraviolet region are 400–800 nm and 190–400 nm, respectively.

It is the measurement of electrons bombarded in and ejected from a sample. This technique provides a tool to study the surfaces up to a depth of 20 Å and is best used in the study of surface chemistry that analyzes friction and wear, catalyst, and adhesion. According to the ejected electron source, it can be classified as X-ray photoelectron spectroscopy (XPS) for chemical analysis. XPS is a powerful analytical tool that provides the quantitative, chemical, and spatially resolved analysis of surfaces of thin films. It is a surface-specific technique and ideal to examine the cause of any type of particular defects in the coatings (Fürbeth and Stratmann, 2001). By combining XPS with the ion beam sputtering, the valuable information about the structure and composition of the film along with the chemistry of the interface region can be obtained (Thomason and Dwight, 1999).

Idla et al. (1997) characterized the synthesized polypyrrole films on mild steel using XPS and AFM. Polypyrrole films were polymerized electrochemically from sodium, p-toluene sulfonate, water, and ethanol solution. XPS detected that a sulfur-rich layer was formed at the interface of the polypyrrole films, which could be due to the accumulation of the *p*-toluene sulfonate on the metal surface. Imine-type nitrogen and high-binding-energy ions were also detected in the polypyrrole films.

Casaletto et al. (2002) prepared hydroxyapatite coatings on polymeric substrates and treated the samples of poly(methyl methacrylate) (PMMA) and poly(2-hydroxyethyl methacrylate) (PHEMA) using the biomimetic method. The surface chemical composition of the samples was studied using X-ray photoelectron spectroscopy and showed the presence of oxygen and carbon and negligible traces of nitrogen, sodium, and silicon. The XPS data indicated that the biomimetic method can be used successfully to obtain good-quality hydroxyapatite coatings on PMMA and PHEMA.

6.1.2.4 Nuclear Magnetic Resonance

Nuclear magnetic resonance (NMR) spectroscopy is a spectroscopic technique used to observe local magnetic fields around atomic nuclei. The sample is placed in a magnetic field, and the NMR signal is produced by the excitation of the nuclei of the sample with radio waves into nuclear magnetic resonance, which is detected with sensitive radio receivers. The schematic representation of the instrumentation is shown in Figure 6.9 (Zia et al., 2019).

Transition occurs between the nuclear spin energy levels of a molecule in an applied magnetic field. Nuclear quadruple resonance result from the transitions between the nuclear spin energy of a molecule arising from the interaction of the unsymmetrical charge distribution in nuclei with the electric field gradients arising from the bonding and non-bonding electrons in the molecule.

Atomic nuclei also behave as spinning magnets. When in contact with a strong magnetic field, these nuclei absorb the radio frequency signal as a function of field strength. For a particular radio frequency which a nucleus absorbs, the magnetic field strength is a function of its immediate electronic environment and the isotopic nucleus. Therefore, the field strength of hydrogen is different for a CH_2 group and a CH_3 group. The NMR of ethanol, hence, shows three different bands for H of CH_2, CH_3, and OH. The most sensitive nuclei are those that behave like strong magnets,

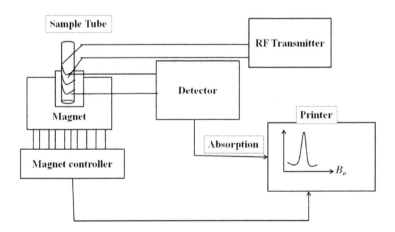

FIGURE 6.9   Instrumentation of NMR (redrawn from Zia et al., 2019).

have high relative isotope abundance, and have nuclear charge distribution approaching spherical distribution.

The important and useful advantages of NMR are polymer characterization and specific identification of the chemical structures of the adjacent groups of a molecule. Hence, it can also be used to distinguish between random and block copolymers and to measure the tacticity of polymer chains. NMR characterizes polymers as monomer sequences (Hirai et al., 1979), stereoregularity (Craver, 1985), and elucidates polymerization mechanisms, and microstructure (Evans et al., 1978). NMR also helps in finding out the formation of covalent bonds Si(Al)-O-Si-C of the alumina and nano-sized silica fillers for polymer reinforcement and scratch-resistant coatings (Bauer et al., 2000). NMR imaging was used to monitor the water uptake of the polyurethane coating (Zhu et al., 2013). Khulbe et al. (2001) used the NMR spectroscopy to determine that poly(2,6-dimethyl-1,4-phenylene oxide) (PPO) radicals in PPO powder and the membranes prepared from it had free radicals that were very much affected by the conditions of the environment.

6.1.2.4.1 Instrumentation

- A sample is placed in the magnetic field. The nuclei excitation produces NMR signal and is detected with the help of sensitive radio receivers.
- Apart from the identification, NMR spectroscopy also provides the information about the dynamics, structure, chemical environment of molecules, and reaction state.
- The most common types of NMR are proton and carbon-13 NMR spectroscopy.

In the field of polymer science, all these spectroscopic techniques are required and researchers use them for pollution monitoring, quality control, analysis of a new product formed, troubleshooting, structural conformation, molecular weight analysis, analysis of functional groups, depth-wise concentration, topographical changes, etc. Due to the high sensitivity of the electronic spectroscopy, it provides conjugated structures that have stronger bands in UV–visible region, in case of more saturated structures.

The spectroscopy based on the dissipating energy behavior of the electronic systems of atoms, molecules, and ions is luminescence spectroscopy. Luminescence is divided into two subdivisions: fluorescence and phosphorescence. Fluorescence formed as an apparent mirror image of the longest wavelength of the absorption spectrum, while in phosphorescence, the energy transition is produced at a lower rate.

In the instrumentation of luminescence spectroscopy, the flexibility of the sample geometry helps in viewing the front surface of the sample, whether it is opaque or geometrically irregular. In case of polymer study, the advantages of luminescence spectroscopy are that it provides measurements for additives in synthetic rubber, monitoring of photooxidation products in highly pure products, and brighteners on synthetic fibers.

Fluorescence spectroscopy helps in studying of aromatic systems, conformational changes as a function of pressure, temperature and kinetics of phase separation in polymer blends and composites and can be investigated in terms of mixing temperature, molecular weight and solvents. Using fluorescence spectroscopy, polymerization reactions can be examined with molecular probes exhibiting viscosity-dependent fluorescence.

6.1.2.5 Mass Spectrometry (MS)

Mass spectrometry requires that the sample under investigation be volatile or produce fragments under high vacuum. In the polymer field, it is principally used to analyze raw materials, residual monomers or solvents, degradation products, and low-molecular-weight oligomers. Instrument capabilities are increasing with the use of high-field magnets and special ionization techniques, so increasing applications to molecular weight determinations are being reported.

6.1.2.5.1 Instrumentation

Ion source: For producing gaseous ions from the substance being studied.

Analyzer: For resolving the ions into their characteristic mass components according to their mass-to-charge ratio.

Detector system: For detecting the ions and recording the relative abundance of each of the resolved ionic species.

In addition, a sample introduction system is necessary to admit the samples to be studied to the ion source while maintaining the high-vacuum requirements (~10^{-6} to 10^{-8} mm of mercury) of the technique, and a computer is required to control the instrument, acquire and manipulate data, and compare spectra to reference libraries. Figure 6.10 shows the schematic representation of mass spectroscopy (Banerjee and Mazumdar, 2012).

6.1.2.6 Electrochemical Impedance Spectroscopy (EIS)

It is a method in which the impedance of an electrochemical system is studied as a function of the frequency of the applied AC wave.

Advantages of EIS: It is particularly suitable for the investigation of corrosion protection by organic coatings for the following reasons. It obtains a more complete view of the process of the deterioration of the paint coating and is able to distinguish the effects that are either concealed or masked in other tests. It obtains precise quantitative data regarding the behavior of the

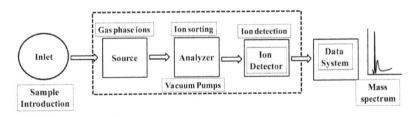

FIGURE 6.10 Instrumentation of mass spectroscopy (redrawn from Banerjee and Mazumdar, 2012)

coatings, given that the data obtained with some of the other tests are rather subjective. It is a faster method for classifying and homologating the protective paint coatings.

EIS is widely used as a standard characterization technique for many material systems and applications (corrosion, plating, batteries, fuel cells, etc.). PC-based modern DSP electronics and software packages now replace lock-in amplifier techniques for implementing EIS. Reinhard et al. (1992) studied the application of corrosion inhibitors in waterborne coatings using EIS. Substitution of mercapto compounds as corrosion inhibitors in acrylate-based dispersion increased with the increase in pH values. EIS also showed that the mercapto compounds were involved as a cross-linking agent during the layer formation.

6.2 APPLICATIONS OF DIFFERENT SPECTROSCOPIC TECHNIQUES IN POLYMERIC COATINGS

Nagaraja et al. (2019) used the different spectroscopic techniques such as FTIR, ^{1}HNMR, and ^{13}CNMR for the structural elucidation of antimicrobial polymeric coatings.

Khulbe and Matsuura (2000) characterized the polymeric membranes using Raman spectroscopy, electron spin resonance, and atomic force microscopy. They found out that after all these three spectroscopic and microscopic techniques, Raman microscopy is best suited for the study of the crystalline structure of the macromolecules and the change in the polymeric structure in membrane. ESR explained the mobility of molecules in membrane polymer matrixes and membrane pores, and atomic force microscope was used for the three-dimensional display of membrane surfaces.

Cao et al. (1998) studied the degradation of polymer coating systems (polyurethane–epoxy-based polymer) by positron annihilation spectroscopy. The Doppler-broadened spectra of positron annihilation were recorded as a function of slow positron implantation energy in a series of polymeric coating systems and then were exposed to UV irradiation for up to 2000 h. The S parameters from the Doppler-broadened energy spectra vs positron energy were calculated using the following equation:

$$S = \frac{\sum_{E_1}^{E_2} N(E_i)}{\sum_{E_i=0}^{\infty} N(E_i)} \tag{6.1}$$

where E_1 and E_2 are 509.41 and 512.59 keV, respectively (the central part). The S parameter showed an increase at very low positron energy (<575 eV) and then a decrease to about 8 keV, followed by a slight increase up to 50 keV. The UV irradiation decreases the S parameter as a function of exposure duration. The significant S parameter decrease was interpreted as the degradation of polymers and the change

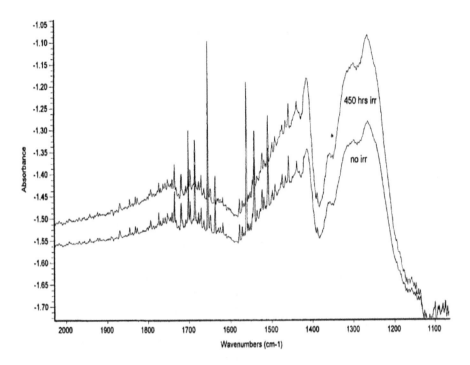

FIGURE 6.11 FTIR absorbance for the 450 h UV irradiated (upper plot) and un-irradiated (lower plot) paint samples (reprinted with permission from Cao et al., 1998).

in the sub-nanometer defect profiles due to UV irradiation. Figure 6.11 shows the FTIR absorbance of the 450-h UV-irradiated (upper plot) and unirradiated (lower plot) paint samples.

Perrin et al. (2015) synthesized and characterized the waterborne corrosion protection coating using polyaniline nanoparticles using EIS, and FTIR analysis showed that polyaniline was doped efficiently with DPA.

Baskar et al. (2014) synthesized poly((E)-(1-(5-(4-(3-(4-chlorophenyl)-3-oxoprop-1-enyl) phenoxy)decyl)-1H-1,2,3-triazol-4-yl)methyl acrylate) and poly((E)-(1-(5-(4-(3-(4-chloro-phenyl)-3-oxoprop-1-enyl)phenoxy)pentyl)-1H-1,2,3-triazol-4-yl) methyl acrylate), cross-linked on photo-irradiation. These polymers were coated on mild steel to protect it from corrosion in 1.0M HCl solution. FTIR analysis showed the presence of photosensitive chalcone moieties. The spectra given by H NMR results confirmed the formation of polymer. Cross-linking of polymer under the influence of photo-illumination was confirmed by absorption spectroscopy. The morphology given by the SEM analysis confirmed that these photo-cross-linking polymers can protect mild steel from corrosion.

Pour-Ali et al. (2014) suggested that the direct addition of polyaniline particles in epoxy matrix resulted in aggregation of the particles. Thus, the polymer was synthesized using a different way. The epoxy matrix was modified using polyethylene glycol (PEG), and polyaniline–camphorsulfonate particles were prepared in the epoxy network. Mild steel was coated using this prepared polymeric coating in order to protect it from corrosion in a 3.5% NaCl medium. The characteristic description of the prepared polymeric coating was done by FTIR and TEM. The anticorrosive nature of the polymeric coating in 3.5% NaCl solution was determined using several techniques such as EIS and salt spray test.

Kocijan et al. (2015) compared and characterized the biocompatible polymer coatings on AISI316L stainless steel. Biofouling reduces the efficiency of materials and contributes to corrosion of metals and alloys. Polydimethylsiloxane (PDMS) and polyethylene glycol (PEG) were used for the surface modification of AISI316L stainless steel.

PDMS induced the hydrophobic nature of the substrate by changing the surface wetting properties. PDMS has a water contact angle greater than 90° and decreases the free energy of the surface. It acted like a water-repellent and protected the surface from corrosion. On the other hand, the water contact angle of PEG was less than 90°, which induced the hydrophilic nature of the system. The free energy of the substrate surface was also increased. PEG has anticorrosive properties, but less than PDMS. X-ray photoelectron spectroscopy determined that PEG coating was slightly thicker than PDMS. Their electrochemical impedance spectroscopy data confirmed that PDMS coatings applied on stainless steel had superior anticorrosion properties than PEG coatings.

Kaur et al. (2018) prepared multilayer coatings by the ex situ addition of cross-linked poly(styrene-co-divinylbenzene) to polystyrene–ethylbenzene solution. The coatings were examined for corrosion protection of mild steel in a simulated water environment equivalent to sea water and acid rain. They used the electrochemical impedance spectroscopy (EIS) technique to determine the corrosion resistance behavior of the prepared polymeric coatings in 3.5 wt % of NaCl aqueous solution. EIS analysis suggested that the sample solution prepared by adding 1% cross-linked polymer in polystyrene–ethylbenzene solution had a better corrosion resistance as compared to the sample solutions prepared by adding 3% and 2% cross-linked polymer in polystyrene–ethylbenzene solution.

6.3 SUMMARY

This chapter briefly described the spectroscopic techniques and their advantages in the field of polymeric coatings. A summary of the works of the researchers using the spectroscopic techniques to characterize and analyze deeply in the field of polymeric coatings has been provided in this chapter.

REFERENCES

Arya, R. K. (2014). Measurement of concentration profiles in thin film binary polymer-solvent coatings using confocal Raman spectroscopy: Free volume model validation. *Drying technology*, 32(8): 992–1002, https://doi.org/10.1080/07373937.2014.880714.

Arya, R. K., Tewari, K., & Shukla, S. (2016). Non-Fickian drying of binary polymeric coatings: Depth profiling study using confocal Raman spectroscopy. *Progress in Organic Coatings*, 95: 8–19, https://doi.org/10.1016/j.porgcoat.2016.02.004.

Arya, R. K., & Vinjamur, M. (2013). Measurement of concentration profiles using confocal Raman spectroscopy in multicomponent polymeric coatings—model validation. *Journal of Applied Polymer Science*, 128(6): 3906–3918, https://doi.org/10.1002/app.38589.

Banerjee, S., & Mazumdar, S. (2012). Electrospray ionization mass spectrometry: A technique to access the information beyond the molecular weight of the analyte. *International Journal of Analytical Chemistry*, 2012, https://doi.org/10.1155/2012/282574.

Barkur, S., Bankapur, A., Pradhan, M., Chidangil, S., Mathur, D., & Ladiwala, U. (2015). Probing differentiation in cancer cell lines by single-cell micro-Raman spectroscopy. *Journal of Biomedical Optics*, 20(8): 085001, https://doi.org/10.1117/1.JBO.20.8.085001.

Baskar, R., Kesavan, D., Gopiraman, M., & Subramanian, K. (2014). Corrosion inhibition of mild steel in 1.0 M hydrochloric acid medium by new photo-cross-linkable polymers. *Progress in Organic Coatings*, 77(4): 836–844, https://doi.org/10.1016/j.porgcoat.2014.01.013.

Bauer, F., Ernst, H., Decker, U., Findeisen, M., Gläsel, H. J., Langguth, H., & Peuker, C. (2000). Preparation of scratch and abrasion resistant polymeric nanocomposites by monomer grafting onto nanoparticles, 1 FTIR and multi-nuclear NMR spectroscopy to the characterization of methacryl grafting. *Macromolecular Chemistry and Physics*, 201(18): 2654–2659, https://doi.org/10.1002/1521-3935.

Bokobza, L. (2017). Spectroscopic techniques for the characterization of polymer nanocomposites: A review. *Polymers*, 10(1): 1–21, https://doi.org/10.3390/polym10010007.

Cao, H., Zhang, R., Sundar, C.S., Yuan, J.P., He, Y., Sandreczki, T.C., Jean, Y.C., Nielsen, B. (1998). Degradation of polymer coating systems studied by positron annihilation spectroscopy. 1. UV irradiation effect. *Macromolecules*, 31(19): 6627–6635, https://doi.org/10.1021/ma9802778.

Casaletto, M., Kaciulis, S., Mattogno, G., Mezzi, A., Ambrosio, L., & Branda, F. (2002). XPS characterization of biocompatible hydroxyapatite–polymer coatings. *Surface and Interface Analysis: An International Journal Devoted to the Development and Application of Techniques for the Analysis of Surfaces, Interfaces and Thin Films*, 34(1): 45–49, https://doi.org/10.1002/sia.1249.

Craver, C. D. (1985). Spectroscopic methods in research and analysis of coatings and plastics. *Paper Presented at the ACS Symp Ser, Appl Polym Sci*, https://doi.org/10.1021/bk-1985-0285.ch029.

Culler, S., McKenzie, M., Fina, L., Ishida, H., & Koenig, J. (1984). Fourier transform diffuse reflectance infrared study of polymer films and coatings: A method for studying polymer surfaces. *Applied Spectroscopy*, 38(6): 791–795, https://doi.org/10.1366/0003702844554512.

Evans, D. L., Weaver, J. L., Mukherji, A. K., & Beatty, C. L. (1978). Compositional determination of styrene-methacrylate copolymers by pyrolysis gas chromatography, proton-nuclear magnetic resonance spectrometry, and carbon analysis. *Analytical Chemistry*, 50(7): 857–860, https://doi.org/10.1021/ac50029a010.

Everall, N. J. (2000). Modeling and measuring the effect of refraction on the depth resolution of confocal Raman microscopy. *Applied Spectroscopy*, 54(6): 773–782, https://doi.org/10.1366/0003702001948439.

Fürbeth, W., & Stratmann, M. (2001). The delamination of polymeric coatings from electrogalvanised steel–a mechanistic approach. Part 1: delamination from a defect with intact zinc layer. *Corrosion Science*, 43(2): 207–227, https://doi.org/10.1016/S0010-938X(00)00047-0.

Hermán, V., Takacs, H., Duclairoir, F., Renault, O., Tortai, J., & Viala, B. (2015). Core double–shell cobalt/graphene/polystyrene magnetic nanocomposites synthesized by in situ sonochemical polymerization. *RSC Advances*, 5(63): 51371–51381, https://doi.org/10.1039/C5RA06847A.

Hirai, H., Koinuma, H., Tanabe, T., & Takeuchi, K. (1979). Polymerization of coordinated monomers. XV. ^{13}C-NMR study on the alternating copolymers of methyl methacrylate with styrene. *Journal of Polymer Science: Polymer Chemistry Edition*, 17(5): 1339–1352. https://doi.org/10.1002/pol.1979.170170509.

Hofmann, S. (1976). Evaluation of concentration-depth profiles by sputtering in SIMS and AES. *Applied Physics*, 9(1): 59–66, https://doi.org/10.1007/BF00901910.

Idla, K., Talo, A., Niemi, H. M., Forsen, O., & Yläsaari, S. (1997). An XPS and AFM study of polypyrrole coating on mild steel. *Surface and Interface Analysis: An International Journal Devoted to the Development and Application of Techniques for the Analysis of Surfaces, Interfaces and Thin Films*, 25(11): 837–854, https://doi.org/10.1002/(SICI)1096-9918(199710).

Kaiser Optical Systems. Inc., Confocal Raman Microscopy, Raman Products Technical Note, No. 1350., from https://image.slideserve.com/696602/raman-microscopy-l.jpg

Kaur, H., Sharma, J., Jindal, D., Arya, R. K., Ahuja, S. K., & Arya, S. B. (2018). Crosslinked polymer doped binary coatings for corrosion protection. *Progress in Organic Coatings*, 125: 32–39, https://doi.org/10.1016/j.porgcoat.2018.08.026.

Khulbe, K., Feng, C., & Tan, J. (2001). *Characterization of Polyphenylene Oxide and Modified Polyphenylene Oxide Membranes Polyphenylene Oxide and Modified Polyphenylene Oxide Membranes*. (pp. 231–303): Springer, https://doi.org/10.1007/978-1-4615-1483-1_8.

Khulbe, K., & Matsuura, T. (2000). Characterization of synthetic membranes by Raman spectroscopy, electron spin resonance, and atomic force microscopy; a review. *Polymer*, 41(5): 1917–1935, https://doi.org/10.1016/S0032–3861(99)00359-6.

Kocijan, A., Conradi, M., Mandrino, D., & Kosec, T. (2015). Comparison and characterization of biocompatible polymer coatings on AISI 316L stainless steel. *Journal of Coatings Technology and Research*, 12(6): 1123–1131, https://doi.org/10.1007/s11998-015-9698–8.

Kumar, D., Singh, B., Bauddh, K., & Korstad, J. (2015). *Bio-Oil and Biodiesel As Biofuels Derived from Microalgal Oil and Their Characterization by Using Instrumental Techniques Algae And Environmental Sustainability*. (pp. 87–95): Springer, https://doi.org/10.1007/978-81-322-2641-3_7.

Nagaraja, A., Puttaiahgowda, Y. M., & Devadiga, D. (2019). Synthesis and fabrication of high-potent antimicrobial polymeric ultrathin coatings. *Journal of Applied Polymer Science*, 136(34): 47893, https://doi.org/10.1002/app.47893.

Perrin, F., Phan, T., & Nguyen, D. (2015). Synthesis and characterization of polyaniline nanoparticles in phosphonic acid amphiphile aqueous micellar solutions for waterborne corrosion protection coatings. *Journal of Polymer Science Part A: Polymer Chemistry*, 53(13): 1606–1616, https://doi.org/10.1002/pola.27602.

Pour-Ali, S., Dehghanian, C., & Kosari, A. (2014). In situ synthesis of polyaniline–camphorsulfonate particles in an epoxy matrix for corrosion protection of mild steel in NaCl solution. *Corrosion Science*, 85: 204–214, https://doi.org/10.1016/j.corsci.2014.04.018.

Rajagopalan, R., & Iroh, J. O. (2003). Characterization of polyaniline–polypyrrole composite coatings on low carbon steel: A XPS and infrared spectroscopy study. *Applied Surface Science*, 218(1–4): 58–69, https://doi.org/10.1016/S0169–4332(03)00579-8.

Reinhard, G., Simon, P., & Rammelt, U. (1992). Application of corrosion inhibitors in water-borne coatings. *Progress in organic Coatings*, 20(3–4): 383–392, https://doi.org/10.1016/0033-0655(92)80026-S.

Saure, R., & Gnielinski, V. (1994). Moisture measurement by FT-IR-spectroscopy. *Drying Technology*, 12(6): 1427–1444, https://doi.org/10.1080/07373939408961014.

Saure, R., Wagner, G., & Schlünder, E.-U. (1998a). Drying of solvent-borne polymeric coatings: I. Modeling the drying process. *Surface and Coatings Technology*, 99(3) : 253–256, https://doi.org/10.1016/S0257–8972(97)00564-1.

Saure, R., Wagner, G., & Schlünder, E.-U. (1998b). Drying of solvent-borne polymeric coatings: II. Experimental results using FTIR spectroscopy. *Surface and Coatings Technology*, 99(3) : 257–265, https://doi.org/10.1016/S0257–8972(97)00565-3.

Schabel, W., Ludwig, I., & Kind, M. (2004). Measurements of concentration profiles in polymeric solvent coatings by means of an inverse confocal micro Raman spectrometer—initial results. *Drying Technology*, 22(1–2): 285–294, https://doi.org/10.1081/DRT-120028234.

Schabel, W., Scharfer, P., & Kind, M. (2003). Measurement and simulation of concentration profiles during drying of thin films with help of confocal-Micro-Raman spectroscopy. *Chemie Ingenieur Technik*, 75(8): 1105–1106, https://doi.org/10.1002/cite.200390274.

Schabel, W., Scharfer, P., Muller, M., Ludwig, I., & Kind, M. (2003). Measurement and simulation of concentration profiles in the drying of binary polymer solutions. *Chemie Ingenieur Technik*, 75(9): 1336–1344, https://doi.org/10.1002/cite.200300055.

Schlotter, N., & Rabolt, J. (1984). Raman spectroscopy in polymeric thin film optical waveguides. 1. Polarized measurements and orientational effects in two-dimensional films. *The Journal of Physical Chemistry*, 88(10): 2062–2067, https://doi.org/10.1021/j150654a025.

Separovic, F., Chau, H., & Burgar, M. (2001). Solid-state NMR study of aging of Colorbond polymer coating. *Polymer*, 42(3): 925–930, https://doi.org/10.1016/S0032-3861(00)00436-5.

Sharma, J., Ahuja, S., & Arya, R. K. (2019). Depth profile study of poly (styrene)–poly (methyl methacrylate)–tetrahydrofuran coatings. *Progress in Organic Coatings*, 134: 297–302, https://doi.org/10.1016/j.porgcoat.2019.05.016.

Sharma, J., Arya, R. K., Ahuja, S., & Bhargava, C. K. (2017). Residual solvent study in polymer– polymer—solvent coatings: Poly (styrene)—poly (methyl methacrylate)—tetrahydrofuran coatings. *Progress in Organic Coatings*, 113: 200–206, https://doi.org/10.1016/j.porgcoat.2017.09.010.

Siebel, D., Schabel, W., & Scharfer, P. (2017). Diffusion in quaternary polymer solutions—model development and validation. *Progress in Organic Coatings*, 110: 187–194, https://doi.org/10.1016/j.porgcoat.2017.05.002.

Siebel, D., Scharfer, P., & Schabel, W. (2015). Determination of concentration-dependent diffusion coefficients in polymer–solvent systems: analysis of concentration profiles measured by Raman spectroscopy during single drying experiments excluding boundary conditions and phase equilibrium. *Macromolecules*, 48(23): 8608–8614, https://doi.org/10.1021/acs.macromol.5b02144.

Suetaka, W. (1983). *Infrared Spectroscopic Investigation of Polymer Coating-Metal Substrate Interaction Adhesion Aspects of Polymeric Coatings* (pp. 225–233): Springer, Boston, MA, USA, https://doi.org/10.1007/978-1-4613-3658-7_11.

Tang, C. Y., Kwon, Y.-N., & Leckie, J. O. (2007). Probing the nano-and micro-scales of reverse osmosis membranes—A comprehensive characterization of physiochemical properties of uncoated and coated membranes by XPS, TEM, ATR-FTIR, and streaming potential measurements. *Journal of Membrane Science*, 287(1): 146–156, https://doi.org/10.1016/j.memsci.2006.10.038.

Thomason, J., & Dwight, D. (1999). The use of XPS for characterisation of glass fibre coatings. composites part A. *Applied Science and Manufacturing*, 30(12): 1401–1413, https://doi.org/10.1016/S1359-835X(99)00042.

Williams, K., Pitt, G., Batchelder, D., & Kip, B. (1994). Confocal Raman microspectroscopy using a stigmatic spectrograph and CCD detector. *Applied Spectroscopy*, 48(2): 232–235, https://doi.org/0003-7028/94/4802-023252.00.

Yeo, B. S., Amstad, E., Schmid, T., Stadler, J., & Zenobi, R. (2009). Nanoscale probing of a polymer-blend thin film with tip-enhanced Raman spectroscopy. *Small*, 5(8): 952–960, https://doi.org/10.1002/smll.200801101.

Zhang, W., Lowe, C., & Smith, R. (2009). Depth profiling of coil coating using step-scan photoacoustic FTIR. *Progress in Organic Coatings*, 65(4): 469–476, https://doi.org/10.1016/j.porgcoat.2009.04.005.

Zhu, H., Huinink, H. P., Adan, O. C., & Kopinga, K. (2013). NMR study of the microstructures and water–polymer interactions in cross-linked polyurethane coatings. *Macromolecules*, 46(15): 6124–6131, https://doi.org/10.1021/ma401256n.

Zia, K., Siddiqui, T., Ali, S., Farooq, I., Zafar, M. S., & Khurshid, Z. (2019). Nuclear magnetic resonance spectroscopy for medical and dental applications: A comprehensive review. *European Journal of Dentistry*, https://doi.org/10.1055/s-0039-1688654.

7 Rheology of Polymer Coatings

M. Gagliardi
NEST, Istituto Nanoscienze-CNR

CONTENTS

7.1 INTRODUCTION

Rheology is the science that studies how a material flows and deforms under the effect of external forces.

Rheological tests identify the flow law and the mechanical constitutive models of raw materials and final products. Such tests are widely used to optimize production processes, being able to predict the long-term behavior of several classes of materials and products.

In the field of polymer coatings, rheological analysis is crucial because the rheological behavior of materials affects all product properties from the production to the storage, application, and durability. Each step among those cited occurs at very different sets of conditions, and then, material rheology should be carefully analyzed to obtain stable and reliable final products.

A polymer coating should be stable in storage conditions, avoiding or significantly limiting flocculation and aggregation of phase changes for a long period and in different environmental conditions. A good coating should be easily applied on its substrate, by means of the technique required for the specific formulation. Then, final macroscopic properties of the surface coating should be smooth and free of wrinkles.

Cited aspects, storage stability, easy application, and smooth surface, determine the quality of the product and are strictly related to rheological analysis, which becomes a key step in product development and monitoring.

In the first part of this chapter, we will explore rheology basics to understand why it is important in coating manufacturing. We will introduce the fundamental terminology and the most common flow rules related to polymeric materials. In the second part, the most important industrial applications of coating rheology will be illustrated to underline the importance of this science in the industrial world.

7.2   FUNDAMENTALS OF RHEOLOGY

Rheology is the science that studies the flow of systems that show a plastic flow rather than elastic deformation under the application of forces. Systems of interest for the rheological analysis are liquid, soft, and solid matter materials.

The main aim of this science is to extend the domains of continuum mechanics (Figure 7.1), analyzing the combination of nonlinear viscous and plastic behaviors in the same system.

Thanks to this comprehensive approach, rheological analysis provides a good prediction of the mechanical behavior of materials at the continuum-scale level, without neglecting material properties at the micro- or nanoscale, such as the molecular size and in-solution molecular architecture and arrangement, or the particle size distribution in a suspension.

Typical fields of interest for rheology are as follows:

* Materials science, in the analysis of polymeric systems, polymer melts, colloids, and solgel systems;
* Paints and coatings development, providing the guidelines for quality check and predicting properties of final products;

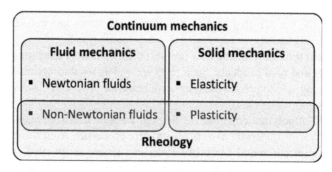

FIGURE 7.1   Schematic representation of the continuum mechanics domain, and the contribution of rheology.

- Food science and cosmetics, improving their stability and helping in the development of packaging and making them easy to use;
- Geology and geophysics, to understand and predict the flow of geological materials at short (e.g., lava) or extended (rheids) timescales;
- Physiology, improving the knowledge of body fluid motion.

Materials with "fluid" characteristics flow under the effect of external loads. A load generates a stress state within the fluid. Such stress is defined as the force per area. As in solid mechanics, different kinds of stresses, such as torsional or shear, that are applied to fluids exist. Consequently, materials can respond with different behavior under the effect of such stresses. Material response depends on some fundamental mechanical properties of the system, mainly depending on stress and strain tensors. It means that each kind of stress induces a different deformation. The identification of a relationship between external loads, internal stresses, and deformation is the core of rheology. Theoretical rheology aims at mathematically explain the relationship between external forces acting on the system and generated internal stresses and strain gradients. Thus, the main outcome of rheological laws is to identify the velocity of the flow arising under the effect of external loads.

Before going in depth, there are some key terms and concepts to clarify.

7.2.1 Introductory Key Terms

In this paragraph, some introductory key terms are defined. Their physical significance will be better clarified in the following sections.

The term *rheology*, or the study of flow and deformation of matter, derives from the Greek words "rheo" (to flow) and "logia" (the study of). It fully identifies the domain of this science, related to the flow rules of solid-like liquids (or liquid-like solids).

Rheological tests are performed to identify the rheological law of a material. The experimental characterization of the rheological behavior of a material is generally indicated as *rheometry*.

In the characterization of a rheological system, the most important material parameter to characterize is called viscosity. The *viscosity* is the resistance of the material to flow or, in other terms, the ratio between the shear stress and the shear rate. Shear stress and shear rate will be defined in the sequel.

Rheometry tests are designed to simulate the actual processing and working conditions that the system undergoes. To have a realistic picture of how the system behaves under specific conditions, several kinds of mechanical loadings can be provided, and the deep knowledge of stress states in rheological systems is of vital importance to maximize the results of rheological characterization. Concerning these amounts, we can identify:

- The *shear stress*, or the force applied to a body in tangential direction;
- The *shear strain*, or the deformation arising after the application of a shear stress;
- The *shear rate*, or the rate of (temporal) change in strain.

Results of rheometry tests are plots of a calculated physical quantity as a function of the applied external load. The most important plots obtained in rheometry are as follows:

- The *flow curve* that is the plot of the shear stress *vs.* the shear rate.
- The *viscosity curve* that is the plot of the viscosity *vs.* the shear rate.

7.2.2 RHEOLOGICAL MODELS

Rheological models are mathematical laws used to describe the viscoelastic behavior of rheological systems. Such mathematical models are developed to simplify the interpretation of experimental data from rheometry tests but also to calculate rheological parameters characteristic of the studied system. Some mathematical models are empirical or semiempirical and are based on the coupling of elastic and viscous elements, to catch both material characteristics.

However, before going in detail, it is fundamental to understand the basic classification of Newtonian or non-Newtonian fluids.

7.2.2.1 Newtonian Fluids (Viscous Liquids)

Isaac Newton formulated the first hypothesis around the resistance of a fluid to motion given by its deformation (Philosophiæ Naturalis Principia Mathematica, 1686).

In his work, Newton stated that "the resistance which arises from the lack of slipperiness of the parts of the fluid, other things being equal, is proportional to the velocity with which the parts of the liquid are separated from one another." The hypothesized "lack of slipperiness" was attributed to the internal friction generated in the fluid. The internal friction was named, for the first time, as viscous friction.

About 150 years later, George Stokes perfected the Newton's theory, studying the forces involved in a system composed of two sliding plates separated by a fluid thin layer. In this experiment, Stokes discovered that the force needed to maintain the motion of the plates is linearly proportional to the differential plate velocity and inversely proportional to the fluid thickness, the viscosity, and the surface separating plates. Moreover, the stress profile across the fluid layer linearly changed, being not dependent on the spatial coordinate.

Observations arising from these experiments indicated that forces are proportional to the rate of change of the velocity in the fluid across all the thickness. Fluids with this linear behavior are known as *Newtonian fluids*.

In other terms, a Newtonian fluid is characterized by a linear dependency of the shear rate upon the shear stress. The mathematical interpretation of a Newtonian fluid behavior indicates that there is a *constant* parameter, the so-called *shear viscosity*, able to relate shear stress and shear strain.

Upon this definition, applied forces in Newtonian fluids are proportional to the rates of velocity vector change in the fluid. Consequently, the viscous stress tensors and strain rate are related by a constant viscosity tensor.

Considering a single element of fluid affected by forces exerted by the motion of surrounding elements (mainly, viscous forces), this fluid element gradually gets deformed over time. Acting forces can be mathematically approximated with a first-order law by a viscous stress tensor τ. At the same time, the deformation of the fluid element with time can be mathematically approximated with a first-order time-dependent strain tensor γ. The strain tensor has a first time derivative called strain rate tensor $\dot{\gamma}$. This tensor describes how the deformation of the fluid element changes with time. Elements of the tensor $\dot{\gamma}$ can be also expressed in terms of velocity gradient Δv.

All the elements of cited tensors depend on the spatial coordinates. Thus, tensors are expressed by 3×3 matrices in the selected coordinate system.

In Newtonian fluids, the stress tensor and the velocity gradient tensor are related by Eq. (7.1):

$$\tau = \mu(\nabla v) \qquad (7.1)$$

μ is a fourth-order tensor, and components do not depend on the velocity or stress state in the fluid.

A particular case of this class of fluids is the isotropic Newtonian fluids. In isotropic Newtonian fluids, mechanical properties do not vary along any direction. In this case, the viscosity tensor is represented by two real coefficients: the resistance to continuous shear deformation and the resistance to continuous expansion/compression.

In anisotropic Newtonian fluids, the tensor μ relates the stress tensor to the deformation rate tensor composed of elements containing spatial derivatives of deformation rate.

7.2.2.2 Non-Newtonian Fluids

Unlike Newtonian fluids, a non-Newtonian fluid does not follow the linear Newton's law of viscosity, and viscosity depends on stress or, most commonly, on shear rate or shear rate history.

In non-Newtonian systems, the relation between the shear stress and the shear rate is not linear and the value of the internal shear stress for a null shear rate could be nonzero.

In this class of systems, the viscosity can be time-dependent. For this reason, it is common to consider the viscosity parameter not adequate for the description of the fluid behavior of non-Newtonian fluid. Generally, such fluids are described through an *apparent viscosity*.

Due to the complexity of their behavior, with several nonlinearities in the constitutive law, the rheological characterization of non-Newtonian fluids is performed in particular conditions, such as in extensional rheometry (applied external loads are purely extensional, without shear) or in oscillatory shear regime (applied external loads follow a sinusoidal time-dependent law).

Non-Newtonian fluids can be classified as follows:

- Systems with non-Newtonian (nonlinear) viscosity;
- Systems with a time-dependent viscosity.

Most common systems with non-Newtonian viscosity are viscoplastic (or Bingham) fluids, shear-thinning (or pseudoplastic) fluids, shear-thickening (or dilatant) fluids, and generalized non-Newtonian fluids.

Viscoplastic fluids have a linear shear stress–shear rate relationship, but the intercept of the curve is not zero. This particular behavior indicates that a finite yield stress is required to activate the fluid flow.

The main characteristic of *shear-thinning fluids* is a decreasing apparent viscosity with increased stress. This nonlinear behavior indicates that the more is the applied external force, the more is the material flow. A shear-thinning behavior is required in paints that should easily flow during their application and limit dropping in the drying phase.

The opposite behavior of shear thinning is the shear thickening. In *shear-thickening fluids*, an increasing apparent viscosity with increased stress is detected. In other words, shear-thickening materials increase their resistance to flow under high stresses, assuming a solid-like behavior.

Finally, in *generalized Newtonian fluids* the viscosity is constant, the shear stress is a discrete or punctual function of the shear strain rate, and the internal stress state does not depend on the shear strain history.

The shear stress *vs.* shear strain rate curves of the described classes of non-Newtonian fluids are illustrated in Figure 7.2.

Systems with time-dependent viscosity are thixotropic fluids and rheopectic fluids.

In thixotropic systems, the viscosity decreases with the duration of the external load application. According to this behavior, thixotropic materials tend to become more fluid after a prolonged applied shear stress (or shear strain).

On the contrary, in rheopectic systems, viscosity increases with the duration of the external load; thus, they become less fluid after prolonged applied loads.

The viscosity *vs.* shear strain rate curves of the described classes of non-Newtonian fluids are illustrated in Figure 7.3.

7.2.3 RHEOMETRY MEASUREMENTS

Experimental apparatus used to evaluate rheological properties of materials are called rheometers. Rheometers can simulate mechanical loads that trigger the rheological

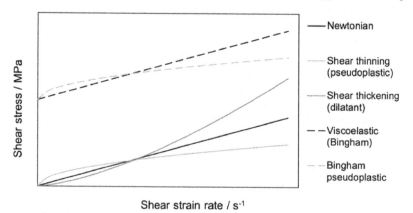

FIGURE 7.2 Shear stress *vs.* shear strain rate functions of non-Newtonian fluids, compared to those of linear Newtonian fluids.

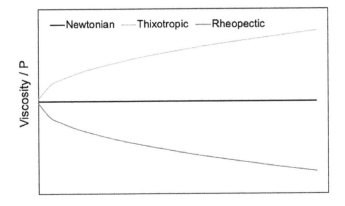

Shear strain rate / s⁻¹

FIGURE 7.3   Viscosity *vs.* shear strain rate functions of non-Newtonian fluids, compared to those of linear Newtonian fluids.

flow of actual systems. The loading scheme is generally obtained by applying constant or time-dependent shear rate functions.

Rheometry measurements enable the calculation of functions and parameters, such as apparent viscosity, shear stress, and complex shear moduli, as a function of the shear rate, the time, or the frequency of load application.

Experimental data from rheometry are used to define the appropriate mathematical models, or the constitutive law of the system, that correlate the shear stress and the shear strain (or the shear strain rate). Constitutive laws are mathematical equations providing the material response, in terms of internal stress state, upon the application of the external applied shear.

The calculated stress tensor depends on the direction and orientation of the applied force with respect to the area where the external load is applied. At the same time, the velocity gradient needs the specification of the direction along which the velocity varies.

In steady shear flow analysis (Figure 7.4), the material is allocated between two flat plates, one moving (top plate) and one fixed (bottom plate). The fluid is sheared by the moving top plate in the direction of the motion.

Indicating with D_x, the displacement of the top plate in the x-direction, and with δ, the distance between plates, the shear strain in the xy plane γ_{xy} is given by:

$$\gamma_{xy} = \frac{D_x}{\delta} \tag{7.2}$$

If the velocity of the top plate in the x-direction is u_x, the velocity gradient, or the shear rate, $\dot{\gamma}$, is given by:

$$\dot{\gamma} = \frac{dv_x}{dy} = \frac{u_x}{\delta} \tag{7.3}$$

The velocity profile v_x across the fluid layer thickness derives from the stretching of the fluid caused by the relative motion of plates. Considering a linear velocity profile,

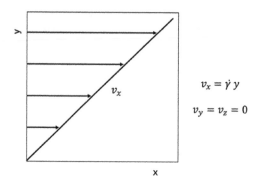

FIGURE 7.4 Steady shear flow; v_x is the velocity profile across the sample thickness that is constant with time.

as that depicted in Figure 7.4, the velocity profile function assumes the values $v_x=v_0$ at $x=0$ (interface with the resting plate) and $v_x=D_R v_0$ at $x=\delta$ (interface with the moving plate), where D_R is the final displacement of the top plate.

In this test, velocity profiles in directions y and z are null.

The deformation of the material between points $x=0$ and $x=\delta$ is indicated with γ_{xx} and is calculated as:

$$\gamma_{xx} = \frac{(D_R v_0 - v_0).\Delta t}{\delta} \tag{7.4}$$

while the deformation in the y-direction is given by $\gamma_{yy}=-\gamma_{xx}$.

The deformation γ is not a scalar. This amount is composed of multiple components in the Cartesian coordinate, and then, it is described by the tensor γ:

$$\gamma = \gamma_{ij} = \begin{pmatrix} \gamma_{xx} & \gamma_{xy} & \gamma_{xz} \\ \gamma_{yx} & \gamma_{yy} & \gamma_{yz} \\ \gamma_{zx} & \gamma_{zy} & \gamma_{zz} \end{pmatrix} \tag{7.5}$$

Consequently, the shear strain rate tensor $\dot{\gamma}$ is the time derivative of γ:

$$\dot{\gamma} = \frac{\partial \gamma_{ij}}{\partial t} = \begin{pmatrix} \dot{\gamma}_{xx} & \dot{\gamma}_{xy} & \dot{\gamma}_{xz} \\ \dot{\gamma}_{yx} & \dot{\gamma}_{yy} & \dot{\gamma}_{yz} \\ \dot{\gamma}_{zx} & \dot{\gamma}_{zy} & \dot{\gamma}_{zz} \end{pmatrix} \tag{7.6}$$

Under the effect of the shear strain caused by the relative plate motion, the material generates a stress with the same direction of the relative plate motion. This stress state, commonly called shear stress, is a tensor, and its components are indicated with τ_{ij}. The stress state generated by the applied deformation is also called deviatoric stress.

As previously stated, shear stress and velocity gradient are correlated through a material function called viscosity η. The mathematical equation describing this correlation is:

$$\tau_{ij} = -\eta \cdot \dot{\gamma}_{ij} \qquad (7.7)$$

In addition to shear stress components, the overall stress state also comprises some additional components orthogonal to the surface over which the external load is applied. Normal components of the stress are given by the hydrostatic pressure and the fluid motion. Such components are two, indicated as N_1 and N_2, and are generally calculated as the difference between tangential stresses:

$$N_1 = \tau_{xx} - \tau_{yy} \qquad (7.8)$$

$$N_2 = \tau_{yy} - \tau_{zz} \qquad (7.9)$$

Components N_1 and N_2 are also indicated as the primary and the secondary normal stress difference, respectively.

Normal stress differences N_1 and N_2 are related to the shear strain rate by means of the following equations:

$$N_1 = -\psi_1 \cdot \dot{\gamma}^2 \qquad (7.10)$$

$$N_2 = -\psi_2 \cdot \dot{\gamma}^2 \qquad (7.11)$$

where ψ_1 and ψ_2 are called primary and secondary normal stress coefficient, respectively.

Material parameters η, ψ_1, and ψ_2 are a function of the shear rate.

Normal stress coefficients are related to the square of the shear strain rate in order to prevent changes in the direction of the shear even for negative y, or by changing the direction of the shear.

In dynamic rheological measures in oscillatory regime (Figure 7.5), the imposed shear varies with time, generally following a sinusoidal law. The oscillatory regime allows calculating the phase angle between the imposed shear strain rate and the measured shear stress.

In this test, the linear viscoelastic region is characterized by a linear proportionality between shear stress and shear strain, and the stress response is a perfect

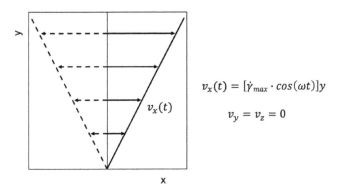

$$v_x(t) = [\dot{\gamma}_{max} \cdot \cos(\omega t)]y$$

$$v_y = v_z = 0$$

FIGURE 7.5 Dynamic (oscillatory) shear flow; v_x is the velocity profile across the sample thickness that depends on time.

sinusoidal function, such as the imposed shear strain function. Also in this test, velocity profiles in directions y and z are null.

In dynamic tests, the oscillatory velocity field is given by:

$$v_x(t) = \left[\dot{\gamma}_{max} \cdot \cos(\omega t)\right] y \tag{7.12}$$

where $\dot{\gamma}_{max}$ is the maximum velocity gradient and ω is the frequency of the sinusoidal applied shear strain.

The ratio between $\dot{\gamma}_{max}$ and ω represents the maximum value of the shear strain, γ_{max}. According to the previous definition of $v_x(t)$, we can describe the shear stress as:

$$\tau_{xy} = \tau_{max} \cdot \cos(\omega t - \varphi) \tag{7.13}$$

where τ_{max} is the maximum value of the shear stress registered during the test and φ is the phase shift between the imposed strain and the measured stress.

The measured oscillatory stress is a complex function in which one term is in phase with the velocity and the other term is out of phase (orthogonal). Combining Eqs. (7.7) and (7.13), the shear stress becomes:

$$\tau_{xy} = -\eta' \cdot \dot{\gamma}_{max} \cdot \cos(\omega t) - \eta'' \cdot \dot{\gamma}_{max} \cdot \sin(\omega t) \tag{7.14}$$

In this equation, η' and η'' are two components of the dynamic viscosities, used to define the complex viscosity η^* as:

$$\eta^* = \eta' - i\eta'' \tag{7.15}$$

The complex viscosity η^* represents the ratio between temporal values of the shear stress and the shear strain:

$$\eta^* = \frac{\tau_{xy}(t)}{\dot{\gamma}_{xy}(t)} \tag{7.16}$$

When the shear strain rate approaches zero and the frequency of the applied shear is low, η' is close to values of viscosity calculated in steady shear flow regime and it is referred to the dynamic viscosity.

Eq. (7.14) can be rewritten as a function of the complex shear modulus as:

$$\tau_{xy} = -G' \cdot \gamma_{max} . \cos(\omega t) - G'' \cdot \gamma_{max} \cdot \sin(\omega t) \tag{7.17}$$

In this equation, G' and G'' are, respectively, the storage modulus and the loss modulus, identified for the calculation of the complex shear modulus G^*:

$$G^* - G' - iG'' \tag{7.18}$$

Complex components of the modulus G^* are related to the dynamic viscosity coefficients through the frequency of the applied shear strain:

$$G' = \omega\eta' \tag{7.19}$$

$$G'' = \omega\eta'' \tag{7.20}$$

To understand the meaning of such parameters, we need to analyze their relationship with the stress response after the application of the external strain loading.

According to Eq. (7.15), G' is linearly correlated with the stress in phase with the applied strain; thus, it is related to the recoverable elasticity of the material. On the other hand, G'' is related to the orthogonal response of the stress to the applied strain, resulting in the dissipative energy losses in the material. According to these definitions, G' approaches the shear modulus of an ideal rubbery material, in which the stress response is perfectly in phase with the applied shear strain. On the contrary, G'' approaches the shear modulus of a purely viscous material, in which there are no elastic recovery phenomena and the applied energy is completely dissipated by viscous flow.

Viscoelastic dynamic moduli G' and G'' are generally dependent on the frequency of the applied strain, in particular in macromolecular materials. Some materials tested under a frequency scan show a point called *crossover*. At the crossover, G' and G'' assume the same value. In materials with this crossover, one modulus dominates the other at low frequencies and vice versa at high frequencies.

7.2.4 CONSTITUTIVE RHEOLOGICAL MODELS

Constitutive rheological models are defined for each class of materials.

For Newtonian fluids that are characterized by only one parameter, the constitutive equation calculated by means of rheological tests can describe and predict the flow behavior in all laminar flows. In this particular case, the shear stress is given by:

$$\tau = -\eta \cdot \dot{\gamma} \tag{7.21}$$

Here, τ and $\dot{\gamma}$ are both tensors (note that the conventional bold symbol is used).

According to the definition given in Eq. (7.7), the tensor τ, also called deviatoric stress tensor, is given by:

$$\tau = \tau_{ij} = \begin{pmatrix} \tau_{xx} & \tau_{xy} & \tau_{xz} \\ \tau_{yx} & \tau_{yy} & \tau_{yz} \\ \tau_{zx} & \tau_{zy} & \tau_{zz} \end{pmatrix} \tag{7.22}$$

Now, we introduce the definition of the strain rate tensor $\dot{\gamma}$, according to the definition given in Eq. (7.1):

$$\dot{\gamma} = \nabla V + (\nabla V)^T \tag{7.23}$$

In this equation, ∇ is the gradient operator, ∇v is the velocity gradient tensor, and the superscript T is the transpose operator. The velocity gradient reads:

$$\nabla V = \begin{pmatrix} \dfrac{\partial v_x}{\partial x} & \dfrac{\partial v_x}{\partial y} & \dfrac{\partial v_x}{\partial z} \\[2ex] \dfrac{\partial v_y}{\partial x} & \dfrac{\partial v_y}{\partial y} & \dfrac{\partial v_y}{\partial z} \\[2ex] \dfrac{\partial v_z}{\partial x} & \dfrac{\partial v_z}{\partial y} & \dfrac{\partial v_z}{\partial z} \end{pmatrix} \tag{7.24}$$

According to Eq. (7.23), components of the strain rate tensor are defined as:

$$\dot{\gamma}_{ij} = \frac{\partial v_i}{\partial x_j} + \frac{\partial v_j}{\partial x_i} \tag{7.25}$$

where v_i and v_j are the Cartesian components of the velocity vector $\mathbf{v}$ and x_i and x_j are the components of the position vector $\mathbf{x}$.

Substituting the definition given in Eq. (7.25) in Eq. (7.7) for a steady shear flow (see Figure 7.4), we obtain:

$$\tau_{xy} = -\eta \dot{\gamma}_{xy} = -\eta \frac{\partial v_x}{\partial y} \tag{7.26}$$

in which the component v_y is null in this kind of test.

Once the parameter η is known for the fluid of interest, Eq. (7.26) can be used to predict every kind of laminar flow in systems with a Newtonian behavior.

The Newtonian model can be generalized for non-Newtonian fluids, introducing the viscosity parameter as a function of the shear rate. To do this, it is necessary to calculate the invariants of the deformation rate tensor. The second invariant of this tensor is the only one that takes both hydrostatic pressure and shear stresses into consideration. In fact, the first invariant is null for incompressible fluids, while the third is null in shear-governed flows.

The second invariant of the shear rate tensor is calculated as:

$$J_2 = \sum_i \sum_j \dot{\gamma}_{ij} \cdot \dot{\gamma}_{ji} \tag{7.27}$$

Viscosity is calculated as the square root of the halved second invariant, in order to have a compliant magnitude of this amount, resulting in:

$$\eta = \sqrt{\frac{1}{2} J_2} = \sqrt{\frac{1}{2} \sum_i \sum_j \dot{\gamma}_{ij} \cdot \dot{\gamma}_{ji}} \tag{7.28}$$

Substituting Eq. (7.28) in Eq. (7.26), it is possible to calculate the laminar flow in generalized Newtonian systems.

The generalized Newton model can be difficult to apply in some cases from a mathematical point of view, such as in transient flows or in fluids with not negligible normal stresses. Then, some semiempirical generalized flow laws were developed with a limited number of parameters, generally calculated by fitting of experimental data.

One of the most used generalized flow models for the prediction of the viscosity is:

$$\eta = K\dot{\gamma}_{xy}^{(n-1)} \tag{7.29}$$

where K and n are two fitting parameters.

This power law is simple to manage, thanks to the low number of parameters involved, and reliable enough to analyze the behavior of a large number of systems, in particular in the inventory of polymer solutions. On the other hand, this model fails for low-flow-rate regimes. In low-flow regimes, the Carreau model can be useful:

$$\frac{\eta - \eta_\infty}{\eta_0 - \eta_\infty} = \left[1 + (\lambda\dot{\gamma}_{xy})^2\right]^{\frac{n-1}{2}} \tag{7.30}$$

In this equation, η_∞ and η_0 are viscosities at high and low shear rate, respectively; λ is a fitting parameter calculated at the shear rate of interest; and n is the corresponding parameter reported in Eq. (7.25).

7.3   PARAMETERS AFFECTING RHEOLOGICAL PROPERTIES

Rheological properties are significantly affected by specific external or internal conditions. In addition to the kind of external applied load and the loading frequency, the most relevant properties affecting rheology are as follows:

- Temperature;
- Micro- or nanostructure of the system.

7.3.1   EFFECT OF TEMPERATURE

Temperature is probably the most important parameter affecting rheological properties of systems. Viscosity widely depends on the temperature, and this dependence is generally expressed as a separate function from the effect of other parameters (e.g., the shear rate or the shear frequency):

$$\eta(\dot{\gamma}, T) = \eta(\dot{\gamma}) \cdot f(T) \tag{7.31}$$

The simpler form of the function $f(T)$ commonly used is:

$$f(T) = \exp[-a \cdot \Delta T] \tag{7.32}$$

In this equation, the parameter a is called temperature sensitivity of the viscosity, while the temperature difference is calculated between the working temperature and the temperature at which viscosity is known.

7.3.2 STRUCTURE OF THE SYSTEM

The structure of the system is particularly important in macromolecular formulations, e.g., in polymer solutions or melts.

Basically, intramolecular and intermolecular interactions determine mechanical properties of a polymer at the macromolecular scale.

Polymeric chains are obtained by covalently attaching a large number of repeating units. Intramolecular forces acting in polymers are prevalently related to the C–C bond (except, of course, in the presence of heteroatoms in the main polymer chain).

The energy of the C–C bond is around 350 kJ/mol, leading to a theoretical stiffness of 300–400 GPa. In the real case, polymer stiffness is considerably lower than the theoretical value. The difference between theoretical and actual stiffness is generally attributed to the macromolecular misaligning. This observation leads to consider also the intermolecular interactions as an important factor in polymer strength. Intermolecular interactions are mainly accounted to be van der Waals forces, and their entity is inversely proportional to the sixth power of the distance between atoms. Thus, in the presence of solvents or plasticizers, or when heated, the distance between two macromolecules increases, and the free volume increases; thus, intermolecular interactions, and consequently stiffness, decrease.

The degree of polymerization, or the number of repeating units per macromolecule, is another factor affecting macromolecular interactions.

Polymers with a low degree of polymerization are liquid or very soft at room temperature while, increasing the degree of polymerization, they gradually become stiffer at the same temperature and with the same additives. However, while the stiffness rapidly tends to a plateau, the viscosity steadily increases with molecular weight, generally with a power law (Figure 7.6).

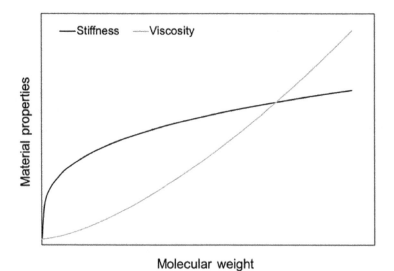

FIGURE 7.6 Variation of stiffness and viscosity in polymers by increasing the molecular weight.

In systems containing macromolecular coils, soft particles, or vesicles, rheological properties are also affected by deformation of these components.

As a general rule, deformable structures dispersed in a liquid phase tend to get deformed in the direction of load application, while non-deformable structures tend to get aligned in the load direction.

In both cases, the most general rule is that systems become shear thinning during load application, while thixotropy dominates when loading is stopped, and systems tend to return thicker with time.

7.4   INDUSTRIAL USES OF RHEOLOGICAL ANALYSIS

Industrial systems are often related to fluid flow and rheology. In this inventory, polymer coatings and polymer-based paints are the major example of industrial rheological systems.

The terms "coating" and "paint" are synonymous. However, coating commonly refers to industrial applications such as in automotive applications, food cans, and papers. On the other hand, paint commonly refers to architectural coatings such as wall and house paints, for both interior and exterior surfaces.

Paints and coatings are used for a variety of applications, from surface protection to aesthetic decorations. From production to final use, paints and coatings formulations, containing solvents, fillers, resins, and pigments, should remain uniformly mixed. These products should remain stable during production and storage and then should be easily applied providing smooth surfaces. For all these reasons, rheological properties of paints and coatings must be carefully analyzed and optimized.

In this section, we will analyze some practical concerns of polymer-based paints and coatings and their rheological characterization, in order to understand how rheology is so important in quality check and product performances.

7.4.1   COMMERCIAL FORMULATIONS OF COATINGS AND PAINTS

Formulations for coatings and paints are colloidal suspensions that solidify to give homogeneous solid films. Conventional formulations contain solvents, plasticizers, inorganic fillers, a polymeric film former, and pigments. More innovative formulations also contain additives to improve some performances, such as stability, surface properties, and durability.

From their production to the final use, such formulations should remain uniformly mixed and stable also after prolonged storage, also in different conditions. The ideal formulation should be pumpable, stirrable, mixable, and dispersible to fulfill all the requirements needed in the whole life cycle.

7.4.1.1   The Importance of Rheology in Paints and Coatings Formulations

Rheological properties of paints and coatings formulations are crucial for their performances and quality; thus, their optimization is vital. Generally, such formulations are developed to be non-Newtonian fluids, in order to have more than one parameter (e.g., temperature) governing their behavior.

Production processes, transportation, and storage, as well as their application on the substrate, imply different shear stress regimes. Thus, formulations have to be optimized to be correctly manipulated, and to ensure correct product performances at all stages.

During the production process, components and products are enlivened across the plant piping, and the flow regime implies moderate-to-high shear stresses.

The storage of final products occurs in a static and shear stress-free manner, in which solid components should maintain their colloidal nature avoiding sedimentation, filler precipitation, or demixing.

Transportation is a low-shear-stress process in which formulations are only subjected to vibrations.

During the final use, formulations should be easily spreadable by brush or roller, or sprayable. Thus, shear stresses related to their application vary on the basis of the selected application process: as in the case of brush spreading (moderate-to-high shear) to spraying (very high shear).

After application and before drying or curing, colloidal formulations undergo surface tension and gravitational forces. During this step, formulations should show a surface-leveling behavior and the internal microstructure should ideally recover after the application of shear stresses. The recovery time should be adequate for deaeration, to give a smooth and droplet-free surface. During solidification, when the coating is almost dried, sagging should be avoided or significantly limited.

Each process is important, and the *perfect* scenario would be to have the optimal viscosity at each stage to meet all requirements.

7.4.1.2 Rheometry of Paints and Coatings

Such very complex systems cannot be rheologically described by a single-point measure, acquired at an arbitrary shear strain rate regime. As a general rule, rheological measurements are difficult to compare to the actual processes that paints and coatings undergo during their overall life cycle. It is due to the variety of external factors occurring in real applications which can be very complex or impossible to "simulate" in a rheometry experimental test.

At the same time, a comprehensive rheological characterization should be performed in both continuous and dynamic regimes, considering different shear strain rates, and thus applying different rheometry techniques. It is because in a real scenario, paints and coatings are subjected to large variations of viscosity.

Viscosity as a function of the shear rate in different processes involved in paints and coatings formulation/storage/application is qualitatively represented in Figure 7.7.

According to Figure 7.7, formulations with high low-shear viscosity are more stable to sedimentation, while formulations with high high-shear viscosity form thicker layers after spreading.

As a general rule, the characterization at low-shear regimes is associated with how the paint or the coating behaves during leveling, sagging, and drying. At low shear regimes, viscous paints are prone to form brush marks after their spreading. Such behavior is due to the high leveling stability of viscous formulations. Too low viscosities at low shear are accounted to paint flow off the surface, resulting in the inapplicability of the product.

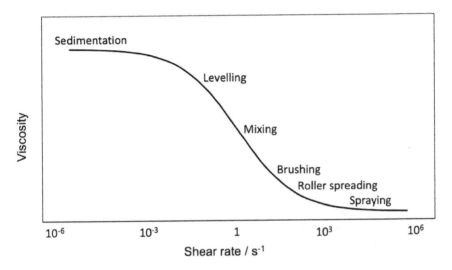

FIGURE 7.7 Qualitative trend of viscosity in paints and coatings formulations, at the different shear rates imposed in processing and application; values range from high to low from sedimentation to spraying.

A preliminary rheological analysis of paints and coatings formulation can be performed in continuous flow, in order to calculate some basic parameters.

Formulations for paints and coatings generally behave as a Bingham fluid. Thus, an important parameter that can be evaluated in continuous flow is the yield stress. Its value can be used to discriminate whether a formulation can be manually spread (e.g., by brush) or needs to be applied with different techniques (e.g., spraying). It is because the yield stress is the value of loading that a formulation can withstand before it starts flowing. This value of stress is commonly indicated as static yield stress. The value of the static yield stress increases with time, during the drying process.

However, yield stress of paints and coatings formulations can be evaluated not only in continuous flow but also by different and more complex rheological tests, e.g., in dynamic conditions. Unlike the static yield stress, indicating the point at which formulations start flowing, the dynamic yield stress is related to the process occurring when a flowing fluid comes to a rest state. The value of the dynamic yield stress can be evaluated from the fitting of experimental flow curves, by using appropriate models (e.g., the power law or the Carreau empirical models).

After the application of the formulation on its substrate, rheology can help to understand the leveling, sagging, and drying processes.

The appropriate rheological characterization to catch aspects related to leveling is obtained with oscillatory measurements, as for all non-Newtonian fluids, coupled to a preshear conditioning. The preshear step is able to simulate the spreading of the formulation over the substrate surface, while the oscillatory low stress applied at a fixed frequency can simulate the rebuilding of the microstructure. The parameter to monitor is viscosity, which should increase with the analysis time. It is expected that the system should show a shear-thinning behavior once the prestress is imposed, to have a simple spreading over the substrate surface, and a shear-thickening behavior once the prestress is released, providing a flow at low-shear regimes, to maximize leveling properties.

In this step of characterization, the G'' modulus is expected to dominate the G' part at low-shear-strain regimes.

After the characterization of parameters related to leveling, a further step of analysis consists in the evaluation of rheological parameters affecting sagging. The formulation should flow at low shear to provide the surface leveling, but this flow should be limited to avoid sagging. This behavior is obtained in materials in which the loss modulus G'' is larger than G' after the preshear condition. It ensures a correct leveling. Then, after a finite time, the paint should show a crossover of G' and G'' moduli, with the G' part dominating after this crossover. This behavior is associated with a *sol-to-gel* transition. After the *sol-to-gel* transition, the liquid formulation becomes more viscous and stops flowing. It means that after a specific timescale, the mechanical behavior of the paint becomes elastically dominated.

Sagging is mainly governed by gravitational loads. The gravitational load that triggers the flow of the applied formulation can be easily calculated. The liquid layer formulation can be considered as spread over an inclined plane (Figure 7.8), and the shear stress distribution profile results in a maximum value at the interface with the substrate, while it is null at the free coating/air interface.

Gravitational effects increase with formulation density, film thickness, and inclination angle. The maximum shear rate, at the substrate/coating interface, increases with a decreasing viscosity. Formulation flow can be minimized by increasing the viscosity at the typical shear stress of the coating/substrate system until the colloidal solution is dried.

Under the value of the apparent yield stress of the system, an optimized formulation behaves like a solid, forming a sort of network given by molecular interactions. An external load larger than the apparent yield stress of the formulation (e.g., the load applied during brushing) can break the network microstructure and can activate the fluid flow. A formulation with this characteristic can prevent sagging but, at the same time, can be easily spread over the substrate.

7.4.2 Thickeners as Rheology Modifiers

Paints and coatings formulations also contain some additives called thickeners. Such additives, also known as rheology modifiers, are used to tailor the viscosity of the final product, to make it more suitable for production, spreading, leveling, and sag resistance.

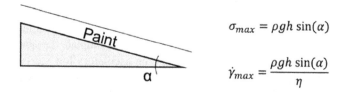

$$\sigma_{max} = \rho g h \sin(\alpha)$$

$$\dot{\gamma}_{max} = \frac{\rho g h \sin(\alpha)}{\eta}$$

FIGURE 7.8 Schematization of paint layer and maximum stress and shear rate acting at the substrate/paint interface.

Thickeners are commonly added in formulation production to improve the grinding process. During the grinding phase, inorganic fillers and pigments are added to the liquid formulation and mixed at high shear stress to maximize their dispersion. Thickeners are added to decrease the size of particle aggregates in the dispersion medium, improving product uniformity and pigment yield.

Thickeners are also used in several mixing phases, in which liquid components are injected or poured in other preparations. In this step, thickeners are useful to facilitate the transfer of liquids from one container to another.

Finally, thickeners are widely used to adapt the product viscosity to the final application procedure. Paints and coatings applied via spraying should be less viscous than those applied by brush or roller. The optimization of formulation viscosity allows optimizing the surface coverage, avoiding using too much material or the formation of uneven coatings.

On the other hand, a too-much use of thickeners can lead to some drawbacks, such as flocculation, "fish-eye" formation, and other film defects.

7.5 CONCLUSIONS

Rheology is the science of fluid flow. Industrial coatings and paints are governed by rheological parameters, and a deep knowledge of formulation rheometry allows optimizing production, storage, and stability of the final products, also providing important information on their application and final properties of coatings, in terms of surface smoothness and homogeneity.

Generally, paints and coatings formulations behave like non-Newtonian fluids, showing different rheological parameters to tune and to be tailored for each specific use. Their complex mechanical behavior can be studied by rheometry, developing *ad hoc* continuous and oscillatory tests to calculate parameters of interest.

Several empirical models can be used to fit experimental data, starting from more complex mathematical models involving stress and strain tensors, and time-dependent viscosity laws.

8 Cure Reactions of Polymer Coatings

Piotr Czub and Anna Sienkiewicz
Cracow University of Technology

CONTENTS

8.1 INTRODUCTION

The basic component of any coating system is the binder, in the case of polymer coatings consisting of an oligomer/polymer and a suitable curing agent or a monomer and an initiator for its polymerization. The cross-linking of the polymer used as the basis for the binder is a necessary step in the formation of the coating film. At the same time, the cross-linking is an equally important as the process responsible for the integrity of the coating (cohesion strength), and even for its adhesion to the substrate or to a subsequent layer in a multilayer coating. The resulting three-dimensional network structure of the cross-linked coating combines the binder with pigments, organic or inorganic fillers and all applied non-reactive additives (e.g. rheological, like a defoamer), thinners, antioxidants, etc. On the other hand, reactive additives (e.g. diluents) taking part in the cross-linking process can be built into the structure of the coating. For example, aliphatic diepoxides, commonly used as the reactive diluents for the bisphenol A-based low-molecular-weight epoxy resins, not only reduce high viscosity of these resins, but they also act like a plasticizer. The cross-linking of the binder provides the coating with the desired mechanical strength, chemical resistance and thermal stability. Depending on the used polymeric binder, the process of cross-linking may occur according to various mechanisms. The most often, this is done by reaction of functional groups present in an oligomer and in a cross-linker. The cross-linking of epoxy, novolac and polyester resins proceeds mainly in this way. In turn, in the case of many important polymer coating materials, such as acrylates, vinyl, alkyd and unsaturated polyester resins, the cross-linking process is carried out by polymerization or copolymerization, usually radical, but also ionic polymerization as well as photopolymerization. A specific process, also involving the reaction of functional groups, is the cross-linking polymerization carried out using of multifunctional monomers (i.e. the synthesis of polyurethanes and polyesters).

8.2 CROSS-LINKING *VIA* FUNCTIONAL GROUP REACTIONS

Monomers used in polyaddition and polycondensation reactions as well as the resulting oligomers or prepolymers contain active functional groups, usually terminal. These end groups can be successfully used to build the polymer chain, but also in construction of the spatial structure in the cross-linking process. The cross-linking density of such structures (which has the greatest impact, among others, on their mechanical strength, water and solvent absorption, and thermal properties) can be controlled by the length of the oligomeric chains of the materials constituting the basis of the binder, as well as the functionality and structure of the cross-linking agents. Depending on the selected type of the polymer binder and the cross-linker, it is possible to prepare and use the one- and two-component coating systems, as will be shown in the examples below.

8.2.1 EPOXY RESINS

The epoxy resins, most commonly used in the commercial coating systems, are the products of the reaction of bisphenols (usually bisphenol A) with epichlorohydrin. As can be seen in Figure 8.1, the oligomeric products of this reaction, which are diglycidyl ethers, contain two epoxy groups at the ends of the chains and a different number (depending on the degree of polymerization) of secondary hydroxyl groups. Both types of functional groups can be used to cross-link the resins, although they differ in their reactivity. However, due to the strained structure of the oxirane rings, they are characterized by exceedingly reactivity towards both nucleophilic and electrophilic species and therefore easily react with a wide range of chemicals that can be used as cross-linking agents. In practice, mainly polyfunctional amines (aromatic, aliphatic or cycloaliphatic), polyaminoamides and carboxylic acid anhydrides are used to cross-link the liquid low-molecular-weight epoxy resins ($n=0.1$–1.2). Generally, aliphatic or cycloaliphatic polyamines, especially with primary groups, however

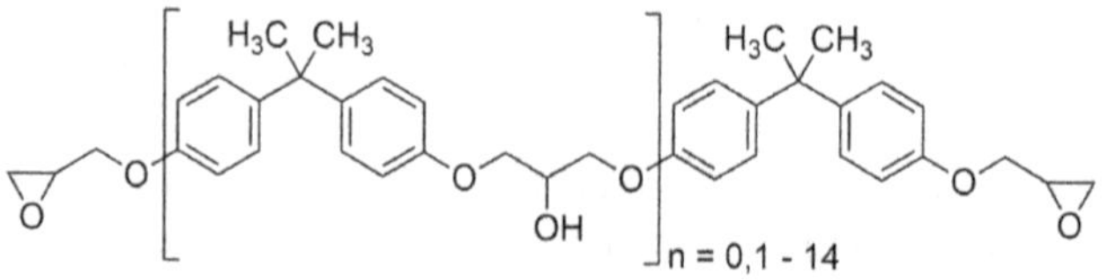

FIGURE 8.1   The chemical structure of the bisphenol A-based epoxy resin.

hazardous to handle, are very convenient cross-linking agents because the curing process using them can be carried out at ambient temperature. After the first step of the reaction of primary amine and epoxy groups, the secondary amine group is created, as shown in Figure 8.2. It reacts with the next epoxy group at the same time cross-linking different oligomeric chains. It should be remembered that cross-linking with very reactive primary amine groups is accompanied by the release of a large amount of heat, and also the volumetric shrinkage of the cured composition appeared. The structure of the industrially most commonly used polyamines and polyaminoamides is shown in Figure 8.3. Aromatic amines require an elevated curing temperature, but provide a higher glass transition temperature value as compared to aliphatic polyamines. Usually, the cross-linking of the composition is carried out in two stages, first at a temperature of up to 60°C and then even up to 120°C–140°C, depending on the type of amine used. Epoxy resins cured with aromatic amines are particularly resistant to thermal ageing, and chemical resistance of such cured resins exceeds that of those cross-linked with aliphatic polyamines or organic carboxylic acid anhydrides. Polyaminoamides, most often obtained from unsaturated fatty acids and polyamines, contain nitrogen-related active hydrogen atoms (in primary and secondary amine groups), thanks to those they can react with epoxy resins like all amines. They are non-volatile and less toxic than amine cross-linking agents (although they may also contain free amines). Polyaminoamides easily mix with the

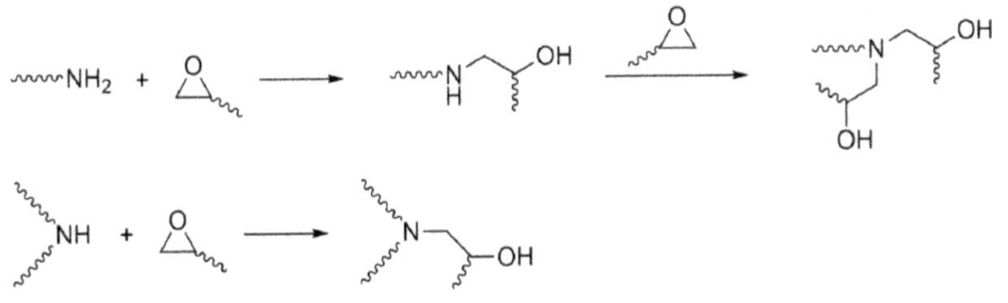

FIGURE 8.2 The cross-linking reaction of epoxy resins using primary and secondary amines.

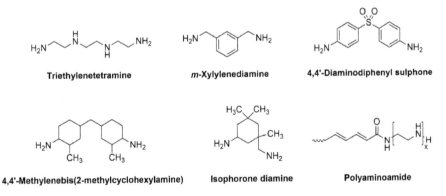

FIGURE 8.3 Examples of aliphatic, aromatic and cycloaliphatic polyamines and polyaminoamides used as curing agents for epoxy resins.

epoxy resins, the curing process can be carried out at room temperature, and shrink-age during cross-linking is very small. They are a kind of internal plasticizers of the cross-linked resin increasing its impact strength. The presence of long fatty acid chains in the vicinity of the amino group gives to polyaminoamides the nature of a surfactant resulting in an increase in a positive effect on the wetting of the substrate as well as the adhesion of the coating film (Czub, Bończa-Tomaszewski, Penczek and Pielichowski, 2002). In turn, lower cross-linking density may, however, cause a decrease in the chemical resistance of the coating.

The above-mentioned secondary hydroxyl groups, present in the structure of the epoxy oligomers, play an important role in the cross-linking with carboxylic acid anhydrides, mostly cyclic anhydrides of diccarboxylic acids and dianhydrides of tet-racarboxylic acids. The cyclic rings of anhydrides first must be opened, and it is real-ized just in the reaction of them with the secondary hydroxyl groups, which is shown in Figure 8.4. Additionally, tertiary amines allow cross-linking of the epoxy res-ins with carboxylic acid anhydrides, acting as the anhydride ring-opening catalysts, which is also shown in Figure 8.4; and they can also limit a possible side reaction that may occur between the epoxide and hydroxyl groups. Examples of the acid anhy-drides used to cure epoxy resins are shown in Figure 8.5. Although the cross-linking with carboxylic acid anhydrides requires heating for a long time at the high tempera-ture (usually above 120°C), the curing process is characterized by a low exothermic thermal effect and a low volumetric shrinkage. Anhydride-cross-linked epoxy resins are free of internal stress and resulting mechanical defects. Therefore, they are char-acterized by high thermal stability and are perfect for the industrial tooling and for insulation materials in the electrical engineering.

Among the group of nucleophilic compounds, the phenolic resins, e.g. phenol-formaldehyde resins, and above all, thermoplastic novolac resins, are also used as cross-linking agents for epoxy resins. As can be seen in Figure 8.6, in addition to the

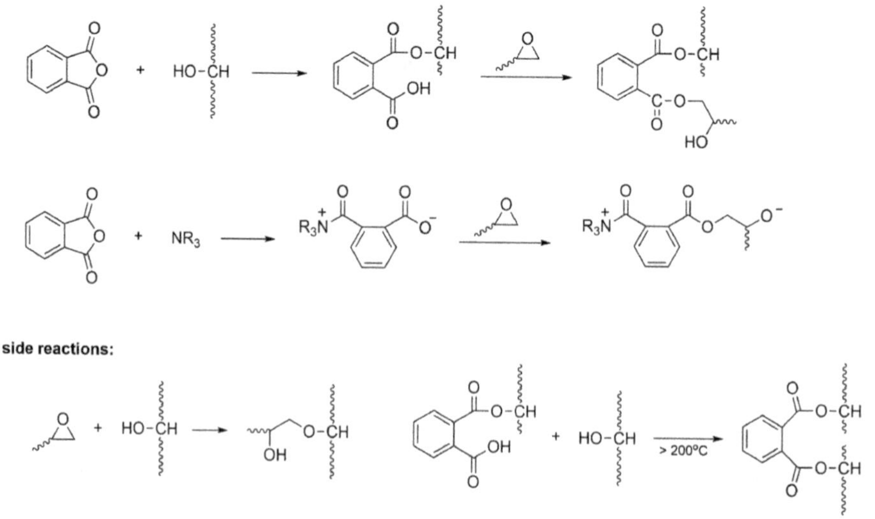

FIGURE 8.4   The cross-linking reaction of epoxy resins using carboxylic acid anhydrides.

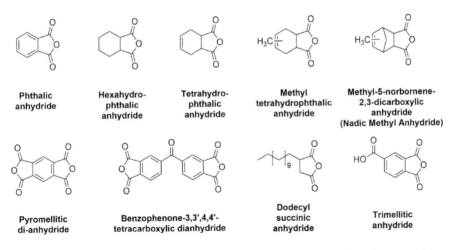

Phthalic anhydride

Hexahydro-phthalic anhydride

Tetrahydro-phthalic anhydride

Methyl tetrahydrophthalic anhydride

Methyl-5-norbornene-2,3-dicarboxylic anhydride (Nadic Methyl Anhydride)

Pyromellitic di-anhydride

Benzophenone-3,3',4,4'-tetracarboxylic dianhydride

Dodecyl succinic anhydride

Trimellitic anhydride

FIGURE 8.5 The structure of exemplary acid anhydrides used for curing of epoxy resins.

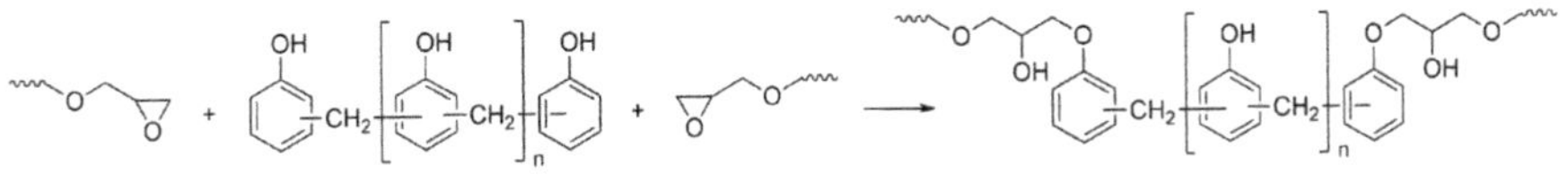

FIGURE 8.6 The cross-linking of epoxy resins with phenolic resins.

main reaction between the phenolic and epoxy groups, other reactions that contribute to cross-linking of the composition take place in parallel. The most important of them is the reaction of epoxy groups with secondary alcohol groups – contained in the initial epoxy resin and formed in the main reaction. An ether bond is formed in this reaction, and a secondary hydroxyl group is formed, which further participates in the cross-linking process. The compositions made of the high-molecular-weight epoxy resins based on bisphenol A cured with the novolacs are the coating materials with excellent adhesion, mechanical strength, chemical resistance and flexibility.

Tertiary amines are commonly used as accelerators of cross-linking of epoxy resins with aromatic amines and carboxylic acid anhydrides. However, tertiary amines can also be used as catalysts in the anionic ring-opening polymerization of epoxides, as shown in Figure 8.7. Hydroxyl groups are necessary to initiate and accelerate polymerization, and they may be present in the resin itself, in an amine catalyst or could be introduced with additives, such as alcohol or phenol. A minimum amount of hydroxyl groups, e.g. those present in moisture from the air, is sufficient to initiate the reaction. The structure of tertiary amines suitable for ring-opening polymerization of epoxy resins is presented in Figure 8.8.

For the high-molecular-weight epoxy resins ($n=4-14$), typical binder for powder coatings, that have oxirane rings only at the ends of long chains, another method of cross-linking must be applied. For such resins, the amount of epoxy groups is too small to cross-link the resin conventionally, and for this purpose, secondary hydroxyl groups present in large amounts in the chains should also be used. Dicyandiamide and 2-methylimidazole are used for this purpose. During the cross-linking process using dicyandiamide, two reactions take place in parallel: epoxy groups with the four active

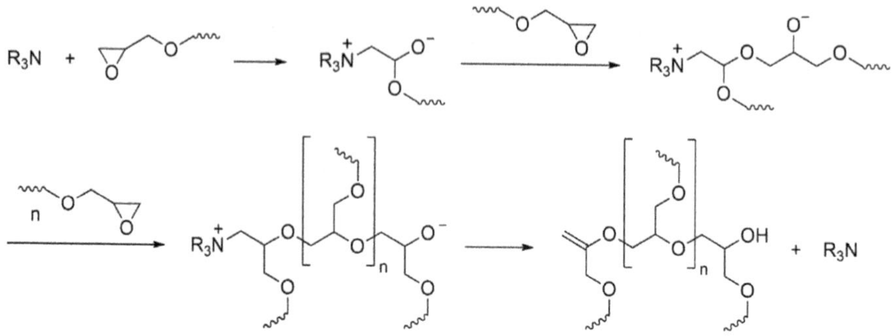

FIGURE 8.7 The mechanism of the tertiary-amine-catalysed polymerization of epoxy resins. (Sufficiently adapted – modified and redrawn based on Czub, Bończa-Tomaszewski, Penczek and Pielichowski, 2002.)

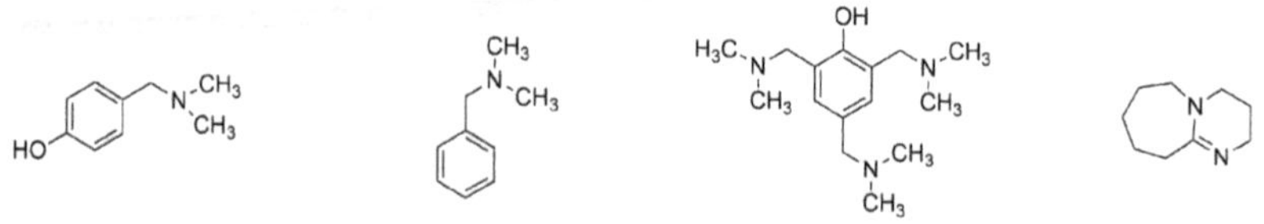

4-((Dimethylamino)methyl)phenol   N,N-Dimethylbenzylamine   2,4,6-tris(dimethylamino-methyl)phenol   1,8-Diazabicyclo[5.4.0] undec-7-ene

FIGURE 8.8 The structure of tertiary amines suitable as catalyst for ring-opening polymerization of epoxy resins.

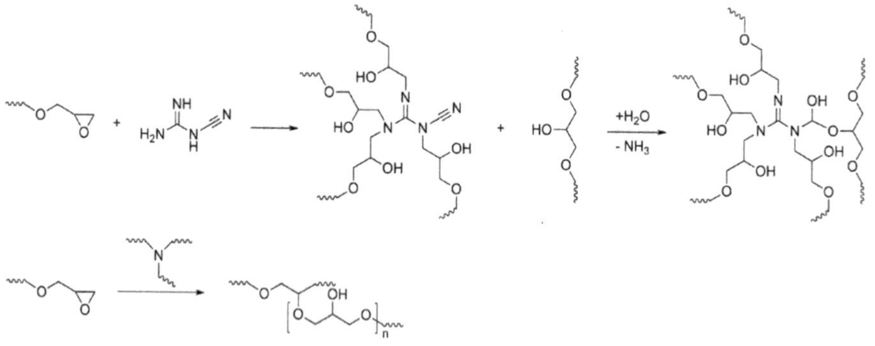

FIGURE 8.9 The cross-linking of high-molecular-weight epoxy resins with dicyandiamide. (Sufficiently adapted – modified and redrawn based on Ellis, 1993.)

protons of each dicyandiamide molecule and homopolymerization of epoxy resin, initiated by the tertiary amine (as shown in Figure 8.9) (Ellis, 1993). The next stage of the process is the joining of polymer network fragments, resulting from the reactions described above, proceeding through the reaction of secondary hydroxyl groups present in the partially hardened resin with the cyano groups in the dicyandiamide. Imidazoles are applied as efficient accelerators for the dicyandiamide (as well as anhydride) cross-linking. Additionally, they also act as catalytic epoxy curing agents at moderate-to-high temperatures. Curing with 2-methylimidazole proceeds according to the mechanism shown in Figure 8.10 (Ellis, 1993), which assumes that the epoxy group first reacts with the hardener's amino group to form an equimolar adduct. Then, the next epoxy group reacts with the resulting adduct, forming a compound containing in its structure a highly reactive alkoxylate ion, initiating rapid anionic polymerization of the epoxy groups. Secondary hydroxyl groups present in the chains of the high-molecular-weight epoxy resins can also be applied to cross-link the resins using blocked isocyanates. After unblocking at elevated temperatures, the released isocyanates typically react with hydroxyl groups, giving the cross-linked product.

8.2.2 Polyurethanes

The basic reaction during polyurethane synthesis occurring between difunctional isocyanates and alcohols (usually diols, glycols, polyesterols or polyetherols) leads to the formation of oligomers or polymers with a linear structure as shown in Figure 8.11. With an excess of isocyanates and at elevated temperatures, the resulting urethane groups (or urea groups in the case of the reaction of isocyanates with amines) can react with isocyanate groups to give allophanates (or biurets). These reactions lead to

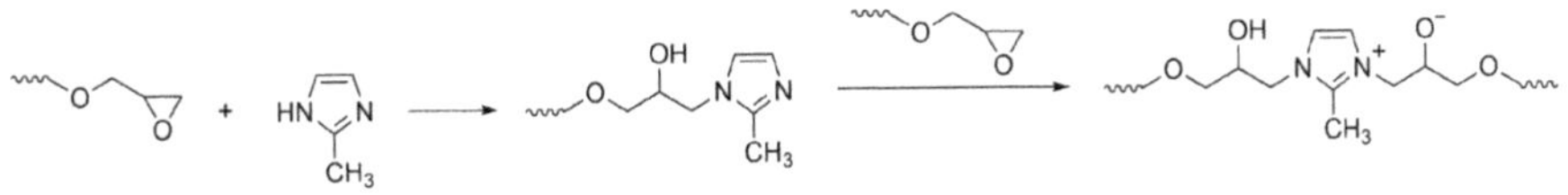

FIGURE 8.10 Ring-opening mechanism of epoxy resin polymerization initiated by 2-methylimidazole. (Sufficiently adapted – modified and redrawn based on Ellis, 1993.)

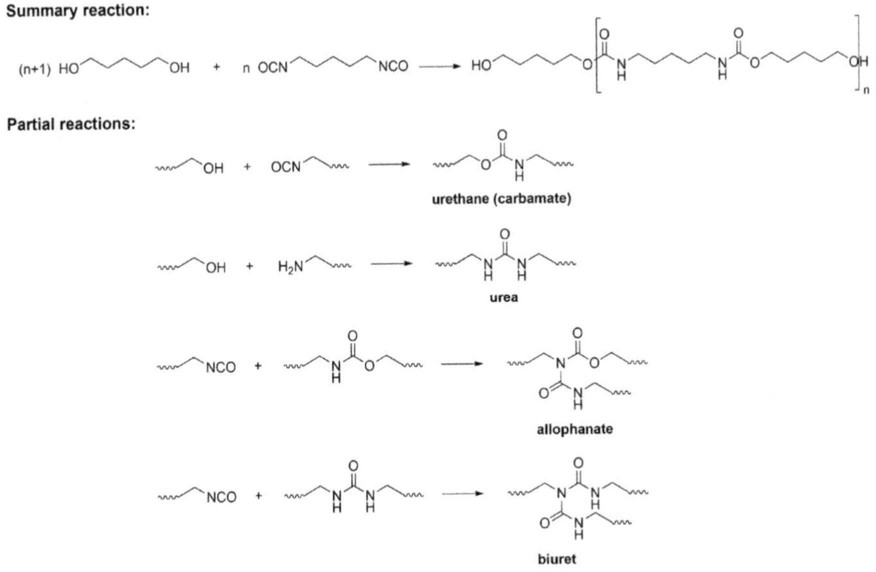

FIGURE 8.11 Reactions occurring during the synthesis of polyurethanes.

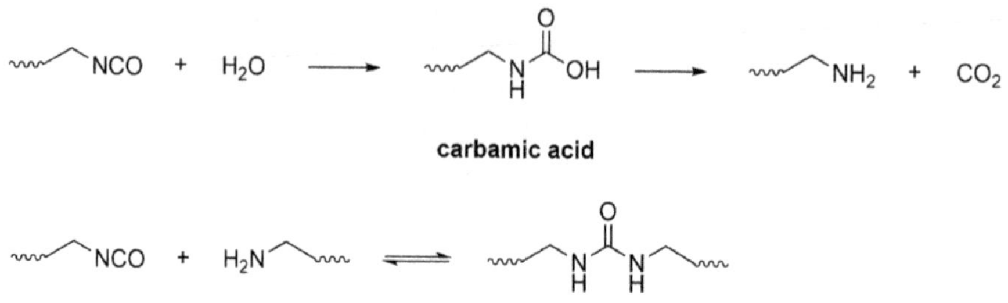

FIGURE 8.12 The reaction of isocyanates with water and subsequent reactions.

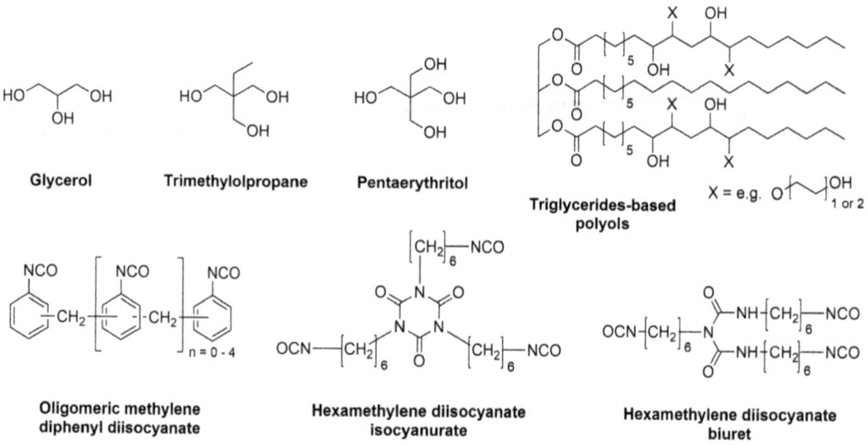

FIGURE 8.13 The multifunctional raw materials for the polyurethane synthesis.

the formation of a branched and cross-linked structure. Nevertheless, the reaction of isocyanate groups with atmospheric moisture is the oldest method of cross-linking of polyurethanes. This reaction leads to formation of an unstable carbamic acid, which quickly dissociates to carbon dioxide and a primary amine, as shown in Figure 8.12. The resulting amine reacts also very quickly with isocyanate groups forming ureas. Analogously, biurets and allophanates are also formed as a result of follow-up reactions. Cross-linking of polyurethanes, in both one-step and two-step syntheses, may be achieved by using reagents with a higher functionality than two. Examples of raw materials enabling the synthesis of cross-linked polyurethanes are shown in Figure 8.13.

8.2.3 PHENOL- AND AMINO-FORMALDEHYDE RESINS

Depending on the conditions of reaction of phenol with formaldehyde, two different products are obtained: resole or novolac resins. Both products, although differing in the structure, are generally oligomers and require cross-linking. The cross-linking makes the coating based on phenol-formaldehyde resins insoluble, tasteless and odourless as well as mechanically strong, and resistant to exposure to high temperature, chemicals (except alkalis), solvents and hot water (Dodiuk and Goodman,

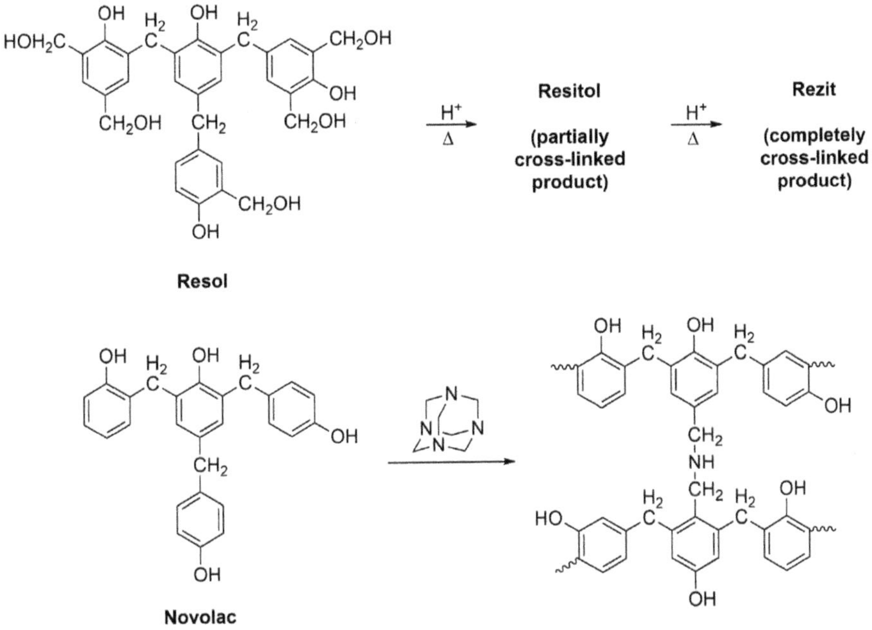

FIGURE 8.14   The structure and cross-linking reaction of phenol-formaldehyde resins.

2014). Resole phenolic resins are methylol-bearing products, while novolacs are non-methylol-bearing products, as can be seen in Figure 8.14. Therefore, the resole resins are ready for self-curing, which is carried out at elevated temperatures with or without an acid catalyst. The creation of the spatial structure occurs at the expense of reactive methylol and hydroxyl groups, in the condensation process leading to creation of methylene bridges. The novolac resins don't have methylol groups, and they are not able to self-cure. Hexamethylenetetramine (urotropin) is typically used for cross-linking of the novolac resins (as shown also in Figure 8.14). It is a condensation product of formaldehyde and ammonia, and therefore, the cured products contain the methylene and dimethylene amino bridges. The novolac resins may be cross-linked also with formaldehyde or methylol-bearing compounds, such as bismethylol cresol and tetramethylol bisphenol A.

Amino resins (the products of formaldehyde condensation mainly with urea or melamine, but also with benzoguanamine or glycoluril) contain, in their structure, methylol groups, which undergo condensation reactions with the liberating water and forming of dimethylene ether linkages (as can be seen in Figure 8.15).

8.2.4   POLYESTER RESINS

Polycondensation of polyols with polycarboxylic acids (or their anhydrides or esters) leads to formation of saturated polyesters terminated with hydroxyl or carboxyl groups, depending on the molar ratio of reagents. Usually, diols (glycols) and dicarboxylic acids (phthalic and terephthalic acids or their anhydrides) are applied; therefore, linear chains of oligomeric/polymeric products are formed, as can be seen in Figure 8.16.

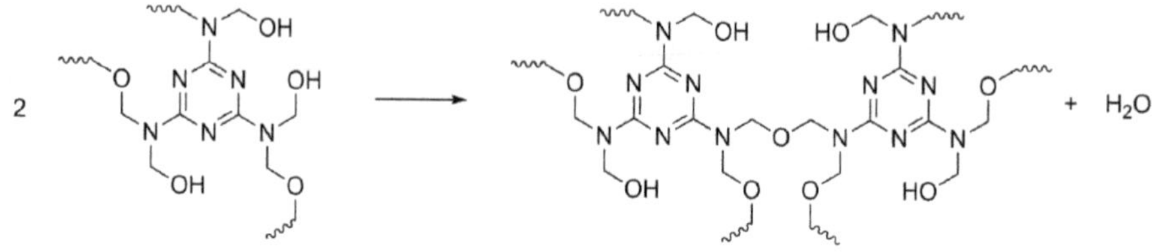

FIGURE 8.15 The cross-linking of a melamine-formaldehyde resin.

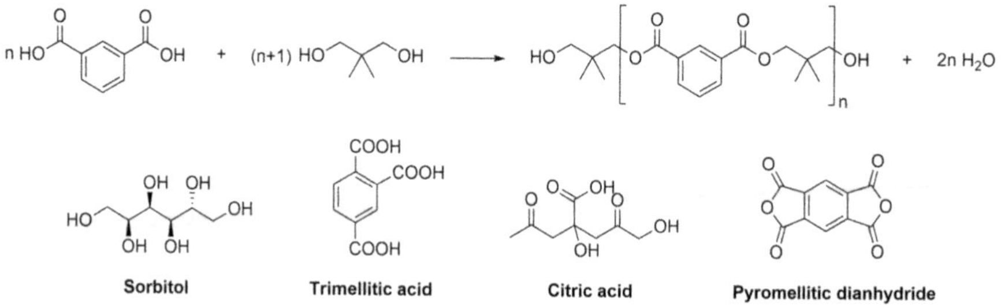

| Sorbitol | Trimellitic acid | Citric acid | Pyromellitic dianhydride |

FIGURE 8.16 The saturated polyester synthesis and commonly used raw materials.

However, reactants with functionality higher than two (shown also in Figure 8.16) are also used, giving branched or cross-linked polyesters, dependent on the poly-condensation conditions. The trifunctional alcohols (glycerol and trimethylpropane), the tetrafunctional pentaerythritol or the hexafunctional sorbitol and trimellitic acid, the tetrafunctional citric acid or pyromellitic dianhydride are most commonly used for this purpose. Also, the functionality of the reagents may depend on the reaction conditions. Thus, glycerol is the bifunctional compound at 180°C, because only the primary hydroxyl groups are active at this temperature. But already at 220°C, the secondary hydroxyl group is involved in the reaction, due to which glycerol becomes a trifunctional compound. The high-molecular-weight polyester resins terminated with carboxyl groups usually were previously cross-linked using 1,3,5-triglycidyl isocyanurate (shown in Figure 8.17), which provides very good flexibility, improved UV-light resistance and high thermal stability for the cured resin. However, due to mutagenic effects, its use is recently restricted for occupational health and safety reasons. It can be successfully replaced by such compounds as tris(oxiranylmethyl) benzene-1,2,4-tricarboxylate and N,N,N′,N′-tetrakis(2-hydroxyethyl)hexanediamide (shown also in Figure 8.17). These compounds also provide very good properties for cross-linked coatings. However, it should be remembered that cross-linking with the first of them may begin already at 40°C, and when curing with the second one, moisture is released, which can cause coating defects. The blocked isocyanates may also be successfully applied. Additionally, oligomeric polyesters terminated with hydroxyl or carboxylic groups can be cross-linked using other resins, such as epox-ides, phenol- and amino-formaldehyde resins or polyisocyanates, selected taking into account their end groups.

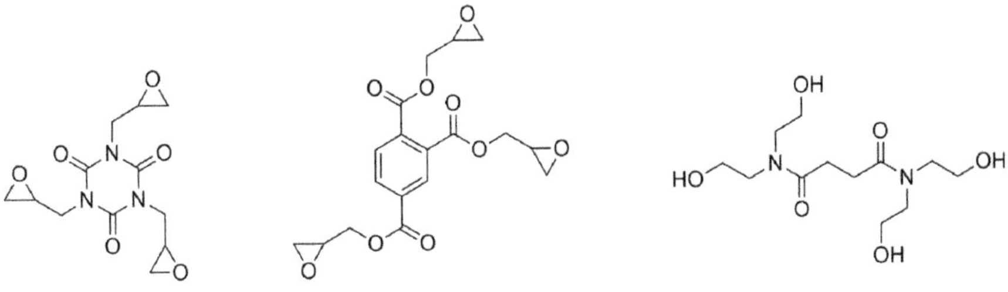

| 1,3,5-triglycidyl isocyanurate | Tris(oxiranylmethyl) benzene-1,2,4-tricarboxylate | N,N,N',N'-tetrakis(2-hydroxyethyl) hexanediamide |

FIGURE 8.17 The cross-linking agents for curing of the high-molecular-weight polyesters.

8.3 CROSS-LINKING *VIA* DOUBLE-BOND POLYMERIZATION

Most of the monomers containing double bonds may be converted into the three-dimensional network polymers. Depending on their structure and reactivity as well as chosen cross-linking conditions and initiator/catalyst systems, this process can be realized according to various reaction mechanisms. The most commonly applied mechanisms in practice are a radical or ionic polymerization/copolymerization and an oxidative cross-linking. The thiol-ene reaction with the use of multifunctional thiols is also a very interesting possibility. The mechanism of a chain-growth polymerization is very well known and can be presented in a general way, taking into account various types of monomer activation, as shown in Figure 8.18. The reaction consists of three stages: activation of the monomer, polymer chain growth (propagation) and termination. The activation of the monomer usually takes place using an initiator (free radical polymerization/copolymerization) or a catalyst (anionic or cationic polymerization), although it is also possible to activate the monomer thermally or by means of UV light, ultrasound, radiation and otherwise. The method of termination of the polymerization depends on its type, and it can take place by recombination, disproportionation and chain transfer (radical polymerization and copolymerization) or the end-group functionalization and the termination through unintentional quenching by impurities, often present in trace amounts in a monomer and a solvent (anionic polymerization), or chain transfer by hydrogen abstraction from the active chain end to a counterion or a monomer (cationic polymerization).

A very important and one of the oldest (thus the longest used) methods of cross-linking is autoxidation of unsaturated bonds. This naturally occurring or catalysed reaction is involved for desired or sometimes undesirable cross-linking of many polymers. It is oxidation that occurs in the presence of oxygen (exactly in air as a source of oxygen). Using an additional oxidizer (such as different peroxides) to speed up the process is also practised. Drying oils, a key component of oil paint and some varnishes, are the first group of coating materials for which cross-linking autoxidation has been applied. Typically, metal coordination complexes are used as catalysts (named as siccatives). These are derivatives of such metals with variable oxidation state as lead (withdrawn due to the toxicity), cobalt, manganese, zinc, zirconium or

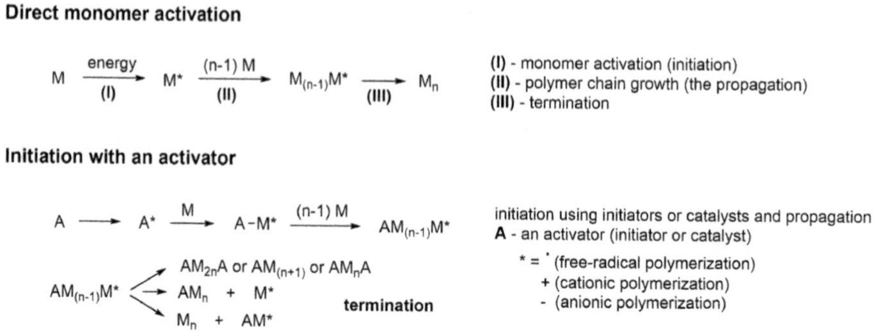

Direct monomer activation

(I) - monomer activation (initiation)
(II) - polymer chain growth (the propagation)
(III) - termination

Initiation with an activator

initiation using initiators or catalysts and propagation
A - an activator (initiator or catalyst)

* = ˙ (free-radical polymerization)
 + (cationic polymerization)
 − (anionic polymerization)

FIGURE 8.18 The mechanism of a chain-growth polymerization.

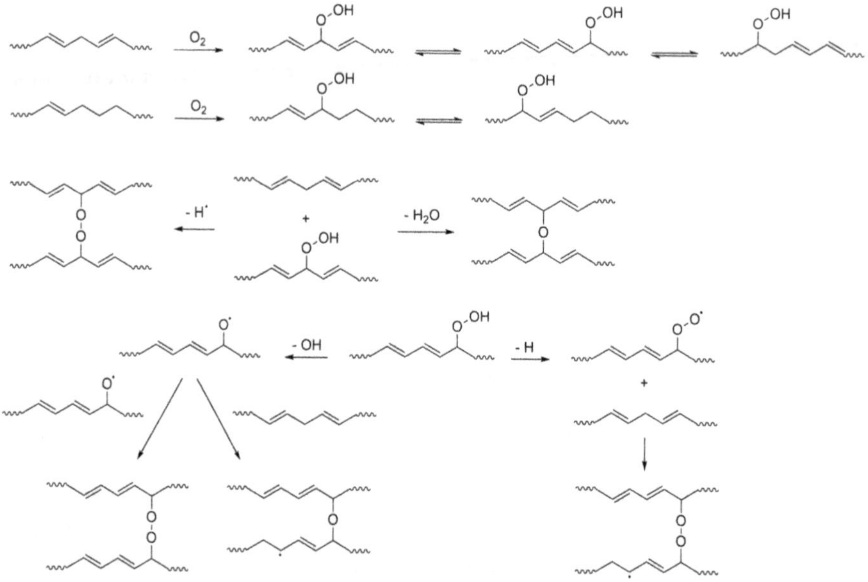

FIGURE 8.19 The mechanism of drying oils' cross-linking *via* autoxidation.

iron and lipophilic carboxylic acids, such as naphthenic acid or octanoic acid, to make the carboxylates oil-soluble. The mechanism of cross-linking *via* autoxidation is shown in Figure 8.19.

The thiol–ene reaction is included in the click chemistry reaction groups, and its advantages are as follows: mild operating conditions and at the same high rate, very high reagent conversion and yield of the reaction, stereoselectivity, lack of by-products and, therefore, high purity of the main product. Two possible mechanisms of the thiol–ene reaction are considered: a free radical addition (initiated by light, heat or radical initiators) and a catalysed (a base or a nucleophilic catalyst) Michael addition. The thiol–ene reaction is a very convenient method of introducing branching into the linear molecule as well as joining the molecules together (as can be seen in Figure 8.20). This reaction is very willingly used in the synthesis of hyperbranched

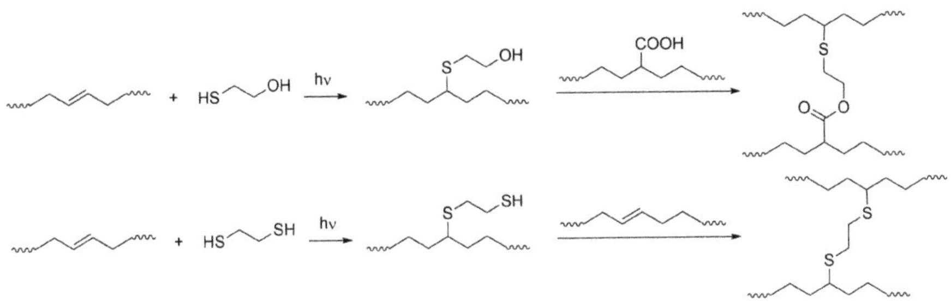

FIGURE 8.20 Examples of thiol–ene reactions leading to cross-linking.

polymers and dendrimers, but it can also be useful for cross-linking polymers with unsaturated bonds located along the chain.

8.3.1 ACRYLATES

Acrylates are the biggest group of monomers with unsaturated bonds practically applied for polymer coatings. They are polymerized with a chain-growth mechanism under free radical or ionic conditions. Using the most often-applied acrylic and methacrylic monomers (as shown in Figure 8.21), polymers with a linear structure are obtained. However, the multifunctional acrylic monomers (as can be seen in Figure 8.21) can be used to obtain highly cross-linked polymers with increased toughness, modulus and solvent resistance. The formation of the linear macromolecules from acrylic monomers takes place involving vinyl groups, and unreacted carboxyl (acrylate) groups remain in the polymer and can be used for following reactions, such as cross-linking. Another possibility is to use acrylic monomers with additional functionality (as shown also in Figure 8.22). In this way, pendant groups terminated with functional groups capable of further reactions can be introduced into the linear macromolecule. The presence of side groups terminated with reactive functional groups, in turn, allows their use to connect the polymer chains with each other, and thus cross-linking polyacrylates.

8.3.2 VINYL RESINS

The most important representatives of this group of polymer coating materials are poly(vinyl chloride) and poly(vinyl acetate) prepared by the free radical chain-growth polymerization of their corresponding monomers. Because of its insolubility and brittleness, poly(vinyl chloride) as a homopolymer is rarely used as a binder for coatings, and more often, copolymers of vinyl chloride and vinyl acetate or vinyl chloride with other monomers are applied. The incorporation of a small amount of maleic anhydride promotes adhesion of coatings and introduces in a form of difunctional monomer a low level of cross-linking. Other co-monomers used with vinyl chloride are presented in Figure 8.23. Vinyl chloride is also copolymerized with maleates (e.g. dibutyl maleate) and vinyl isobutyl ether. Acrylates are often applied as co-monomers for the vinyl monomers, and diacrylates or triacrylates (as shown in Figure 8.23) may also play the role of cross-linking agents. Vinyl chloride and

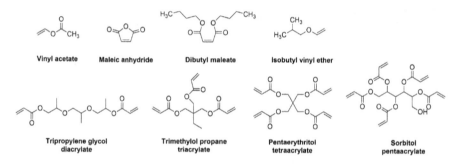

Ethyl acrylate   **2-Ethyl hexyl acrylate**   **Methyl methacrylate**   **n-Butyl methacrylate**

FIGURE 8.21   The most important acrylic and methacrylic monomers.

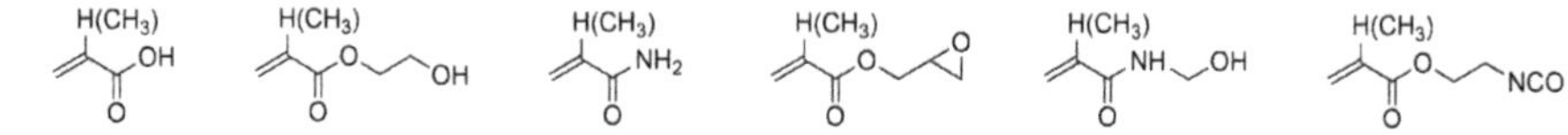

(Meth)acrylic acid   Hydroxethyl   (Meth)acrylamide   Glycidyl   N-(Hydroxymethyl)   (Meth)acryloyloxy
                       (meth)acrylate   (meth)acrylate   (meth)acrylamide   ethyl isocyanate

FIGURE 8.22   Examples of acrylic monomers with additional functionality.

Vinyl acetate   Maleic anhydride   Dibutyl maleate   Isobutyl vinyl ether

Tripropylene glycol   Trimethylol propane   Pentaerythritol   Sorbitol
diacrylate   triacrylate   tetraacrylate   pentaacrylate

FIGURE 8.23   Co-monomers industrially used with vinyl chloride.

vinyl acetate terpolymers with dicarboxylic acids may be cross-linked by involving the carboxyl groups within their structure. Terpolymers of vinyl chloride/acetate and monomers containing hydroxyl groups (e.g. hydroxyethyl- and hydroxypropylacrylate/methacrylate) can be cross-linked with polyisocyanate or melamine resins.

8.3.3   ALKYD RESINS

Alkyd resins, as the polyesters synthesized by polycondensation of polyols (e.g. glycerine, trimethylolpropane, glycols or pentaerythritol) and dicarboxylic acids (most often isophthalic acid) or carboxylic acid anhydrides (mainly phthalic or maleic anhydride), are terminated with carboxylic or hydroxyl groups. These terminal acid and hydroxyl functionalities may be involved in the cross-linking of saturated alkyd resins (named also as "non-drying" resins). Alkyd resins are modified by the addition of fatty acids as monofunctional monomers in the process of synthesis. But more often, they are synthesized using products of plant or vegetable oil alcoholysis. The transesterification of the triglycerides using glycerol leads to a mixture of mono- and diglyceride derivatives, applied next in the typical polycondensation process with dicarboxylic acids or carboxylic acid dianhydrides. On the one hand, it is an economically profitable, and on the other hand, a bio-based process leads to building into the resin structure fatty acid residues containing various amounts of unsaturated bonds (depending on the origin of the oil used). Due to the presence of unsaturated bonds in the branches of the resin's main chains, it is possible to cross-link resins according to the autoxidation mechanism. Analogously to drying oils, the cross-linking of alkyd resins obtained in the glyceride process (named as "drying" resins) is usually accelerated using catalysts (siccatives/driers, e.g. Co(II) naphthenate or octoate as the primary catalyst and Ca(II) or Zn(II) octoates as the auxiliary catalysts). The mechanism of the cobalt-catalysed cross-linking (drying) of alkyd resins by the autoxidation is shown in Figure 8.24. Autoxidation is a typical free radical chain process, which can be divided into three stages: (1) the diene (or triene) autoxidation by hydrogen abstraction from the centre methylene groups which leads to formation of the hydroperoxide, (2) the decomposition of the resulting hydroperoxide to radicals in a redox process and (3) the recombination of radicals or the addition of radicals to double bonds in another unsaturated side chain generating the cross-linked structure.

8.3.4 UNSATURATED POLYESTER RESINS

Unsaturated polyesters are polymeric products formed in the condensation process of polyhydric alcohols (usually glycols, predominately 1,2-propylene, ethylene, diethylene or 1,3-butylene glycols) and unsaturated acid anhydrides (mainly maleic anhydrides) or dibasic acids (fumaric or itaconic acids). To reduce the number of double bonds in the polymer chain, unsaturated polyesters are usually synthesized with the addition of saturated dibasic acid (preliminary phthalic anhydride or adipic and sebacic acids), which further modifies the properties of the resin. The cross-linking of unsaturated polyesters takes place by means of a copolymerization reaction with vinyl, allyl or acrylic monomers (as shown in Figure 8.25). This process is very complex. Considering the cross-linking of the polyester with styrene, in addition to proper connecting of the polyester chains through styrene molecules (intermolecular cross-linking), the following reactions take place: intermolecular cross-linking of polyester without linking through styrene monomer, intramolecular cross-linking of the polyester molecules linking through styrene, chain branching on the polyester molecules and homopolymerization of styrene. Of course, chain branching and homopolymerization of styrene do not contribute towards network formation. Styrene is still the most commonly used monomer, also acting as a reactive polyester diluent, due to

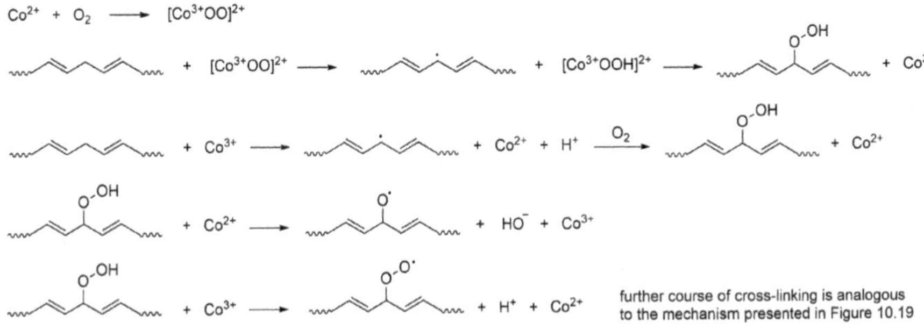

FIGURE 8.24 The mechanism of alkyd cross-linking *via* the cobalt-catalysed autoxidation.

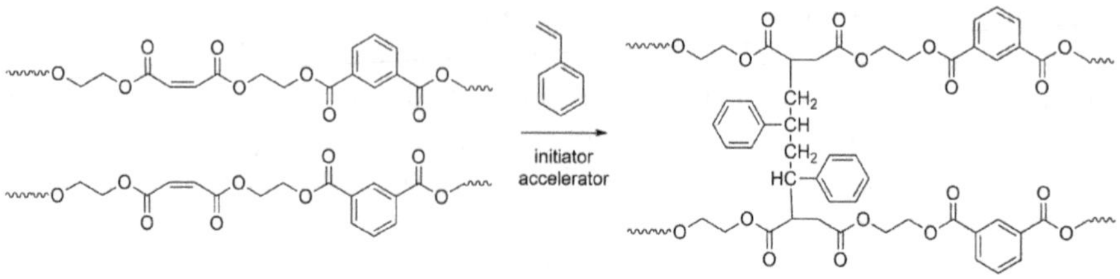

FIGURE 8.25 The cross-linking of unsaturated polyesters *via* the copolymerization with styrene.

both the low price of styrene and the good quality of the cured product. The presence of styrene ensures the good solubility and low viscosity of polyester solutions obtained from both aromatic and aliphatic acids. Moreover, styrene has a high reactivity and copolymerizes with the unsaturated polyesters at elevated or room temperature in the presence of radical catalyst systems. The disadvantages of styrene as a monomer are listed as follows: high volatility and flammability (flash point at 31°C) combined with toxic properties and a tendency to the oxygen inhibition. Additionally, the presence of even traces of polystyrene in the monomer creates the risk of gelation of the polyester during its dissolution at high temperatures. Therefore, styrene currently is being replaced by other commercially important monomers including vinyl esters, dimethacrylates and divinylbenzene (as shown in Figure 8.26). Vinyltoluene, 2-hydroxyethyl methacrylate and glycerol monoallyl ether, which exhibit lower volatility and toxicity in comparison with styrene, may also be used. The relatively low price of vinyltoluene and the good quality of cross-linked resins, especially their low water absorption, determine the use of vinyltoluene in industrial practice. The cross-linking through free radical copolymerization of unsaturated polyesters and monomers is initiated usually by the organic peroxides or hydroperoxides (predominately by dibenzoyl peroxide, di-*t*-butyl peroxide or methyl ethyl ketone peroxide) and azo compounds (mainly by azo-di-isobutyronitrile) (as shown in Figure 8.27). Together with the peroxide initiators, accelerators of their decomposition are commonly used to ensure room temperature or the low temperature of cross-linking of polyesters at a reasonable rate. Accelerators have a reducing ability, and therefore, peroxide–accelerator systems are called the redox systems. Various compounds used as accelerators act selectively only on certain groups of peroxides. Thus, the decomposition of ketone peroxides and alkyl hydroperoxides is accelerated by heavy metal salts (usually by cobalt naphthenate and cobalt octoate) and that of diacyl peroxides by tertiary aromatic amines (e.g. N,N-dimethylaniline or N,N-dimethyl-*p*-toluidine). Another group of additives, called promoters, is also used for cross-linking of unsaturated polyesters. They accelerate the action of, e.g., N,N-dimethylaniline on the redox system hydroperoxide–cobalt salt, resulting in a synergistic effect and faster gelation of the polyester resin. They are usually compounds with reducing properties (e.g. acetyl acetone, ethyl acetoacetate, amides of acetoacetic acid or tertiary aromatic amines), and they are called promoters due to the fact that they only accelerate the action of the hydroperoxide–cobalt system without the reacting with hydroperoxides. The above-mentioned accelerators and promoters are presented in Figure 8.28.

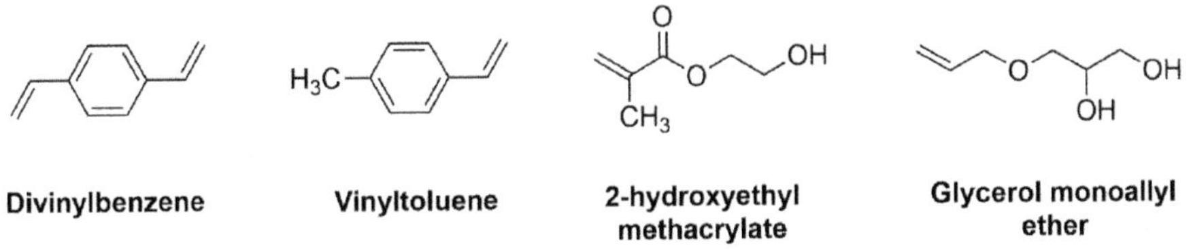

Divinylbenzene   **Vinyltoluene**   **2-hydroxyethyl methacrylate**   **Glycerol monoallyl ether**

FIGURE 8.26 The monomers used for the cross-linking of unsaturated polyester resins.

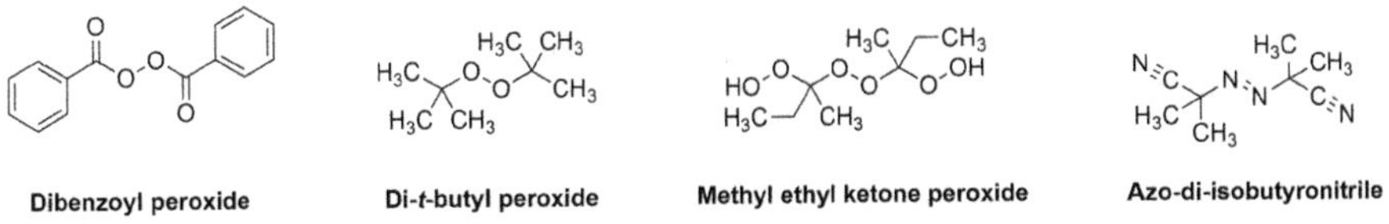

Dibenzoyl peroxide   Di-*t*-butyl peroxide   Methyl ethyl ketone peroxide   Azo-di-isobutyronitrile

FIGURE 8.27 Initiators for the copolymerization of unsaturated polyesters with vinyl monomers.

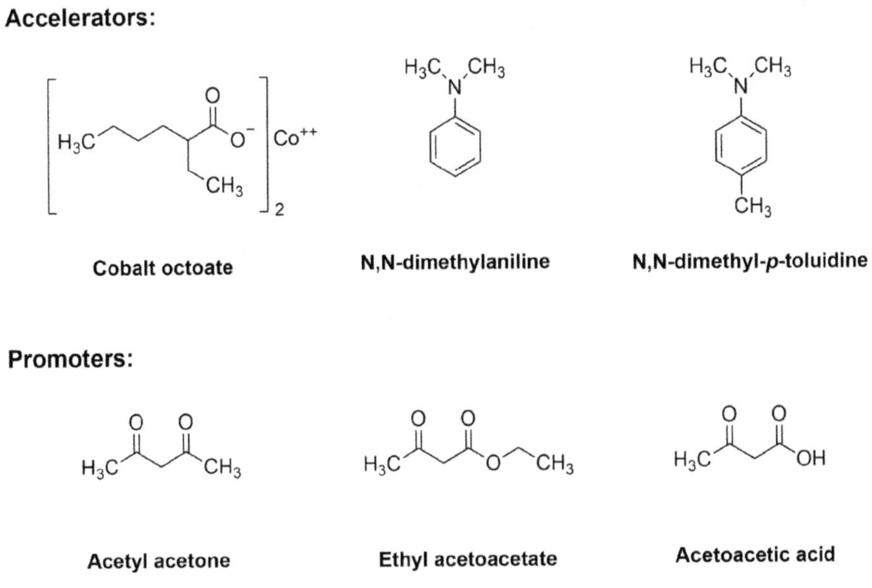

Accelerators:

Cobalt octoate   **N,N-dimethylaniline**   **N,N-dimethyl-*p*-toluidine**

Promoters:

Acetyl acetone   **Ethyl acetoacetate**   **Acetoacetic acid**

FIGURE 8.28 Accelerators and promoters used for the cross-linking of unsaturated polyesters.

8.4 PHOTO-CROSS-LINKING

The use of UV radiation for the cross-linking of polymer coatings is one of the possible methods of the binder curing upon exposure to radiation. At the same time, it is currently the most economically viable method for industrial use in a wide range. It does not require large expenditures on the exposure equipment (and on protective measures, as in the case of the electron beam cross-linking). It is also the method that ensures the better control of the process of radiation exposure of the coating and is safer for both the coating material and the substrate material than radiation with high-energy electrons. The process of photo-cross-linking may consist of irradiation-initiated radical or ionic polymerization of monomers, leading to the formation of a three-dimensional structure of the polymer, or it may consist of cross-linking of polymer chains after formation of free radicals on them. In the second case, the cross-linking process occurs by recombination of free radicals and is analogous to the autoxidizing cross-linking process (as can be seen in Figure 8.19). Irradiation of the coating with high-energy electrons allows supply of energy large enough to break the covalent bond and generate free radicals. In turn, UV radiation cross-linking usually requires the use of photo-initiators (or sometimes a sensitizer, either a photo-initiator together with a sensitizer) and the polymerization process takes place with a typical chain-growth mechanism consisting of the following steps: initiation, propagation and termination. Photo-initiators of different structure may be applied to generate radicals. Taking into account the method of the free radical generation, photo-initiators can be basically divided into two groups: type I – compounds that produce radicals by intramolecular photocleavage (homolytic cleavage, homolysis); and type II – compounds that produce radicals by the intermolecular abstraction of hydrogen. The process of the free radical generation from photo-initiators belonging to group I is unimolecular, and for compounds from group II, it is bimolecular and requires using a co-initiator. Benzoin and its derivatives (e.g. benzoin alkyl ethers, benzoin esters), ketones and aromatic hydroxyl ketones (in which the bond at carbon α in relation to the carbonyl group breaks – α-cleavage) as well as oxosulphones and oxosulphides (in which the bond at carbon β in relation to the carbonyl group breaks – β-cleavage) are commonly used as photo-initiators from group I. Aromatic ketones (benzophenone, Michler's ketone, thioxanthone or benzyl- or anthraquinone) are predominately used as photo-initiators from group II with an H-donor (tertiary amines, alcohols or thiols, e.g. 2-sulfanyl benzoxazole). Examples of photo-initiators from both groups are shown in Figure 8.29. The photo-initiated free radical polymerization or copolymerization is very widely applied to the UV cross-linking of acrylates/methacrylates, acrylated epoxy and urethane systems and unsaturated polyester resins. Ionic photo-initiators, mainly dedicated to cationic polymerization, are widely used to cross-link important binders for polymer coating systems such as cycloaliphatic epoxides, bisphenol A-based epoxy resins, vinyl ethers, lactones, sulphides

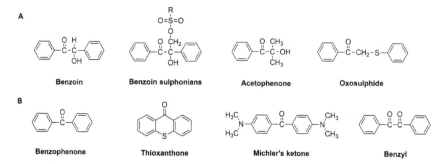

FIGURE 8.29   The structure of photo-initiators generating free radicals: type I (A) and type II (B).

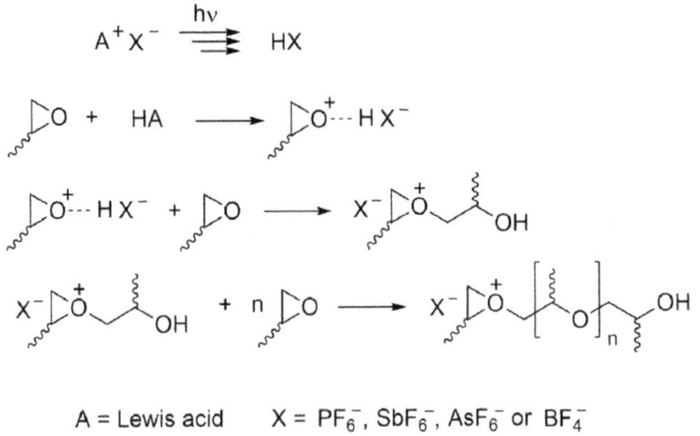

$$A^+X^- \overset{h\nu}{\rightleftharpoons} HX$$

A = Lewis acid   X = PF_6^-, SbF_6^-, AsF_6^- or BF_4^-

FIGURE 8.30   The mechanism of the cationic ring-opening polymerization of epoxides.

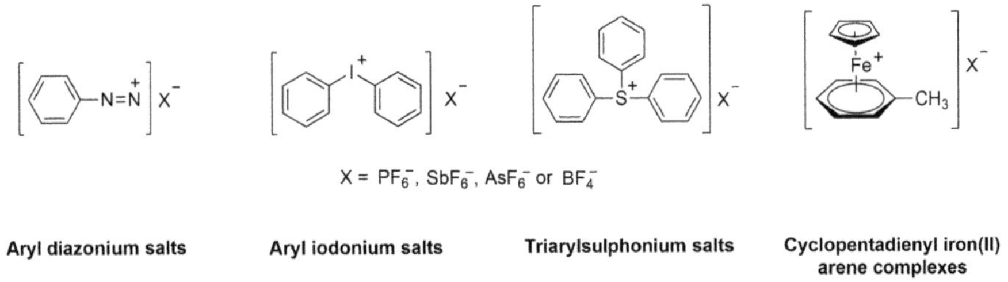

X = PF_6^-, SbF_6^-, AsF_6^- or BF_4^-

Aryl diazonium salts   **Aryl iodonium salts**   **Triarylsulphonium salts**   **Cyclopentadienyl iron(II) arene complexes**

FIGURE 8.31   Examples of cationic photo-initiators.

and cyclic ethers *via* ring-opening polymerization. The mechanism of the epoxide cross-linking using cationic photo-initiators is shown in Figure 8.30. The most commonly used cationic photo-initiators are the salts containing aryl diiodonium, aryl diazonium and triarylsulphonium cations as well as the following weakly nucleophilic or non-nucleophilic anions: BF_4^-, BF_6^-, AsF_6^- or SbF_6^- (as can be seen in Figure 3.31). In contrast to photo-initiated free radical polymerization, the ionic ring-opening polymerization is not inhibited by oxygen and also polymerization doesn't stop when the radiation is removed, but continues for an appreciable time. Cationic photopolymerization is a fast process, and the photo-initiators used in it are characterized by high photodissociation efficiency.

BIBLIOGRAPHY

Czub, P., Bończa-Tomaszewski, Z., Penczek, P., and, Pielichowski J. (2002). *Chemistry and Technology of Epoxy Resins*, IV ed., Warsaw: WNT.

Dodiuk H. and Goodman S.H., (editors) (2014). *Handbook of Thermoset Plastics*, III ed. Oxford: William Andrew and Elsevier.

Ellis B. (1993). *Chemistry and Technology of Epoxy Resins*. Dordrecht: Springer Science and Business Media.

Kłosowka-Wołkowicz Z., Penczek P., Królikowski W., Czub P., Pielichowski J., and Ostrysz R. (2010). *Unsaturated Polyester Resins*, III ed. Warsaw: Wydawnictwa-Naukowo Techniczne.

Lambourne R. and Strivens T.A., (editors) (1999), *Paint and Surface Coatings. Theory and Practice*, II ed. Cambridge: Woodhead Publishing Ltd.

Marrion A., (editor) (2004). *The Chemistry and Physics of Coatings*, II ed. Cambridge: The Royal Society of Chemistry.

Pascault J-P. and Williams R.J.J., (editors) (2010). *Epoxy Polymers. New Materials and Innovations*, Weinheim: Wiley-VCH Verlag GmbH & Co. KGaA.

Penczek P., Czub P., and Pielichowski J. (2005). Unsaturated polyester resins: chemistry and technology recent research and development, in *Crosslinking in Materials Science*, Heidelberg: Springer-Verlag, https://link.springer.com/book/10.1007/b136215#toc

Ratna D. (2009). *Handbook of Thermoset Resins*. Shropshire: iSmithers.

Stoye D. and Freitag W., (editors) (1998). *Paints, Coatings and Solvents*, II ed., Weinheim: Wiley-VCH.

Tracton A.A., (editor) (2006). *Coatings Technology Handbook*, III ed., Boca Raton, FL: Taylor & Francis Group.

Wicks Jr. Z.W., Jones F.N., Pappas S.P., and Wicks D.A. (2007). *Organic Coatings. Science and Technology*, III ed. Hoboken, NJ: John Wiley & Sons, Inc.

De With G. (2018). *Polymer Coatings. A Guide to Chemistry, Characterization, and Selected Applications*. Weinheim: Wiley-VCH Verlag GmbH & Co. KGaA.

9 Thermal and Mechanical Properties of Polymer Coatings

Emerson Escobar Nunez
Universidad Autónoma de Occidente

Kian Bashandeh and Andreas A. Polycarpou
Texas A&M University

CONTENTS

9.1 INTRODUCTION

Tribology (which is the study of friction, wear, and lubrication) has emerged as one of the fields that contribute to sustainable development through environmentally friendly solutions. These solutions are aimed to reduce energy consumption by lowering power losses during sliding applications. Power losses at sliding interfaces can be diminished by reducing the abrasive and adhesive nature of hard metal asperity interactions. One way to reduce friction losses is using high-performance polymer coatings (HPPCs) as interacting (sliding) surfaces (tribopairs).

It has been shown that HPPCs based on polyether ether ketone (PEEK), polytetrafluoroethylene (PTFE), and aromatic thermosetting copolyester (ATSP) blended with solid lubricants can withstand high bearing loads and provide self-lubrication.

This is due to their ability to form tribofilms by transferring material to the hard metal asperities of the counterpart, thus reducing the friction and wear at the sliding interface [1,2]. HPPCs possess the advantage of deposition onto metallic substrates in the tens of microns range (10–40 m), and it has been shown that this thickness range is sufficient to form a well-adhered and robust transfer film [3–5]. Micromechanical properties of HPPCs are key and important factors, not only to sustain the transfer film at the contact interface of polymer/metal tribopairs, but also to maintain their structural stability, including under harsh operating conditions and adverse environments. Furthermore, realistic tribological experiments are typically more complex to conduct compared to the measurement of micromechanical properties, and such property measurements could be correlated to tribological performance and thus become a predictive tool as well.

Among the advantages that make high-performance polymers (HPPs) and HPPCs stand out in different tribological applications, compared to typical metallic tribopairs one can mention: lightweight, self-lubricity, chemical inertness and stability at different temperatures, etc. For instance in low earth orbit space environments, when a spacecraft passes through the earth's shade side, it suffers thermal cycling due to cryogenic changes in temperature, which can lower the self-lubricating capability of HPPs and their thermal and structural stability. Also, under low-earth-orbit conditions, components made from HPPs are exposed to degradation caused by atomic oxygen, ultraviolet radiation, and high vacuum, which can negatively impact the tribological, thermomechanical, and optical properties and also erode the polymer surface, thus reducing their useful lifespan [6–9]. Moreover, advanced materials are used in the oil and gas industry for tilting pad bearings in electrical submersible pumps, where tribological problems are caused when the oil film breaks down causing seizure of the hydrodynamic bearings by the action of rock debris in the pump seals [1]. In this regard, HPPs and HPPCs are an attractive solution in this industrial sector, which in the absence of a lubricant can provide self-lubricity and avoid scuffing or galling of the sliding components [10]. In applications such as extended reach drilling, the drilling fluid temperature can be as high as 200°C due to its interaction with solid rocks, which highlights the importance of tribological and micromechanical properties of HPPCs at elevated temperatures [11]. In the current chapter, the micromechanical properties of HPPCs will be addressed at different temperature regimes, with focus on micro-/nanoindentation and micro-/nanoscratch techniques.

9.2 HIGH-PERFORMANCE POLYMER (HPP) BLENDS

HPPs are materials at the top of the polymer pyramid: Among these materials, PEEK, PTFE, and polyimide (PI) are recognized as the most attractive thermoplastic HPPs for different demanding industrial applications. Some of the properties that make them stand out in different technological applications are (1) high load-bearing capacity, (2) high operating temperature, (3) great dimensional stability or small changes in volume by the influence of temperature and pressure, and (4) acceptable values of coefficient of friction (COF) and wear rates [12–21]. ATSP is another HPP that possesses good mechanical properties and chemical resistance, combined with low wear rates and low COF, when blended with PTFE [22–24]. Polymer bulk-on-polymer coating sliding has proven to be highly beneficial for bearing applications. The ATSP-based coatings were shown to work as a pre-deposited transfer layer that decreases friction and wear, stabilize the contact, and protect the metallic surface [25].

PEEK, PTFE, and ATSP in powder form can be blended with solid lubricants and applied as coatings by electrostatic deposition on traditional engineering substrates, such as gray cast iron, sintered iron, bronze, and aluminum [3,26]. A literature review shows that in the last decade, a great effort has been made to study and understand the tribological behavior of HPPs and HPPCs under different conditions of normal load, sliding velocity, temperature, and pressure (using different atmospheres). Overall, these studies report that engineering substrates coated with HPPCs such as PEEK, PTFE, and ATSP are very promising in different technical and industrial applications. Specifically, under boundary/mixed lubrication regime, or in the absence of a lubricant, metallic asperity interactions can be avoided by the formation of transfer films from polymer to metal, at the sliding interface. The quality of third bodies and their ability to form transfer films have a close relation to their micromechanical properties [27].

9.3   THERMAL PROPERTIES OF POLYMER COATINGS

Mechanical and micromechanical properties of HPPs and HPPCs are highly influenced by temperature and strain rates [28,29]. At microscale, temperature gradients affect the mobility of polymer chains, which modifies bulk mechanical properties such as the elastic modulus and yield strength [29]. In semicrystalline thermoplastics, while initial deformation is mainly controlled by lamellar structures (also known as spherulites), final deformation is governed by amorphous polymer structures. This final deformation is more sensitive when the temperature approaches the glass transition temperature (Tg) of the material [30].

Differential scanning calorimetry (DSC) and thermogravimetric analysis (TGA) are techniques used to study the thermal stability of HPPs and HPPCs. DSC is employed to study the crystallinity and glass transition temperatures, and TGA is used to measure the weight loss and degree of degradation with thermal cycles [31]. TGA and DSC are usually complemented with dynamic mechanical analysis (DMA), to investigate the temperature changes of storage modulus and loss factor (tan), which provide information about the rigidity of the polymer and the relaxation temperatures, respectively [32]. For instance, TGA measurements performed on ATSP powder showed a weight loss less than 3%, indicating that monomer residue fraction was very small and that the polymerization of ATSP powder was completed (see Figure 9.1) [22]. In the same study, DMA analysis showed that ATSP/PTFE blends of higher ratio displayed both a larger storage modulus and a larger-loss-factor glass transitions, compared to pure PTFE, as can be observed in Figure 9.2a and b. It was proven by pin-on-disk tribological experiments that higher ATSP/PTFE ratios lead to lower wear rates.

In another study, the tribological behavior of gray cast iron disks coated with ATSP/PTFE blends (i.e., 5% PTFE) was evaluated under cryogenic conditions [33]. To understand the thermal stability, DMA tests were performed on the ATSP coating

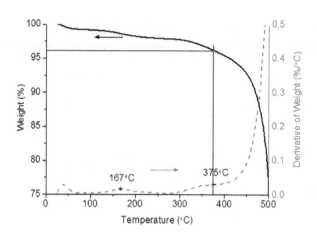

FIGURE 9.1 TGA scan on ATSP powder. (Reprinted from [22] with permission from John Wiley & Sons.)

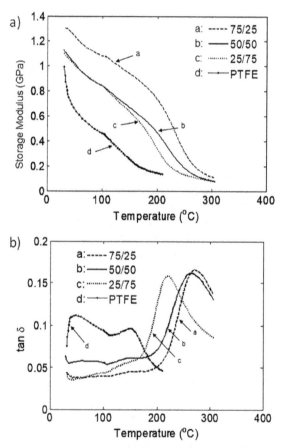

FIGURE 9.2 DMA experimental results for different ATSP/PTFE compositions: (a) flexural modulus; (b) loss factor. (Reprinted from [22] with permission from John Wiley & Sons.)

from room temperature to −140°C. Experimental results showed a peak in the loss modulus at −90°C and an increment in the storage modulus from 1.4 to 14.2 GPa when the temperature reduced from room temperature to −140°C. It was shown that ATSP becomes stiffer at cryogenic temperatures, which makes it more prone to form cracks under sliding conditions. Unlike ATSP/PTFE, PEEK/PTFE coating is less sensitive to form cracks due to its lower storage modulus under the same cryogenic and tribological conditions [33].

9.4 MICRO-/NANOINDENTATION AND MICRO-/NANOSCRATCH TECHNIQUES

Technological developments of high-resolution control and force sensors allowed the development of instrumented micro-/nanoindentation (referred to as nanoindentation) as a widely used technique to measure the micromechanical properties of coatings. This technique has been proven to be effective to measure micromechanical properties due to its ability to penetrate and plastically deform materials on a very small scale [34–36]. Pioneering nanoindentation methods to measure the hardness and reduced elastic modulus have not only been reported, but have broadly been cited over the last three decades [37–39]. Oliver and Pharr proposed a method for determining hardness and reduced elastic modulus, which became a widely used method [32]. In this regard, Oliver and Pharr nanoindentation method expands and relies on Doerner and Nix's initial idea that the contact area remains unchanged through the unloading step [37].

Based on the previous assumption, the resulting unloading curve can be considered linear. Nonetheless, careful analysis of the unloading curves shows a non-linear behavior, or even more, this behavior can barely be considered linear, even in the initial part of the unloading curve. Therefore, assuming that the contact area during unloading remains unchanged is not fully correct. Based on the previous remarks and despite the curvy nature of the unloading curves, Oliver and Pharr claimed that the unloading portion could be well approximated by using a power law curve fitting. Figure 9.3a and b shows a graphical representation of an indentation section and the normal load versus indenter vertical displacement, respectively. From these figures, the following equations describe the physical parameters involved during an indentation:

$$h = h_c + h_s \tag{9.1}$$

where h is the total vertical displacement, h_c is the vertical distance where the contact is made (referred to as contact depth), and h_s is the displacement of the surface at the boundary of the contact. Also, from the load–displacement data on the unloading curve, the reduced elastic modulus can be obtained using the following analytical expression:

$$S = \frac{dp}{dh} = \frac{2\beta}{\sqrt{\pi}} E_r \sqrt{A} \tag{9.2}$$

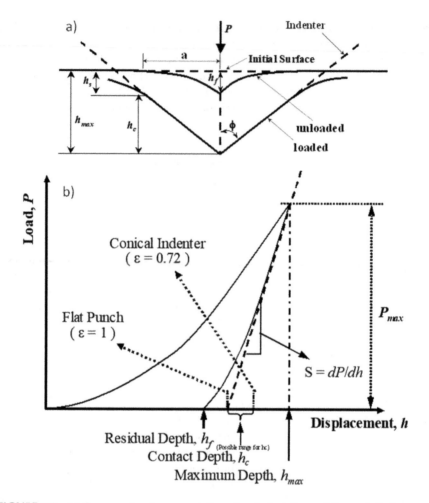

FIGURE 9.3 (a) Cross-sectional representation of an indentation and the analytical parameters involved. (b) Schematic representation of load versus indentation displacement and the analytical parameters considered to obtain the mechanical properties.

where S represents the measured contact stiffness of the initial portion of the unloading curve, β is a constant that depends on the geometry of the indenter, E_r is the reduced elastic modulus, and A is the projected area of the elastic contact. The area function of the indenter is one of the most difficult parameters to measure. It requires a relation between the contact area and the contact depth (h_c) for the specific geometry of the indenter [34]. For instance, for a Berkovich tip, considering the fact that the three-sided pyramid is not geometrically perfect and the top of the tip possesses some roundness, the area function takes the following analytical expression:

$$A_c(h_c) = C_o h_c^{\,2} + C_1 h_c + C_2 h_c^{\,1/2} + C_3 h_c^{\,1/4} + C_4 h_c^{\,1/8} + C_5 h_c^{\,1/16} + \ldots\ldots C_8 h_c^{\,1/128} \quad (9.3)$$

where in equation (3) the contact depth h_c can be expressed as:

$$hc = h_{max} - \varepsilon \frac{P_{max}}{S} \qquad (9.4)$$

The coefficients C_i in Eq. (9.3) can be obtained using a standard or reference material with known elastic modulus (such as fused quartz), and then the analytical contact area for each corresponding contact depth can be obtained, and the C_i coefficients can be found by linear regression. In Eq. (9.4), h_{max} is the maximum displacement and ε is a constant depending on the tip geometry. Referring to Figure 9.3b, $\varepsilon = 1$ for a flat punch, 0.72 for a conical indenter, and 0.75 for a Berkovich pyramidal indenter. (In this case, the top of the tip can be approximated to a sphere at shallow depths.) Usually, tip shape calibration needs to be performed before performing indentation experiments at different contact depths. In this regard, material selection for tip calibration and the range of contact depth are important parameters affecting tip calibration.

Mechanical properties and the nature of the calibration material should be similar to the material under analysis. For instance, in the specific case of polymer coatings, bismaleimide polymer has been reported as a calibration material to obtain the micromechanical properties of coatings in the range of tens of microns thickness [40]. Since the thickness of HPPCs is in the range of tens of microns, Berkovich indenter (as seen in Figure 9.4 a and b) is preferred to perform indentation on these thick coatings (shown in Figure 9.5 a and b). Reduced elastic modulus and hardness of PEEK/PTFE and PTFE/MoS$_2$ coatings measured at room temperature using a Berkovich tip can be seen in Figure 9.6 [41]. From this figure, it can be observed that the PTFE/MoS$_2$ coating is stiffer than the PEEK/PTFE coating, and the loading curves employed to obtain their properties will be explained later. Loading curves for polymer coatings should be especially designed to avoid viscoelastic creep effects when measuring stiffness on the unloading curve.

In miniature devices and thin films, the measurement of wear and scratching resistance is of importance, taking into account that the methods to quantify wear at the macro-level would no longer be considered valid at small scale. Micro-/nanoscratch

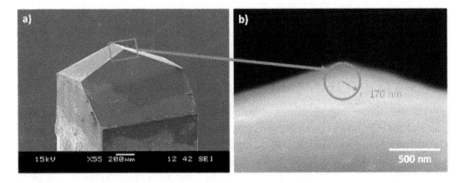

FIGURE 9.4 (a) Typical scanning electron microscopy image of a Berkovich tip used in the high-load indentation measurements of polymer coatings and (b) a zoom-in of the end of the tip radius of curvature of approximately 170 nm. (Reprinted from [40] with permission from Elsevier.)

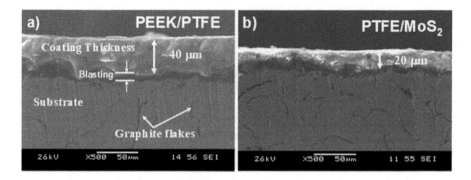

FIGURE 9.5 Scanning electron microscopy images of typical polymer coatings in the tens of micron thickness range: (a) PEEK/PTFE and (b) PTFE/MoS₂. (Reprinted from [5] with permission from Elsevier.)

is another widely used technique to assess the resistance of coatings to shear stress and delamination and to measure their scratch hardness, adhesion, and shear strength on a substrate. Micromechanical properties such as the yield strength (σ_y) of the coatings can be approximated using Tabor's empirical model. Tabor correlated σ_y and H for ductile materials experimentally and observed that hardness $H \sim 3\sigma_y$. This relationship can be reasonably employed to obtain the σ_y for thin crystalline coatings [42]. Moreover, substrate effects are somewhat delayed when probing thin films, compared to indentation [43–46]. In this technique, the indenter moves along the surface of the coating over microns distance according to a specific normal load profile: By measuring the normal and tangential loads, as well as the residual cross-sectional area, a better understanding of the microstructural effects such as grain size, grain boundaries, and surface roughness over the scratch hardness of coatings can be obtained [47].

While the scratch resistance of polymer coatings is influenced by the viscoelastic behavior of the coating material, its cross-linking density, glass transition temperature T_g, and stress gradient induced during the scratch are governed by the normal and lateral loads, sliding speed, and geometry of the indenter [48]. A typical conospherical indenter tip used to perform nanoscratch experiments is shown in Figure 9.7. This tip is usually loaded with a high-load transducer in cases where delamination of coatings needs to be induced [49].

9.5 MICROMECHANICAL PROPERTIES OF HPPCS

HPPCs and their composites have been used in a wide variety of applications composed of rubbing and bearing components made of similar and dissimilar tribopairs. The tribological properties are of paramount importance in achieving high load-bearing capacity in such applications. Since the tribological performance is mainly determined by the mechanical properties of the topmost surface layers, it is imperative to measure the mechanical properties of the top layers of the sliding materials (versus the bulk properties). The conventional techniques for the determination of mechanical properties of bulk polymers, such as bending or tensile tests, are not suitable for the characterization of surface properties as in coatings, since the bulk

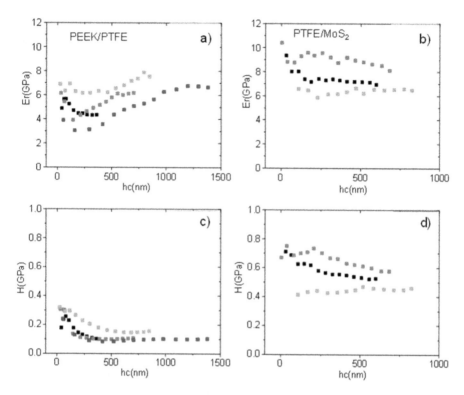

FIGURE 9.6 Reduced elastic modulus (E_r) and hardness (H) of (a) and (c) PEEK/PTFE polymer coatings; (b) and (d) PTFE/MoS$_2$ polymeric coatings. (Adapted from [41].)

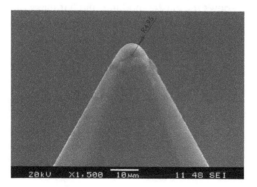

FIGURE 9.7 Typical conospherical tip with a 60° angle used to perform microscratch experiments on coatings. (Reprinted from [49] with permission from Elsevier.)

properties could be different, compared to the properties of the surface layers [50]. Nanoindentation has been proven as a valuable tool in the measurement of mechanical properties of the surface and subsurface regions in the scale of tens of microns to submicron levels. Hardness and reduced elastic modulus are the two primary properties obtained through nanoindentation. As previously mentioned, care must be taken in interpreting the indentation responses of polymeric materials for a number of reasons. The hardness and elastic modulus of the surface coating/film could vary compared to the bulk due to a phenomenon known as indentation size effect. Here, the hardening effect is either due to the higher modulus of the surface than bulk, or due to the uncertainties that pertain to shallow-depth indentations [51], and the indentation size effects are excluded.

The role of substrate should be minimized to ensure the measurement of the properties of the coating (and surface layers) without significant substrate effect [52]. Polymers are viscoelastic and, therefore, exhibit a time–temperature-dependent behavior. During indentation under constant load, or even during unloading, a continuous increase in the contact depth could occur due to creep behavior. Such behavior needs to be eliminated by the proper selection of the load function [51]. The mechanical behavior could also depend on the testing conditions such as strain amplitude, strain rate, and temperature, which must be properly determined before the experiments [53].

9.5.1 Substrate Effects

Minimizing the substrate effect is imperative to determine the intrinsic properties of the coatings. To determine the behavior of coating/substrate system, indentations at several loads are required to create different contact depths [54]. Assuming the true coating properties are obtained at shallow depths, the role of the substrate begins when the elastic modulus starts to increase (in the case of a polymer coating on a harder metal substrate) from the plateau region. Two PTFE-based coatings, namely resin/PTFE/MoS$_2$ and PTFE/pyrrolidone-2, were deposited on gray cast iron substrate [40]. The thicknesses of the coatings were 20±5 and 23±5 μm, respectively. Micromechanical properties were measured using an instrumented nanoindenter at room temperature.

Figure 9.8 a and b shows the elastic modulus and hardness values for the two coatings extracted from load–displacement curves at different contact depths. Indentations

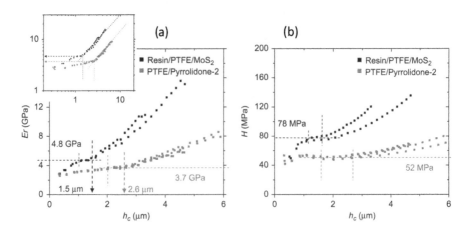

FIGURE 9.8   (a) Reduced elastic modulus and (b) hardness from partial unloading indentations for resin/PTFE/MoS$_2$ and PTFE/pyrrolidone-2 coatings. Initial plateau regions show mechanical properties without substrate effects. (Reprinted from [40] with permission from Elsevier.)

up to 600 mN were performed to obtain contact depths up to 6 μm. The resin/PTFE/ MoS$_2$ showed a higher modulus and hardness compared to PTFE/pyrrolidone-2. The plateau region where the elastic modulus and hardness become independent of the substrate was found to be at about 10% of the coating thickness (i.e., 8% and 9% for resin/PTFE/MoS$_2$ and PTFE/pyrrolidone-2, respectively). The so-called 10% rule (where the indentation depth is 10% of the coating thickness) is generally believed to be more effective for hardness, but still is a good approximation for reduced modulus determination [55]. In a similar study by Tayebi et al. [56], 20% threshold was reported for soft metal films on hard substrates. The lower values for soft polymers indicate the role of substrate effect at lower contact depths compared to soft metals. A slight increase in elastic modulus and hardness at shallow indentation depth was observed, which was attributed to the more pronounced effects of surface roughness, tip imperfections, and stage vibration at shallow-depth indentation.

9.5.2 CORRELATION OF MICROMECHANICAL PROPERTIES WITH MICROSTRUCTURE

Performing a number of indentations over different locations on the surface and contact depths can provide useful information about the uniformity and homogeneity of the coating microstructure through the thickness and surface area [57]. Indentations were performed on PTFE- and PEEK-based coatings to examine their differences in micromechanical properties and their correlation with microstructure [40]. Figure 9.9 a–d shows the load–displacement curves and the extracted reduced modulus and hardness values for indentations at different contact depths and different locations on PTFE/pyrrolidone-2 and PEEK/PTFE coatings. The repeatability (represented by small error bars or less dispersion) of the data over the surface of PTFE/pyrrolidone-2 coating and the smooth increase in the loading curves through the thickness suggested an amorphous-like microstructure with uniform mechanical properties and structure. On the other hand, the PEEK/PTFE coating represented disordered load–displacement curves with different slopes and large error bars for the extracted values of reduced modulus and hardness. This implies different mechanical properties over the surface and heterogeneous microstructure through the thickness, suggesting a semicrystalline (amorphous and crystalline) structure for the coating.

9.5.3 CREEP EFFECTS

The time-dependent viscoelastic creep behavior of polymers brings challenges in performing indentation. Creep usually manifests itself by a round-shaped "nose" at the start of the unloading segment. Therefore, the measurement of mechanical properties would be impossible due to the negative slope of the unloading curve, which is used to determine the properties, such as reduced modulus and hardness. A typical procedure to eliminate/reduce the creep effect is to use a trapezoidal load function with a holding segment at the maximum load to hold the indenter tip for a sufficient time until the material reaches mechanical equilibrium, before the unloading segment begins [58,59]. Indentations were performed on PTFE/pyrrolidone-2 coatings using a triangular load function without hold time and a trapezoidal load function with 5 s of holding time [40]. The creep behavior is readily observed on the curves by

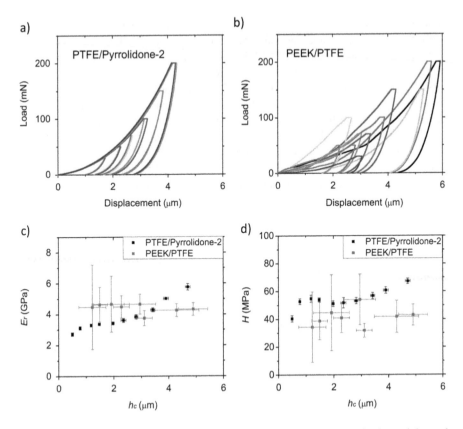

FIGURE 9.9   (a) and (b) Load–displacement curves; (c) and (d) reduced elastic modulus and hardness, respectively, for PTFE/pyrrolidone-2 and PEEK/PTFE coatings. (Adapted from [40].)

the nose shape of the curves and the continuous increase in the displacement after the load started to decrease during unloading. It was shown that the 5-s hold time could eliminate the creep effect.

9.5.4   Thermomechanical Effects

Understanding the mechanical behavior of polymers at elevated temperatures is of great importance to obtain reliable performance for applications that demand operation at elevated service temperatures. High-temperature indentation has been used and is a powerful tool for the testing and assessment of thermomechanical behavior at nanoscale/micron scale. Thermal drift is a common challenge for the indentation of materials at high temperatures, and it needs to be minimized. In case of polymers, the challenge is further compounded due to the viscoelastic nature of polymers exhibiting temperature- and time-dependent mechanical behavior [60]. Nanoindentation experiments were performed on two advanced polymer coatings, namely ATSP and PEEK, from room temperature to elevated temperatures of up to 260°C [61]. A continuous decrease in hardness was measured for both the coatings with an increase

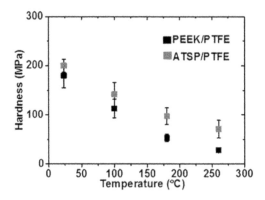

FIGURE 9.10 Nanoindentation hardness for PEEK/PTFE and ATSP/PTFE coatings at different temperatures. (Reprinted from [61] with permission from Elsevier.)

in temperature, as shown in Figure 9.10. A similar trend for hardness and modulus was observed for other polymers such as polyamide 6 [62]. The viscous/relaxation and elastic terms of the two coatings were calculated through curve fitting the quadratic Maxwell model by the unloading data from the nanoindentation experiments at different temperatures. The authors concluded that the obtained parameters were characteristic of the polymers and could describe the viscoelastic behavior very well.

9.5.5 MICRO-/NANOSCRATCH EXPERIMENTS

Scratch testing is frequently used in the assessment of material properties and performance in a number of ways, including the following [44,45,49,63,64]:

a. Single-line constant-load scratch experiments to measure the scratch COF, the in situ penetration depth (which includes both the elastic and plastic components), and scratch recovery by post-scanning the surface to measure the residual depth. These measurements can also be used to measure the scratch hardness [44].
b. Multi-scratch (repeated) single-line constant-load experiments to measure the COF over repeated scratching of the coating and also to assess the durability of the coating.
c. Multi-scratch under planar conditions to obtain "wear" patterns on the surface and thus assess the long-term micro-/nanowear of the coating.
d. Ramp loading scratch (starting from zero normal load and increasing to a maximum load to cause coating delamination/distraction). Such measurements can be used to measure the (1) yield strength of the coating [45], (2) normal strength of the coating on the substrate, (3) shear strength of the coating on the substrate [49], and (4) maximum pressure to delaminate the coating [64].

Microscratch experiments were performed on ATSP and PEEK coatings with a diamond conospherical indenter probe using a ramp load function with a maximum force

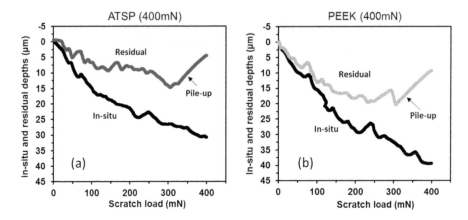

FIGURE 9.11 Microscratch experiments on (a) ATSP- and (b) PEEK-based coatings, with in situ and residual depth plots using a maximum ramp force of 400 mN and a scratch distance of 400 μm. (Reprinted from [26] with permission from Elsevier.)

of 400 mN [26]. Figure 9.11 a and b shows the measured in situ and residual depths versus scratch load from the microscratch experiments. Compared to PEEK coating, the ATSP coating exhibited a lower in situ scratch depth and larger recovery revealed by a smaller residual depth. The recovery rate was calculated by dividing the elastic recovery with the in situ scratch depth, and it provides useful information about the coating's resistance to damage and plastic deformation. The ATSP coating had a 25% higher elastic recovery compared to PEEK coating, which was in good agreement with macro-tribological experiments under three-body abrasive conditions [26]. Scratch testing was also used to estimate the interphase thickness in carbon-fiber-reinforced PEEK matrix. The methodology relied on the measurement of scratch COF through the scratch path from matrix to the fiber region, and the interphase thickness was estimated by variations in the COF [65].

9.6 CORRELATION BETWEEN MECHANICAL PROPERTIES AND TRIBOLOGICAL PERFORMANCE OF HPPCs

Our previous work has shown the advantages of micro-/nanoindentation and scratch experiments in the characterization of the tribological behavior of polymer-based coatings. The studies were performed to understand the role of micromechanical properties in the friction and wear performance [5,26,39,66,67]. For instance, studies on the tribological performance of PTFE/pyrrolidone-2 and PEEK/PTFE coatings showed a stable and low friction for PTFE/pyrrolidone-2, whereas PEEK/PTFE exhibited a significantly higher fluctuation in COF with an order of magnitude higher wear rate [59]. The difference in the tribological behavior was attributed to the differences in the mechanical properties of the coatings and the repeatability of load–displacement curves during the indentation measurements. Consistent mechanical properties on the surface and through the thickness, as well as uniform microstructure, resulted in the stable friction behavior and long-term reliability of PTFE/pyrrolidone-2 coating. In Ref. [61], an elevated temperature friction model was developed based on high-temperature nanoindentation experiments. Nanoindentation experiments were performed (see Figure 9.10) at similar temperatures to macro-scale ball-on-disk tribological experiments for ATSP- and PEEK-based coatings. Compared to PEEK coating, ATSP coating showed a higher hardness and indentation recovery rate at all temperatures, which explained the better wear resistance of ATSP coating [68].

9.7 SUMMARY

The mechanical and thermal properties of different advanced polymer-based coatings were reviewed, and the following conclusions could be drawn:

1. HPPCs are attractive tribopairs for engineering applications where the sliding components are exposed to high normal loads, where retardation or isolation of adhesion between metal asperities is achieved by transfer of polymer films. Transfer of polymer films is formed by third bodies and is governed by the ability to keep the films at the sliding interface and maintain their micromechanical properties unchanged with an increase in interfacial temperature.
2. Nanoindentation and nanoscratch techniques are widely used to measure the micromechanical properties of polymer coatings. The former provides the in situ reduced elastic modulus and hardness of the coating at the location of the indent, whereas the latter provides an average hardness, yield strength, and delamination strength of the coating over the plastically deformed scratch distance.
3. By comparison of micromechanical properties of PEEK/PTFE versus PTFE/pyrrolidone-2, it was observed that the former displayed a higher variation of the measured reduced elastic modulus and hardness compared to the latter and these non-uniform properties were also evidenced on the load–displacement curves. This variation was attributed to the non-homogenous microstructure of PEEK/PTFE coating compared to PTFE/pyrrolidone-2 coating.
4. Analysis of micromechanical properties at varying temperatures (from room temperature to 260°C) shows relaxation of the polymer coating with an increase in temperature for both PEEK- and ATSP-based coatings. Unlike PEEK coatings that displayed an abrupt change in the viscoelastic parameters as a function of temperature, ATSP coatings showed a higher in situ hardness and indentation recovery at all temperatures. As a result, ATSP coatings proved to have better wear resistance during tribological testing.

REFERENCES

1. Nunez EE, Gheisari R, Polycarpou AA. 2019. Tribology review of blended bulk polymers and their coatings for high-load bearing applications. *Tribol. Int.* 129: 92–91.
2. Nunez EE, Polycarpou AA. 2015. The effect of surface roughness on the transfer of polymer films under unlubricated testing conditions. *Wear.* 326–327: 74–83.

3. Dascalescu D, Polychronopoulou K, Polycarpou AA. 2009. The significance of tribo-chemistry on the performance of PTFE-based coatings in CO_2 refrigerant environment. *Surf. Coating. Technol.* 204: 319–29.

4. Nunez EE, Demas NG, Polychronopoulou K, Polycarpou AA. 2010. Comparative scuff-ing performance and chemical analysis of metallic surfaces for air-conditioning compres-sors in the presence of environmentally friendly CO_2 refrigerant. *Wear.* 268: 668–676.

5. Nunez EE, Yeo SM, Polychronopoulou K, Polycarpou AA. 2011. Tribological study of high bearing blended polymer-based coatings for air-conditioning and refrigeration compressors. *Surf. Coat. Technol.* 205: 2994–3005.

6. Lv M, Wang Q, Wang T, Liang Y. 2015. Effects of atomic oxygen exposure on the tribo-logical performance of ZrO_2-reinforced polyimide nanocomposites for low earth orbit space applications. *Comp. Part B.* 77: 215–222.

7. Park SY, Choi HS, Choi WJ, Kwon H. 2012. Effect of vacuum thermal cyclic exposures on unidirectional carbon fiber/epoxy composites for low earth orbit space applications. *Comp. Part B.* 43: 726–38.

8. Han JH, Kim CG. 2006. Low earth orbit space environment simulation and its effects on graphite/epoxy composites. *Compos. Struct.* 72: 218–26.

9. Grossman E, Gouzman I. 2003. Space environment effects on polymers in low earth orbit. *Nucl. Instrum. Meth. Phys. Res. B.* 208: 48–57.

10. Lan P, Meyer JL, Vaezian B, Polycarpou AA. 2016. Advanced polymeric coatings for tilting pad bearings with application in the oil and gas industry. *Wear.* 354–355: 10–20.

11. Lan P, Polycarpou AA. 2018. High temperature and high pressure tribological experi-ments of advanced polymeric coatings in the presence of drilling mud for oil & gas applications. *Tribol. Int.* 120: 218–225.

12. Friedrich K. 2018. Polymer composites for tribological applications. *Adv. Ind. Eng. Polym. Res.* 1: 3–39.

13. Cannaday ML, Polycarpou AA. 2005. Tribology of unfilled and filled polymeric sur-faces in refrigerant environment for compressor applications. *Tribol. Lett.* 19: 249–262.

14. Lin L, Pei X, Bennewitz R, Schlarb AK. 2018. Friction and wear of PEEK in continuous sliding and unidirectional scratch tests. *Tribol. Int.* 122: 108–93.

15. Schroeder R, Torres FW, Binder C, Klein AN, De Mello JDB. 2013. Failure mode in sliding wear of PEEK based composites. *Wear.* 301: 717–726.

16. Li G, Qi H, Zhang G, Zhao F, Wang T, Wang Q. 2017. Significant friction and wear reduction by assembling two individual PEEK composites with specific functionalities. *Mater. Des.* 96: 152–159.

17. Rodriguez V, Sukumaran J, Schlarb AK, De Baets P. 2016. Reciprocating sliding wear behavior of PEEK-based hybrid composites. *Wear.* 362–363: 161–169.

18. Li J, Qu J, Zhang J. 2015. Wear properties of brass and PTFE–matrix composite in traveling wave ultrasonic motors. *Wear.* 338–339: 385–393.

19. Song F, Wang Q, Wang T. 2016. The effects of crystallinity on the mechanical proper-ties and the limiting PV (pressure×velocity) value of PTFE. *Tribol. Int.* 93: 1–10.

20. Liu H, Wang T, Wang Q. 2012. Tribological properties of thermosetting polyimide/TiO2 nanocomposites under dry sliding and water-lubricated conditions. *J. Macromol.* 51: 2284–96.

21. Zhao G, Hussainova I, Antonov M, Wang Q, Wang T, Yung DL. 2015. Effect of tem-perature on sliding and erosive wear of fiber reinforced polyimide hybrids. *Tribol. Int.* 82: 525–533.

22. Zhang J, Demas N.G, Polycarpou A.A, Economy J. 2008. A new family of low wear low coefficient of friction polymer blends based on polytetrafluoroethylene and aromatic thermosetting polyester. *Polymer Adv Technol.* 19: 905–912.

23. Zhang J, Polycarpou AA, Economy J. 2010. An improved tribological polymer coating system for metal surfaces. *Tribol. Lett.* 38: 355–365.

24. Demas NG, Zhang J, Polycarpou AA, Economy J. 2008. Tribological characterization of aromatic thermosetting copolyester–PTFE blends in air conditioning compressor environment. *Tribol. Lett.* 29: 253–258.
25. Bashandeh K, Lan P, Polycarpou AA. 2019. Tribological performance improvement of polyamide against steel using polymer coating. *Tribol. Trans.* 62: 1051–1062.
26. Lan P, Polychronopoulou K, Zhang Y, Polycarpou AA. 2017. Three-body abrasive wear by (silica) sand of advanced polymeric coatings for tilting pad bearings. *Wear.* 382: 40–50.
27. Yeo SM, Polycarpou AA. 2014. Fretting experiments of advanced polymeric coatings and the effect of transfer films on their tribological behavior. *Tribol. Int.* 79: 16–25.
28. Jaekel DJ, MacDonald DW, Kurtz SM. 2011. Characterization of PEEK biomaterials using the small punch test. *J. Mech. Behav. Biomed. Mater.* 4: 1275–1282.
29. Rae PJ, Brown EN, Orler EB. 2007. The mechanical properties of poly(ether-ether-ketone) (PEEK) with emphasis on the large compressive strain response. *Polymer.* 48: 598–615.
30. Gonzalez DG, Zaera R, Arias A. 2017. A hyperelastic-thermoviscoplastic constitutive model for semi-crystalline polymers: Application to PEEK under dynamic loading conditions. *Int. J. Plast.* 88: 27–52.
31. Gupta R, Shinde S, Yella A, Subramaniam C, Saha SK. 2020. Thermomechanical characterisations of PTFE, PEEK, PEKK as encapsulation materials for medium temperature solar applications. *Energy.* 194: 96921.
32. Wang Q, Zheng F, Wang T. 2016. Tribological properties of polymers PI, PTFE and PEEK at cryogenic temperature in vacuum. *Cryogenics.* 75:19–25.
33. Lan P, Gheisari R, Meyer JL, Polycarpou AA. 2018. Tribological performance of aromatic thermosetting polyester (ATSP) coatings under cryogenic conditions. *Wear.* 398–399: 47–55.
34. Garcia IA, Berasategui EG, Bull SJ, Page TF. 2002. How hard is fullerene-like CNx? some observations from the nanoindentation response of a magnetron-sputtered coating. *Phil. Mag.* 82:2133–2147.
35. Lemoine P, Zhao JF, Quinn JP, McLaughlin JA, Maguire P. 2000. Hardness measurements at shallow depths on ultra-thin amorphous carbon films deposited onto silicon and Al_2O_3–TiC substrates. *Thin. Solid. Films.* 379: 166–172.
36. Charitidis C, Logothetidis S. 2000. Nanomechanical and nanotribological properties of carbon based films. *Thin. Solid. Films.* 482: 120–125.
37. Doerner MF, Nix WD. 1986. A method for interpreting the data from depth-sensing indentation instruments. *J. Mater. Res.* 1: 601–609.
38. Oliver WC, Pharr GM. 1992. An improved technique for determining hardness and elastic modulus using load and displacement sensing indentation experiments. *J. Mater. Res.* 7: 1564–1583.
39. Cheng Y, Cheng C. 1998. Scaling approach to conical indentation in elastic-plastic solids with work hardening. *J. Appl. Phys.* 84: 1284–1291.
40. Yeo SM, Polycarpou AA. 2013. Micromechanical properties of polymeric coatings. *Trib. Int.* 60: 198–208.
41. Nunez EE. 2010. *Scuffing and Wear of Engineering Materials under Different Lubrication Regimes in the Presence of Environmentally Friendly Refrigerants.* PhD Thesis, Champaign, IL: University of Illinois.
42. Tabor D. 1951. *Hardness of Metals.* Oxford: Oxford University Press.
43. ASTM D7027-13. 2013. *Standard Test Method for Evaluation of Scratch Resistance of Polymeric Coatings and Plastics Using an Instrumented Scratch Machine.* West Conshohocken, Pennsylvania: ASTM Intern.
44. Tayebi N, Conry TF, Polycarpou AA. 2003. Determination of hardness from nano-scratch experiments: Corrections for interfacial shear stress and elastic recovery. *J. Mater. Res.* 18: 2150–2162.

45. Lee KM, Yeo CD, Polycarpou AA. 2008. Relationship between scratch hardness and yield strength of elastic perfectly plastic materials using finite element analysis. *J. Mater. Res.* 23: 2229–2237.

46. Chen Y, Bakshi SR, Agarwal A. 2010. Correlation between nanoindentation and nano-scratch properties of carbon nanotube reinforced aluminum composite coatings. *Surf Coating Technol.* 204: 16–17.

47. Pelletier H, Mendibide C, Riche A. 2008. Mechanical characterization of polymeric films using depth-sensing instrument: Correlation between viscoelastic-plastic properties and scratch resistance. *Prog. Org. Coat.* 62: 162–178.

48. Lahijania YZK, Mohseni M, Bastani S. 2014. Characterization of mechanical behavior of UV cured urethane acrylate nanocomposite films loaded with silane treated nano-silica by the aid of nanoindentation and nanoscratch experiments. *Tribol. Int.* 69: 10–18.

49. Lee J, Polychronopoulou K, Cloud AN, Abelson JR, Polycarpou AA. 2014. Shear strength measurements of hafnium diboride thin solid films. *Wear.* 318: 168–176.

50. Sinha SK, Briscoe BJ. 2009. *Polymer Tribology.* Imperial College Press, London.

51. VanLandingham MR, Villarrubia JS, Guthrie WF, Meyers GF. 2001. Nanoindentation of polymers: an overview. *Macromol. Symp.* 167: 15–43.

52. Strojny A, Xia X, Tsou A, Gerberich WW. 1998. Techniques and considerations for nanoindentation measurements of polymer thin film constitutive properties. *J. Adhes. Sci. Technol.* 12: 1299–1321.

53. VanLandingham MR. 2003. Review of instrumented indentation. *J. Res. Natl. Inst. Stand. Technol.* 108: 249–265.

54. Tsui TY, Pharr GM. 1999. Substrate effects on nanoindentation mechanical property measurement of soft films on hard substrates. *J. Mater. Res.* 14: 292–301.

55. Fischer-Cripps AC. 2009. *Handbook of Nanoindentation.* Forestville, Australia: Fischer-Cripps Laboratories Pty Ltd.

56. Tayebi N, Polycarpou AA, Conry TF. 2004. Effect of substrate on determination of hardness of thin films by nanoscratch and nanoindentation techniques. *J. Mater. Res.* 19: 1791–802.

57. Dragatogiannis DA, Koumoulos E, Ellinas K, Tserepi A, Gogolides E, Charitidis CA. 2015. Nanoscale mechanical and tribological properties of plasma nanotextured cop surfaces with hydrophobic coatings. *Plasma Process Polym.* 12: 1271–1283.

58. Briscoe BJ, Fiori L, Pelillo E. 1998. Nano-indentation of polymeric surfaces. *J. Phys. D. Appl. Phys.* 31: 2395–2405.

59. Klapperich C, Komvopoulos K, Pruitt L. 2000. Nanomechanical properties of polymers determined from nanoindentation experiments. *J. Trib.* 123: 624–631.

60. Duan ZC, Hodge AM. 2009. High-temperature nanoindentation: New developments and ongoing challenges. *JOM.* 61: 22–36.

61. Lan P, Zhang Y, Dai W, Polycarpou AA. 2018. A phenomenological elevated temperature friction model for viscoelastic polymer coatings based on nanoindentation. *Tribol. Int.* 99: 299–307.

62. Seltzer R, Kim JK, Mai YW. 2011. Elevated temperature nanoindentation behaviour of polyamide 6. *Polym. Int.* 60: 1753–1761.

63. Lee KM, Yeo CD, Polycarpou AA. 2007. Nanomechanical property and nanowear measurements for sub-10-nm thick films in magnetic storage. *Experimental Mech.* 47: 107–121.

64. Humood M, Qin S, Song Y, Polychronopoulou K, Zhang Y, Grunlan JC, Polycarpou AA. 2017. Influence of graphene reduction and polymer cross-linking on improving the interfacial properties of multilayer thin films. *ACS Appl. Mat. & Interf.* 9(1):907–18.

65. Molazemhosseini A, Tourani H, Naimi-Jamal MR, Khavandi A. 2013. Nanoindentation and nanoscratching responses of PEEK based hybrid composites reinforced with short carbon fibers and nano-silica. *Polym. Test.* 99:525–534.

66. Bashandeh K, Lan P, Meyer JL, Polycarpou AA. 2019. Tribological performance of graphene and PTFE solid lubricants for polymer coatings at elevated temperatures. *Trib. Lett.* 67: 99.
67. Yeo SM, Polycarpou AA. 2012. Tribological performance of PTFE-and PEEK-based coatings under oil-less compressor conditions. *Wear.* 296: 638–647.
68. Lan P, Meyer JL, Economy J, Polycarpou AA. 2016. Unlubricated tribological performance of aromatic thermosetting polyester (ATSP) coatings under different temperature conditions. *Trib. Lett.* 61: 10.

10 Conducting Polymer Coatings for Corrosion Protection

George Kordas
Institute of Nanoscience and Nanotechnology (INN)

CONTENTS

10.1   CONDUCTIVE POLYMERS

The conversion of an organic polymer into a conductive material may be done in two ways: first, creating a polymeric matrix with inclusions such as particles, fibers, and flakes, or synthesizing polymers ensuring charge transport such as hydroquinone and polyvinyl polymers that have the ferrocene or carbazole group [1]. Examples are given in Figure 10.1.

The last category presents the second family of polymers named as intrinsic conductive polymers (ICPs), employed for batteries [2,3], electrochromic devices [4], gas sensors [5], and corrosion protection coatings on metals [5]. The application of the ICPs as anticorrosive coatings is proposed to replace the carcinogenic Cr (VI) coatings, because of their environmentally friendly nature. The conjugated organic polymers with low conductivity range from 10^{-10} up to 10^{-5} S/cm can be converted into the metallic conductive state ($1–10^{14}$ S/cm) via doping. Doping is achieved by controlled addition of chemical quantities less than 10% affecting largely their electrical, electronic, magnetic, optical, and structural properties of the polymer. The process is reversible and can be switched back to initial polymer state normal without damage of the main framework. These processes occur chemically or electrochemically.

This mini-review describes the effects of adding molecules to the conductive polymers and their effect on the properties responsible for their anticorrosion protection.

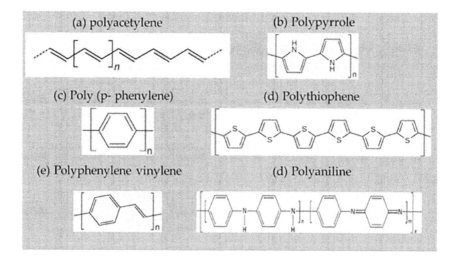

FIGURE 10.1   Chemical types of six widely used endogenously conductive polymers [1].

10.2 SYNTHESIS METHODS

There are two routes for the synthesis of ICPs: first, chemical synthesis; and second, electrochemical synthesis. They both involve polymerization of suitable monomers. In the chemical synthesis methods, the polymer is synthesized and then converted into a conductive state via a doping process. On the contrary, electrochemical synthesis methods synthesize the polymer directly in the conductive state. Characteristic conductive polymers are shown in Figure 10.1 and are (a) polyacetylene, (b) polypyrrole, (c) poly(p-phenylene), (d) polythiophene, (e) polyphenylene vinylene, and (d) polyaniline. Here, we will report only the electrochemical route because of limited space.

10.3 ELECTROCHEMICAL SYNTHESIS (ES) OF CONDUCTIVE POLYMERS

The electrochemical synthesis occurs without the use of organic solvents leading to pure products without the need of purification as must be done with the chemical synthesis methods [7]. The electrochemical production method of conductive polymers produces polymers with extended π-electrons along the chain. For example, polypyrrole can be produced from aqueous sulfuric acid via potentiostatic, potentiodynamic, and galvanostatic techniques. It has been calculated that 2.3 electrons per monomer must be offered to form the polymeric chain for the production of one conductive polypyrrole film on the working electrode (anode) [1]. The suggested mechanism is described in Eq. (10.1):

$$xPy - 2xe^- \rightarrow Py_x + 2xH_3O^+ \tag{10.1}$$

The remaining 0.3 e$^-$ create excess negative load that is the excess electronic density prerequisite for the introduction of an adequate quantity of cations in polymeric chain, as shown in Eq. (10.2):

$$Py_x - 0.3xe^- + 0.3xH_2SO_4 \rightarrow \left[Py_x^{0.3x+} + 0.3x^+ HSO_4^- \right] + 0.3xH_3O^+ \tag{10.2}$$

10.4 CYCLIC VOLTAMMETRY (CV)

CV is a commonly used technique for the production of conductive polymers on metal electrodes [8,9]. Usually, the CV takes place in aqueous solutions of electrolytic substance in electrochemical cell using three electrodes. The main feature of the technique is the linear variation of electrode potential and the recording of the corresponding potential. The linear potential scan is done between two extremes (E_{min} and E_{max}) n times in aqueous solutions by the dynamic releases of H_2 (cathodic) and O_2 (anodic).

In 1–2–1, 1′–2′–1′, etc., are the 1st, 2nd, etc. dynamic scanning cycle (Figure 10.2a), while rising sections correspond to anodic bias and cathodic section in cathodic polarization of the electrode, respectively. The anodic section corresponds to oxidizing reactions and in the cathodic section reduction. Each time, the selected scanning speed $v_{sweep} = dE/dt$ is kept constant during the cycles and takes values between 10^{-1} and 10^6 mV/s depending on the system. These anodic and cathodic runs in potential

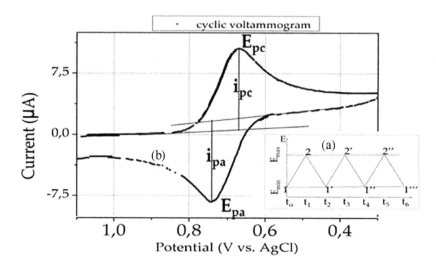

FIGURE 10.2   (a) Linear variation of potential in cyclic voltammetry over time, t_n. (b) A cyclic voltammogram shows the cathodic and anodic peaks corresponding to a reversible reaction.

may be repeated n times. The current at the working electrode is plotted versus the applied voltage to give the cyclic voltammogram trace, which is shown in Figure 10.2b.

10.5   DOPING

The oxidizing and/or reduction reactions that take place during electropolymeriza-tion processes have as product polymers with extended π-electron systems along the chain. The term doping is used because of the doping process relating to semicon-ductor materials. Doping of both semiconductors and conductive polymers affects oxidation state without altering the structure. The higher the concentration of doping agents, the greater will be the mobility of charge [10]. The doping can be n και p as in semiconductors. Na, K, Li, Ca, etc., are electron donors producing n-type CPs, while PF_6, BF_4, etc., are electron receivers producing p-type polymers. One can achieve maximum doping between 0.1 and 0.3 per monomer of polymeric chain. The limit of doping ability amounts to fractions (0.1–0.3) of a charge per monomer unit. A few polymers can be p- and n-doped. The conductivity of a polymer after doping may be increased up to 1000 orders of magnitude, namely as high as 500 S/cm [10]. It is important to note that this process is reversible to a very high degree [10].

Figure 10.3 shows the conducting state of polyaniline represented by emeral-dine salt produced by emeraldine base or leucoemeraldine via $2H^+ + 2X^-$ (X=Cl) or $2e^- + 2X^-$ (X=Cl) doping, respectively, leading in both cases to the production of the same product, emeraldine salt [1].

10.6   CORROSION PROTECTION OF PRODUCTION POLYMERS

The first published work on corrosion protection via conductive polymers was carried out by A. MacDiarmid, H. A. Heeger, and Shirakawa [11]. On metals, workers found

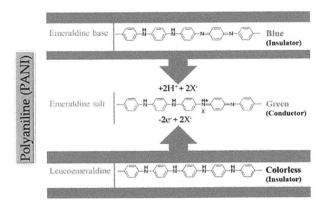

FIGURE 10.3 Production of emeraldine salt by doping emeraldine base or leucoemeraldine via $2H^+ + 2X^-$ (X=Cl) or $2e^- + 2X^-$ (X=Cl), respectively.

protection of metals by applying conductive polymer coatings [12]. Since then, many scientists worked on that subject making progress improving anticorrosion properties of conducive polymers [6]. Others devoted their research to determine the nature of the mechanism responsible for the corrosion protection [12]. Overall, the corrosion protection depends on the method of application of conductive polymer to the metal surface and the nature of the corrosion environment. So, one has to carefully choose the appropriate method of manufacturing for the desired application [13]. A product is available in the market based on conductive polymers for corrosion protection, called Corrpassiv® from the company Ormecon® [14].

10.7   SELF-HEALING EFFECT

As previously mentioned, the process of doping conductive polymers is largely reversible. The conductive polymers employ corrosion inhibitors during the electrochemical polymerization that they are released due to corrosion [13]. When this mechanism is triggered by corrosive conditions, the underlying metal surface is protected.

Barisci and colleagues [15] were the first to show that the release of corrosion inhibitors was a result of the coupling of the galvanic corrosion of metal and the conductive polymers. M. Kendig [16] showed how coating polyanilines could unleash doping agents, when the pH was increasing as a result of oxygen reduction reaction at the surface of the conductive polymers. Their analysis suggests that when covered by metallic substrate worn locally (i.e., hacked), the film is biased cathodic in relation to the anodic polarized metal. The anodic oxidation reaction of the metal causes the reduction of conductive polymers and/or the reduction of oxygen on the surface of the polymer coating. The reduction of these processes is the driving force for the release of counterions that if corrosion occurs, inhibitors will prevent corrosion. As Figure 10.4 shows, the conductive polymer can be represented as a set of machines that are activated by the upward reactions accompanying the corrosion and spray the surface of the metal with compounds that passivate the metal [16].

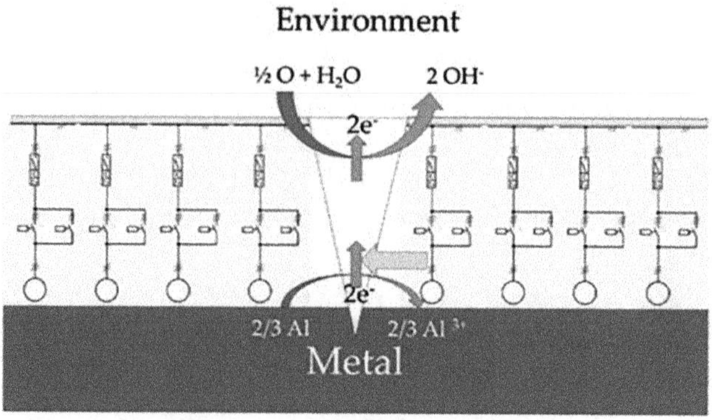

Environment

FIGURE 10.4 Schematic of corrosion inhibitors release mechanism at the point, where the metal has been damaged by the conductive polymer [H15].

10.8 POLYANILINE (PANi)

PANi was discovered in 1934 by form aniline black. PANi exists in nature as a copolymer with polyacetylene or polypyrrole. The polyaniline can be polymerized electrochemical on iron [17], aluminum [18], steel [19], and titanium [20] substrates. The literature shows a variety of electrolytes for the electropolymerization of aniline in oxalic acid [17,18], nitric acid, and sulfuric acid [20].

10.8.1 ELECTROCHEMICAL POLYMERIZATION OF ANILINE

The first study of electrochemical polymerization of aniline to polyaniline was done by Y. Wei and his collaborators [21]. They described polymerization as a "non-classical chain polymerization." The electropolymerization of aniline can be described as a two-molecule reaction in which a radical cation is produced as an intermediate. The enthalpy of reaction equals 121 kJ/mol^{-3}. The reaction shows a product nα equal to 1.2–1.3, where n corresponds to the number of electrons disbursed through polymerization and α is the transfer coefficient [10]. The oxidation of aniline monomer to form a dimer determines the speed of the process. More accurately, the oxidation of monomer to dimer formation is not the more slowly evolving stage of multilateralism but what requires the highest potential; the potential of the system increases by 0.40 V at 0.78 V with respect to calomel electrode. This stage is the literature called induction period; during this stage formed various oxidized monomers and polyaniline in pernigraniline condition (PB) [22].

10.9 POLYPYRROLE (PPy)

Polypyrrole (PPy) is, together with polyaniline, one of the most studied conductive polymers due to the stability it exhibits in atmospheric conditions and its easy composition. Its low oxidation potential offers advantages—as regards its electrochemical composition—in comparison with the synthesis of other conductive polymers in active metal electrodes. In the literature appears to be electropolymerizing in aluminum

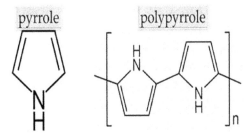

FIGURE 10.5   Pyrrole and polypyrrole molecule structure.

electrodes [23], iron [24], steel [25,26], titanium [25] via potentiostatic [23], galvanostatic [25], and potentiodynamic techniques [26]. At the same time, a number of electrolytes appear to be used to polymerize polypyrrole, such as p-toluenesulfonic acid, oxalic acid, nitric acid, and sulfuric acid [27]. Polypyrrole displays a linear structure. Figure 10.5 shows the structure of pyrrole and polypyrrole.

10.9.1   ELECTROCHEMICAL POLYMERIZATION OF PYRROLE (PY)

Py's electropolymerization mechanism shows several similarities to the anillin counterpart. Initially, the monomer is oxidized to the formation of a root cation, two radical cations react to bilateral formation, the bilateral molecules are further oxidized and react with monomers, or other bilateral to the formation of trimers or tetramers respectively and do it until it's formed polypyrrole. The advantage that pyrrole shows in relation to polyaniline is its ability to polymerization in a wider range of pH and dynamic [16]. F. Beck and R. Michaelis [28] managed to create via electrochemical polymerization of polypyrrole film which shows adhesion power to the iron electrode 11 N/m². The electropolymerized polypyrrole exhibits morphology rich in spheroid aggregates; in the bibliography, the characteristic structure formed by polypyrrole is characterized as the structure of cauliflower. Results from potentiostatic and potentiodynamic techniques deposits of polypyrrole in aluminum electrodes indicate that in both cases, oxidation and passivation of the substrate precedes the oxidation of the monomers of pyrrole and simultaneous its polymerization. The morphology of the oxide layer will determine the morphology of the developing polypyrrole film; its defects and cracks will be the starting points of polymerization [28].

10.10   POLYPYRROLE—PROTECTION AGAINST CORROSION

Polypyrrole coatings are porous, thus allowing easy penetration of electrolyte molecules, especially if they have a small radius such as chlorides, binding and simultaneously releasing counterions. On the other hand, the permeability of the polypyrrole coatings makes the aluminum layer susceptible to the needle erosion as chlorides can easily reach [28]. Monitoring of changes in open-circuit potential over time can help assess the passivation of a polymer-coated metal.

10.10.1 Effect of the Doping on the Corrosion Resistance of Polypyrrole

A number of compounds can be used for doping polypyrrole, such as camphor sulfonic acid (CSA), p-toluenesulfonic acid (p-TSA), phenylphosphonic acid (PPA), oxalic acid (OA), and cerium nitrate ($Ce(NO_3)_3$). All doping agents are known for their inhibitory and anticorrosion effects. For the evaluation of their corrosion properties, potentiodynamic polarization, open-circuit potential, and electrochemical spectroscopy of impedance were made.

Figure 10.6 shows the tungsten electrodeposition curve from p-TSA solution. The first cycle is due to the strong passivation of aluminum expressed with the peak at +1.5 V vs SCE. After the first cycle, the surface is partially passivized by the deposition of the polymer and partially by the development of aluminum oxide where the current density in the next cycles decreases dramatically. The oxidation of pyrrole takes place in close to 0 V vs SCE and is followed by reduction to lower potentials. The peak observed in −0.75 V vs SCE in the first two cycles is attributed to ion exchange reactions.

The image from SEM (Figure 10.7) shows the morphology of (a) pyrrole–p-TSA and (b) pyrrole–PPA coatings. The morphology depends on the kind of different doping ions. In the case of CSA doping, the coatings exhibit more roughness and fewer imperfections compared to p-TSA, PPA, and OA. The coatings formed with the help of nitrates form a surface of similar roughness to the coatings used in p-TSA and PPA with a reduced number of imperfections.

Open-source and direct-current polarization measurements are accomplished in an electrochemical three-electrode cell with working electrode the modified surface of the aluminum alloy with an area of $1\,cm^2$. The opposite electrode is a thin platinum plate, and the reference electrode is the SCE. The DC polarization measurements can be done in the range 1 V vs SCE up to 0 V vs SCE with a scan speed of 1 V/s and where the Evans diagrams can be exported and then the Tafel diagrams can be fitted via the appropriate software.

Figure 10.8a and b shows the continuous-current polarization curves and the open potential curve of pyrrole–$Ce(NO_3)_3$ coating, respectively. The current polarization curves for PPy–cerium coating was measured at different exposure times in the

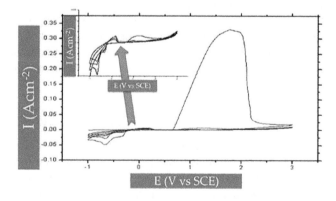

FIGURE 10.6 Electrodeposition diagrams p-TSA for coatings with potential in the range from −1 V to 3 V vs SCE and 30 mV/s scan speed.

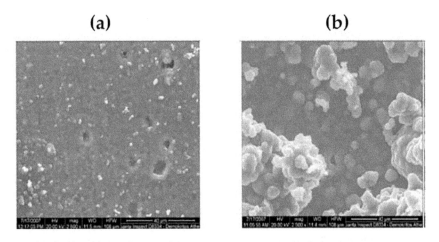

FIGURE 10.7 SEM micrograph for coatings (a) pyrrole p-TSA and (b) pyrrole PPA.

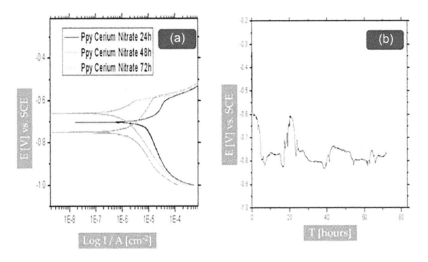

FIGURE 10.8 (a) Direct current polarization curves of the pyrrole Ce $(NO_3)_3$ coating and (b) open-potential curve of the pyrrole Ce $(NO_3)_3$ coating.

corrosive environment. It can easily be concluded that the behavior of PPy-cerium coating depends on the exposure time.

The coating after 48 hours of exposure to the sodium chloride solution has a lower corrosion current (I_{corr}) and greater polarization resistance (R_p) compared to its around exposure to the corrosive environment. In this behavior, there is evidence of recovery with the end of time and improvement of protection, which can be derived either from the formation of aluminum oxides or from the release of the ions of doping. The fact that the free potential is reduced to 48 hours suggests that the improvement is not the result of the formation of the aluminum oxide layer as it would then be accompanied by an increase in the corrosion potential. The release of nitric ions and part of the cations of cerium contributes to downward protection as the intensity of currents in the cathode part of the polarization curve is significantly reduced. Increased protection is also present as there is a decrease in the increase of the current. The open-potential measurement shows that there is a tendency to stabilize the potential after 24 hours of immersion in the corrosive environment. At 72 hours of immersion, the coating shows a barrier effect or upward corrosion inhibition. The gradual movement of the corrosion potential at less electric values is the result of the reduction of the anodic current, thus reducing the dissolution of the metal, which was observed by successive measurements of continuous-current polarization. After the coating switches from the conductive to the non-conductive form, the mains current will decrease and the doping factor which is and corrosion inhibitor will cause the passivation of the surface. The values of corrosion, corrosion current, and polarization resistance for all doped coatings are summarized in Table 10.1.

Figure 10.9a and b shows the Bode diagrams from the measurements of the samples of PPy doped with p-TSA. All samples show a time constant at high frequencies associated with the presence of the coating and another constant at the low frequency values attributed to the aluminum oxide layer or the action of corrosion in longer time of the sample in the corrosive environment. The polypyrrole coatings with anions from p-TSA, phenylphosphonic acid, and oxalic acid showed little protection from corrosion from the first hours of exposure. When the anion doping is

TABLE 10.1
Summary of Values of Corrosion, Corrosion Current, and Polarization Resistance for All Doped Coatings

PPy–Cerium Nitrate	I_{corr} (A)	E_{corr} (mV)	R_p (kΩ/cm^2)
24 h	9.83E-06	−708.73	2.65
48 h	1.84E-06	−748.56	14.19
72 h	9.75E-07	−659.67	26.6
PPy–PPA			
24 h	5.20E-06	−659.08	5.02
48 h	6.69E-06	−786.34	3.90
72 h	7.26E-06	−762.93	3.59
PPy–p-TSA			
24 h	1.62E-05	−723.68	1.61
48 h	4.65E-06	−745.93	5.61
72 h	5.18E-06	−739.01	5.03
PPy–CSA			
24 h	7.14E-06	−707.50	3.65
48 h	5.34E-06	−708.90	4.89
72 h	1.19E-05	−708.16	2.20
PPy–oxalic acid			
24 h	3.64E-06	−766.13	7.17
48 h	4.84E-06	−824.29	5.39
72 h	5.92E-06	−831.71	4.41

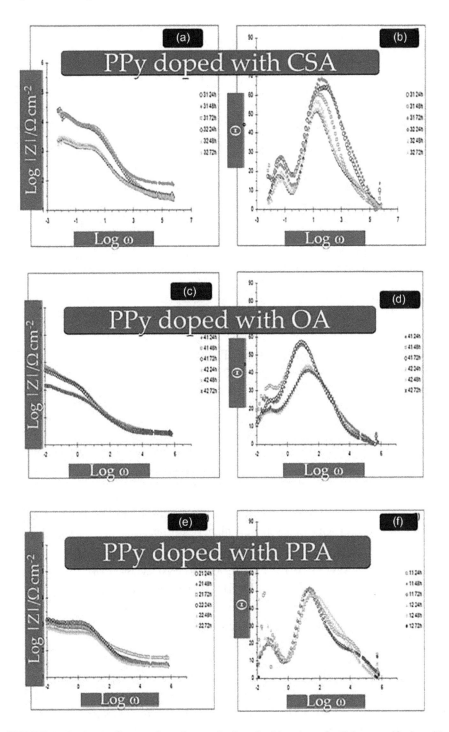

FIGURE 10.9   Bode diagram for polypyrrole doped with anion of p-Toluene sulfonic acid (p-TSA): (a) impedance measurement and (b) phase-angle measurement.

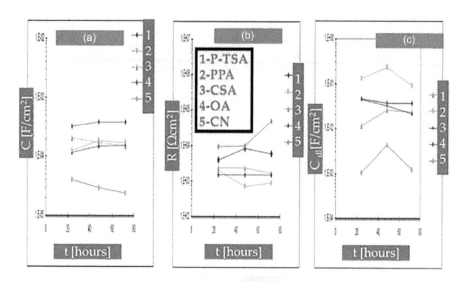

FIGURE 10.10 Calculation of parameters for coatings developed from dopants p-toluene sulfonic acid (1-p-TSA), p-toluene sulfonic acid (2-p-TSA), camphor sulfonic acid (3-CSA), oxalic acid (4-OA) and cerium nitrate (5-CN).

anion from camphor sulfonic acid (CSA; Figure 10.9b and c) and cerium nitrate (CN; Figure 10.9d and e), an improvement is observed of the protective properties of coatings, where in both cases the value of impedance increases with the time of exposure of sample in the sodium chloride solution giving indication of self-healing.

Figure 10.10 shows the calculated capacity at high frequencies, C (Figure 10.9a); double layer, C_{dl} (Figure 10.9c); and low-frequency resistance, R (Figure 10.9b). These results are an indication of better protection against corrosion and indicate that the process of forming a double layer on the surface is more difficult and corrosion is more difficult. The highest resistance values are presented by the coating developed by a cerium nitrate where it shows that the charge transfer process is more difficult to occur. This sample presents the best behavior against the corrosive environment as it has the smallest capacity values and the highest resistances.

10.10.2 IMPROVEMENT OF THE PROTECTION AGAINST EROSION OF POLYPYRROLE COATINGS DEVELOPED BY A SOLUTION OF p-TSA WITH ELECTROCHEMICAL INTEGRATION OF CATIONS EXHIBITING CORROSION-INHIBITING ACTION

Polypyrrole can act as a means of exchanging cations or anions depending on the nature of its doping agent. For this purpose, we developed coatings, e.g., on 2024-T3 aluminum alloy substrates using p-toluenesulfonic acid (p-TSA) as a doping agent. The coatings were developed with cyclic voltammetry in solutions containing nitrates of different cations known for their inhibitory erosion. In the following, we present results using cobalt nitrate, manganese nitrate, and cerium nitrate. Continuous-current polarization measurements were used to evaluate the protective properties of coatings. It was found that in coatings applied the cyclic voltammetry, the cations were incorporated into the coatings offering better protection against corrosion compared to simple coatings with p-TSA. Better protection was also observed from the coatings created by cerium nitrates. Polypyrrole has properties that are interesting in anticorrosion applications. It is very stable in its oxidative state, and the potential it needs for its deposition is quite low. It can be deposited either chemically or electrochemically with the help of quite a large number of electrolytes. With its oxidation, polypyrrole is doped with anions from the electrolytic solution for the purpose of electroneutrality. During successive oxidation and reduction processes of polypyrrole, small doping ions, such as chlorides and nitrates, can be bonded to the polymer grid. Unlike large anions, such as p-TSA, their movement is quite limited. The properties of ion interchange of polypyrrole depend significantly on its anion doping.

10.10.3 Development of Polypyrrole Coatings from a Solution of p-TSA

An electrochemical cell of three electrodes can be used for the electropolymerization of pyrrole. The aluminum alloy 2024-T3 that was first cleaned in a caustic soda solution and then in nitric acid is the working electrode, and a platinum plate is the opposite electrode. The measurements were recorded using the saturated calomel electrode (SCE) as reference electrode. Deposition solutions are aqueous with concentrations of 0.2 m for organic acid and 0.1 m for pyrrole. The cyclic voltammetry was chosen for the deposition of polymers in the range from −1 V to 3 V vs SCE for 4 cycles with a scan rate of 30 mV/s.

Figure 10.11 a shows the voltammogram of pyrrole from a solution of para-toluene. The first cycle is due to the strong passivation of aluminum expressed with the peak at + 1.5 V vs SCE. After the first cycle, the surface is partially passivized by the deposition of the polymer and partially by the development of the aluminum oxide layer,

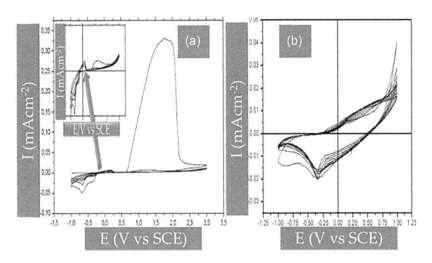

FIGURE 10.11 (a) Electrodeposition of polypyrrole (0.1 m) from solution with p-TSA (0.2 m), scanning speed 30 mV/s, in magnification are cycles from 2 to 4 and (b) Cyclic voltammogram of p-TSA pyrrole coating in a solution 0.1 M P-TSA.

and the density of oxidative currents in the next cycles decreases dramatically. The oxidation of pyrrole takes place in positively dynamically close to 0 V vs SCE and is followed by reduction in lower potentials. The peak observed in −0.75 V vs SCE in the first two cycles is attributed to ion exchange reactions.

After the creation of the coating, a 1-cm^2 circular surface was used for the process in the range from −1 V to +1 V vs SCE with a scanning speed of 20 mV/s for 10 cycles in a solution with a concentration of 0.1 m p-TSA without the presence of pyrrole monomers. Figure 10.11b shows the voltammogram. There is quite good stability and repeatability of cycles from the first to the last, giving indications for a coating quite homogeneous and well attached to the substrate. In the area of anodic currents, there is a wide peak with little intensity due to either the introduction of anion from the solution to the coating, the peak intensity is small due to the small extent of the phenomenon due to the large volume of or in the extraction of cations from the coating (protons). In the cathode area, a peak of −0.4 V vs SCE appears, and it appears to be due to feedback reactions. The reaction may be related to the introduction of cations, the extraction of cations, or both phenomena together. For such a system, the cation extraction is quite difficult as the size of anion doping is quite large and cannot be easily moved and removed from the polymer grid, the prevailing reaction is the introduction of proton in polymer.

10.11 ELECTROCHEMICAL INCORPORATION OF CERIUM, MANGANESE, AND COBALT IONS INTO POLYPYRROLE COATING DEVELOPED FROM A p-TSA SOLUTION

Figure 10.12 shows the third of the 10 cycles from the tungsten chloride solutions to cerium nitrate, manganese nitrate, and cobalt nitrate concentrations of 0.1 m. Compared to the voltammogram in the solution with p-TSA, it is obvious that the ion import and export process is significantly obstructed due to the size of the anion. All coatings have good stability as the intensity of currents in each cycle does not change until the last. No acute peak due to aluminum oxides showing good adhesion of polymer to substrate. The voltammograms from nitric cobalt and manganese nitrate look

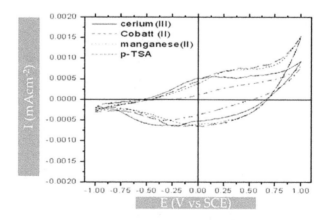

FIGURE 10.12 Cyclic voltammogram (3rd cycle) of pyrrole p-TSA coating in 0.1 M cerium, manganese nitrate, cobalt nitrate and p-TSA.

similar though cerium nitrate is different. For nitric cobalt and manganese nitrate, the upward curve shows a wide peak at $+270\,mV$ vs SCE. Oxidation either in cation extraction or in anion intake. Assuming that only anion intake can be done, the apex can be attributed to the introduction of nitrate groups without significant influence from the cations. However, by using the cerium nitrate the voltammogram differs considerably both in the position of the peaks and in their intensity. It is an indication that cations strongly affect the response of the coating and the process of compensating the load; moreover, the polymer is doped with a large enough anion that is difficult to move. It is easier for a small cation to move in the polymer grid than a large anion. ICP/AES measurements were made to confirm the incorporation of cations in the coating. For this purpose, after the introduction of cations, the coating was rinsed with distilled water to remove the naturally adsorbed ions; then, the dry coating in the air was removed from the substrate and dissolved in aqueous solution. Each of the samples showed the characteristic emission of the cation.

10.12 CONTINUOUS-CURRENT POLARIZATION MEASUREMENTS FOR PYRROLE COATING WITH p-TSA

Figure 10.13 shows the continuous-current polarization curves at different exposure times in sodium chloride 0.5 m for the coating of pyrrole doped with p-TSA without cations with corrosion-inhibiting action. After 24 hours of exposure to corrosive environment, the corrosion current increases and the polarization resistance decreases compared to the 6-hour exposure of the coating (inserted table in Figure 10.13). But after 72 hours, the coating works differently, reducing the corrosion current and increasing polarization resistance. One reason this phenomenon may occur is the release of anion when erosion has begun. From these results, it can be concluded that polypyrrole can be classified as a receptor ion with the export capacity when it is needed.

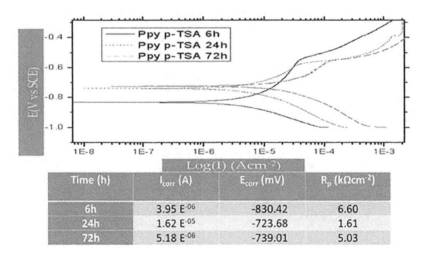

Time (h)	I_{corr} (A)	E_{corr} (mV)	R_p (kΩcm^{-2})
6h	3.95 E^{-06}	-830.42	6.60
24h	1.62 E^{-05}	-723.68	1.61
72h	5.18 E^{-06}	-739.01	5.03

FIGURE 10.13 Continuous current polarization curves of the pyrrole coating p-TSA for 6, 24 and 72 hours of exposure to sodium chloride (inserted table I_{corr}, E_{corr} and R_p after 6, 24 and 72 hours in NaCl solution for p-TSA polypyrrole).

10.13 CONTINUOUS-CURRENT POLARIZATION MEASUREMENTS FOR THE PYRROLE COATING WITH p-TSA AND ELECTROCHEMICALLY INTEGRATED IONS CERIUM, MANGANESE, AND COBALT

Figure 10.14 shows the continuous-current polarization curves for the pyrrole-doped p-TSA after 24 hours of exposure to sodium chloride solution. The introduction of cations was done by a linear increase in the potential without altering the doping ions in the polymer grid. The import process did not cause any major changes in the polymer structure or conductivity.

All enriched with cationic coatings have moved their streams to lower prices (inserted table in Figure 10.14) by about one size class compared to simple coatings. By studying the anodic and cathodic areas of the curves, there is information about the further action of inhibitors. Cerium shows a dominant decrease in cathode currents compared to other inhibitors. From the literature, the action of cerium as a cathodic corrosion inhibitor is known, the cerium acts with the hydroxyl ions of the environment forming an insoluble hydroxide layer of cerium specific areas with a local increase in pH protecting the substrate from further corrosion. Reduction of cathodic currents was observed with the introduction of other cations. The anodic area of the curves was also moved to lower current values, with coatings containing manganese having the largest decrease in anodic currents and higher polarization resistance, confirming its nature as an effective corrosion inhibitor for aluminum alloys. This study shows that manganese ions affect the anodic corrosion currents. The incorporation of cobalt ions into the coatings successfully acted against corrosion by increasing the polarization resistance and reducing the corrosion current compared to the unmodified coating, but its effect was not as effective as the manganese and cerium ions. Cobalt influenced anode and cathode corrosion currents.

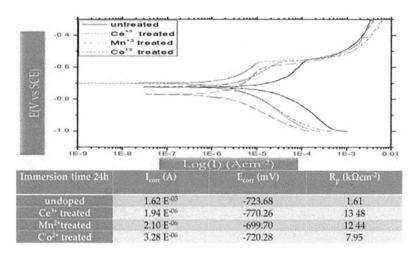

Immersion time 24h	I_{corr} (A)	E_{corr} (mV)	R_p (kΩcm^{-2})
undoped	1.62 E^{-05}	-723.68	1.61
Ce^{3+} treated	1.94 E^{-06}	-770.26	13 48
Mn^{2+}treated	2.10 E^{-06}	-699.70	12 44
Co^{2+} treated	3.28 E^{-06}	-720.28	7.95

FIGURE 10.14 Continuous current polarization curves for 24 hours of NaCl exposure for coatings with Ce^{3+}, Mn^{2+}, Co^{2+} and simple coating (inserted table I_{corr}, E_{corr} and R_p after 24 hours in NaCl solution for coatings with Ce^{3+}, Mn^{2+}, Co^{2+} and undoped coating).

Manganese and cerium ions provided the coatings with greater polarization resistance than cobalt ions, but all enriched with cationic coatings showed a significant improvement compared to the undoped polypyrrole coating with p-TSA on 24 hours of immersion in the sodium chloride solution.

10.14 POLYPYRROLE COATINGS DEVELOPED BY CERIUM NITRATE SOLUTION

Figure 10.15 shows the electrodeposition diagram of the polypyrrole. The direct-current polarization curves of the coating after exposure to 0.5 M concentration sodium chloride include the corrosion potential and polarization resistance. In the 24 hours of immersion in the sodium chloride solution, the polypyrrole coating developed by a nitrate-based solution showed a greater corrosion potential (Figure 10.15) than the polypyrrole coating with p-TSA and cerium ions. The explanation that can be given by the fact that the deposition of the first system was in the presence of cerium ions, when the intensity of the capacity from −1 V vs SCE was increased to a more positive dynamic the dissolution of aluminum is significantly impeded by the cerium ions and the application of the corresponding hydroxide layer of cerium, at the same time pyrrole uses nitrates to equalize the charge. When the polymerization of pyrrole is made from a solution with p-TSA, the above phenomena are not observed as the anions are mainly used for their introduction into the pyrrole matrix. As shown by the graphs, the coating with anion and p-TSA has a lower current of corrosion and the distance of the potential corrosion with the potential of localized corrosion is greater than the nitrate coating. The cathodic current is also smaller, indicating that the cathodic corrosion inhibition works best for these coatings because of cerium ions. Furthermore, by studying the anodic and cathodic parts of the curves separately, we can conclude that the system with the ions p-TSA and cerium maintains its properties for a long time. This behavior may be due to

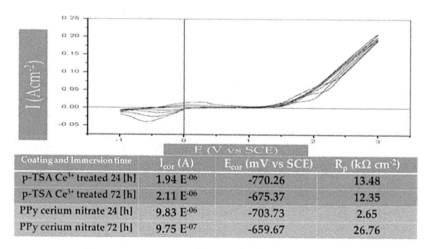

Coating and Immersion time	I_{cor} (A)	E_{cor} (mV vs SCE)	R_p (kΩ cm^{-2})
p-TSA Ce^{3+} treated 24 [h]	1.94 E^{-06}	-770.26	13.48
p-TSA Ce^{3+} treated 72 [h]	2.11 E^{-06}	-675.37	12.35
PPy cerium nitrate 24 [h]	9.83 E^{-06}	-703.73	2.65
PPy cerium nitrate 72 [h]	9.75 E^{-07}	-659.67	26.76

FIGURE 10.15 Electrodeposition of polypyrrole (0.1 m) from solution with Ce $(NO_3)_3$ (0.2 m), scan speed 30 mV/s. The inserted table gives the current values of I_{corr}, E_{corr} and R_p after 24 and 72 hours in NaCl solution for coatings p-TSA and Ce^{3+}, Ce $(NO_3)_3$.

the presence of both ions as both are corrosion inhibitors. For this coating, there is an increase in the potential of localized corrosion to stay longer in the corrosive environment offering significantly better protection from the coating developed by a solution of nitrate. Conductive polymers can act as a reservoir of corrosion inhibitors, and when one inhibitor is exhausted then the other is released, due to the size of the anion of p-TSA is likely to be released last and cerium ions used first to suspend corrosion.

When polypyrrole is formed from a solution of cerium nitrate, the polymer is doped with nitrates to equalize the charge after the oxidation of the pyrrole. The cerium can be on the coating only as a naturally adsorbed. In this way, its action is limited due to its small concentration compared to the coating containing the ions p-TSA and cerium.

10.15   PANi–PPy COMPOSITE COATINGS

Recently, some researchers have started to deal with the simultaneous electropolymerization of monomers anillin and pyrrole and other monomers [29,30], as the composite coatings are expected to exhibit improved properties in comparison with simple polyaniline or polypyrrole coatings. There are several reports in the world bibliography about composite polyaniline–polypyrrole coatings presenting improved corrosion-resistant properties as a result of the combined properties of individual polymers [29,30].

10.15.1   Electrochemical Polymerization of Composite Coatings

Copolymers or complex polyaniline–polypyrrole coatings appear deposited in aluminum electrodes, steel, gold, and iron, via potentiometer, galvanized, and potentiodynamic methods [31–33]. Via cyclic voltammetry, composite coatings are produced, through a two-step process (substrate pre-treatment and electrochemical deposition). Their substantially simple production process makes them suitable for industrial production. Through cyclic voltammetry, they can be synthesized within a few cycles (number of cycles ≥ 5) at different scanning rates. In the literature appear to be synthesized in speeds between 1 [9] και 120 mV/s [33] (Figure 10.16).

J. Danilidis and associates [9] studied the effect of the rate of conductive polymer coating deposition speed from isotonic amounts of anillin–pyrrole–oxalic acid in its morphology and anticorrosion properties. At low speeds (1, 15 mV/s), the coating consists mainly of polypyrrole, while the speed of the texture of the coating is not dominated by any of the two polymers. It seems that the relationship between the deposition rate and the observed maximum of the deposition currents is linear and is consistent with the Randles–Sevcik equation [34]. The rate of scanning of the potential plays an important role in the coating morphology as well as the corrosion properties of the polymer coating. The composite coatings that appear in the bibliography are strongly attached to the substrate and exhibit a uniform structure [9].

10.15.2   Composite Coatings—Protection against Corrosion

The Tafel curves received after 6, 24, and 72 hours of NaCl (3.5% w/w) of copolymers pyrrole and anillin (aqueous polymerization solution: 0.2 m oxalic acid, 0.1 m anillin,

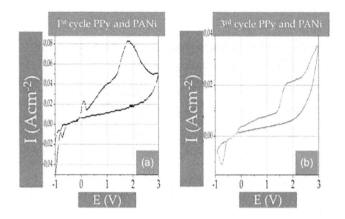

FIGURE 10.16 Cyclic voltammogram (30 mV/s) corresponding to simultaneous polymerization of pyrrole and anillin electrode Al2024-T3: (a) 1st cycle and (b) 3rd cycle.

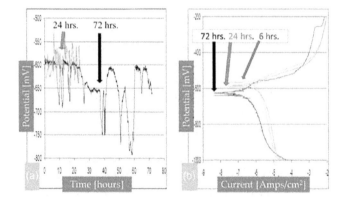

FIGURE 10.17 (a) OCP and (b) Tafel polyaniline–polypyrrole copolymer taken 6 (dark grey), 24 (light grey) and 72 (black) hours after exposure to NaCl (w/w) solution [19].

and 0.1 m pyrrole) are represented in Figure 10.17. We observe the recovery occurring during the transition from the first 6 hours of exposure to 3 days (72 hours) to the anticorrosion behavior of the system [18]. Table 10.2 lists OCP values, corrosion rates (corrosion currents), and polarization resistance as determined by fitting in upward and downward curves for 6, 24, and 72 hours of exposure. OCP appears to be shifted toward more negative values as exposure time rises, due to the significant decrease in cathode current (especially in the case of 72 hours of corrosion). The morphology of the coatings depends on the anillin x wt. % - pyrrole (1-x) wt.% composition [9].

The cathodic current appears to be decreasing from 6 to 72 hours of exposure approximately of a magnitude, while at the same time the polarization resistance increases dramatically, indicating a strong recovery of the corrosion-resistant performance of the coating. Composite coatings bring about a decrease in anodic and cathodic currents compared to bare aluminum, offering corrosion protection. Finally, the improvement of the corrosion performance of coatings with time is attributed to

TABLE 10.2
OCP Values, Corrosion Rates, Polarization Resistance and
Corrosion Current of the Coating

Exposure Time (h)	E_{corr} (mV)	I_{corr} (A/cm²)	R_p (Ω/cm²)
6	−569.9	1.74 5E^{-6}	14,984.00
24	−591.4	9.52 E^{-7}	27,394.00
72	−627.4	2.35 5E^{-7}	111,042.00

the activation of a self-healing mechanism. The mechanism may be triggered by changes in potential on the surface of the coating, pushing the layer to release oxalic ions which react with aluminum create a protective passive layer [9].

10.16 CONCLUSIONS

Polymers created from a solution of cerium nitrate exhibit a strong barrier effect. They have a reduced cathodic and anodic current, established in the corrosion current, and increased polarization resistance after 48 and 72 hours of exposure to the corrosive environment, giving clues to this – treatment (self-healing). The release of nitric ions and part of the cations of cerium, which are naturally adsorbed to the polypyrrole coating, contributes to cathodic protection as the intensity of currents in the cathodic part of the polarization curve is significantly reduced. Anodic protection is also present as there is cathodic and anodic current after 72 hours of exposure to sodium chloride solution. The open-potential measurement shows that there is a tendency to stabilize the potential after 24 hours of immersion in the corrosive environment. Using for the formation of the polymer solution of p-TSA coatings show barrier protection for a short immersion time and signs of possible this – treatment in long exposure intervals. The barrier protection that exists in the first hours begins to collapse after 24 hours of exposure to the corrosive environment, as it is concluded comparing the results obtained after 48 hours of exposure. Increasing the anodic current shows signs that the coating is starting to degrade. The anticorrosion properties of the coatings with PPA are similar to those with cerium nitrate. With p-TSA, OA and CSA are created coatings are created with stability that can passivate the surface at the end of time, with little reduction of their protective properties for the last two in long exposure times. For the coating doped with the anion of p-TSA stability is observed at the end of time may be explained by the nature of doping. Compared to nitrates, the anions of p-TSA are larger in size, making the process of releasing them more difficult, so the time of protection of coatings increases. For the coating of a CSA solution, the continuous-current polarization measurements indicate that the coating's behavior is stable after 48 hours of stay in the corrosive environment, with an increase in the corrosion current after 72 hours. It is important to stress that after 72 hours, the corrosion current did not increase significantly despite the ripples of open potential. Quite stable behavior of coatings is observed in the case of OA without significant alteration of the corrosion properties in long periods of stay in sodium chloride solution. The corrosion current rises and the polarization resistance decreases to 72 hours of exposure. For all coatings, it was found that the release of ion doping is not done immediately by dipping the coating into the sodium chloride solution, but due to the different water uptake process, as shown by the OCP measurements in the first hours of immersion in the sodium chloride solution, the release of the anion is different.

The recovery of the properties of the coatings after corrosion with a tendency to re-stabilize the OCP in combination with the direct-current polarization curves recommends self-healing. This conclusion opens new research in this area.

ACKNOWLEDGMENTS

The author thanks the European Union for funding this work under **FP6 IP Program** MULTIPROTECT (Advanced environmentally friendly multifunctional corrosion protection by nanotechnology).

REFERENCES

1. Rupprecht, L. Eds. 1999. *Conductive Polymers and Plastics in Industrial Applications.* Plastics Design Library, Norwich.
2. Mirmohseni, A., Solhjo, R. 2003. Preparation and characterization of aqueous polyaniline battery using a modified polyaniline electrode. *J. Eur. Polym.* 39: 219–223.
3. Wang, J. Yang, J. Xie, J. Xu, N. 2002. A Novel conductive polymer–sulfur composite cathode material for rechargeable lithium batteries. *Adv. Mater.* 14: 963–965.
4. Boehme, JL., Mudigonda, DSK., Ferraris, J. 2001. Electrochromic properties of laminate devices fabricated from polyaniline, poly(ethylenedioxythiophene), and poly(N-methylpyrrole). *Chem. Mater.* 13 (12): 4469–4472.
5. Shaolin, M., Huaiguo, X., Bidong, Q. 1991. Bioelectrochemical responses of the polyaniline glucose oxidase electrode. *J. Electroanal. Chem. Interf. Electroch.* 304(1–2): 7–16.
6. Herrasti, P., Recio, FJ., Ocón, P., Fatás, E. 2005. Effect of the polymer layers and bilayers on the corrosion behaviour of mild steel: Comparison with polymers containing Zn microparticles. *Prog Org Coat.* 54 (4): 285–291.
7. Erdem, E., SaCak, M., Karakiqla, M. 1995. Synthesis and properties of oxalic acid -doped PANi. *Polymer Int.* 39: 153–159
8. Rajagopalan, R., Iroh, J. 2001. Development of polyaniline–polypyrrole composite coatings on steel by aqueous electrochemical process. *Electrochimica Acta*, 46: 2443–2455
9. Tsirimpis, A.I., Kartsonakis, I., Danilidis, I., Liatsi, P., Kordas, G. 2010. Synthesis of conductive polymeric composite coatings for corrosion protection applications. *Prog. Organ. Coat.* 67(4): 389–397.
10. Syed, S., Dinesan, M. 1991. Review: Polyaniline – A novel polymeric material. *Talanta.* 8: 815–837.
11. Shirakawa, H., Louis, HJ., MacDiarmid, AG., Chiang, CK., Heeger AJ. 1995. Synthesis of electrically conducting organic polymers: Halogen derivatives of polyacetylene, $(CH)_x$. *Chem. Comm.* 24: 2473–2670.
12. Walker, DA., D'Silva, C. 2014. Electrochemical and physical properties of N-substituted arylmethylene pyrrole polymers and N-alkylmethine pyrrole copolymers. *Electroch. Acta* 116: 175–182.

13. Weng, CJ., Chang, CH., I-Li Lin, IL., Chen, PH. 2012. Advanced anticorrosion coating materials prepared from fluoro-polyaniline-silica composites with synergistic effect of superhydrophobicity and redox catalytic capability. *Surf. Coat. Techn.* 207: 42–49.

14. Wessling, B., Posdorfer, J. 1999. Corrosion prevention with an organic metal (polyaniline): Corrosion test results. *Electroch. Acta.* 44: 2139–2147.

15. Barisci, J., Spinks, G. 1998. Conducting polymers as a basis for responsive materials systems. *J. Intel Mater Syst. Str.* 9: 723–731.

16. Kendig, M., Hon, M., Warren, L. 2003. 'Smart' corrosion inhibiting coatings. *Progr. Org. Coat.* 47: 183–189.

17. Sazou, D. 2001. Electrodeposition of ring-substituted polyanilines on Fe surfaces from aqueous oxalic acid solutions and corrosion protection of Fe. *Synth. Metals.* 118:133–147.

18. Huerta-Vilca, D., de Moraes, DSR. 2004. Anodic treatment of aluminum in nitric acid containing aniline, previous to deposition of polyaniline and its role on corrosion. *Synth. Metals.* 140: 23–27

19. Galkowski, M., Malik, M. 2003. Protection of steel against corrosion in aggressive medium by surface modification with multilayer polyaniline based composite film. *J. Electroch. Soc.* 150: B249–B253.

20. Zhong, L., Zhu, H. 2006. A passivation mechanism of doped polyaniline on 410 stainless steel in deuterated H_2SO_4 solution. *Electroch. Acta*, 51: 5494–5501

21. Wei, Y., Yang, G. 1990. Polymerization of aniline and alkyl ring-substituted anilines in the presence of aromatic additives. *J. Phys. Chem.* 94: 7716–7721.

22. Wei, Y., Hsueh, F. 1994. Monitoring the chemical polymerization of aniline by open-circuit-potential measurements. *Polymer.* 35: 3572–3575.

23. Saidman, S., Quinzani, O. 2004. Characterisation of polypyrrole electrosynthesised on aluminium. *Electroch. Acta* 50: 127–134.

24. De Bruyne, A., Delplancke, J. 1998. Comparison between polypyrrole films obtained on mild steel by electropolymerization from oxalic acid and sodium sulphate aqueous solutions. *Surf. Coat. Technol.* 99: 118–124.

25. Iroh, J., Voevodin, N. 2003. Electrochemical synthesis: A novel technique for processing multi-functional coatings. *Progr. Org. Coat.* 47: 365–375.

26. Richardson, J., Woo, G. 1970. A study of the pitting corrosion of Al by scanning electron microscopy. *Corr. Sci.* 10: 313–323.

27. Sakmeche, N., Lacaze, P. 2000. Usefulness of aqueous anionic micellar media for electrodeposition of poly-(3,4-ethylenedioxythiophene) films on iron, mild steel and aluminium. *Electroch. Acta* 45: 1921–1931.

28. Beck, F., Michaelis, R. 1994. Filmforming electropolymerization of pyrrole on iron in aqueous oxalic acid. *Electroch. Acta* 39: 229–234.

29. Yagan, A., Ozc, N. 2007. Inhibition of corrosion of mild steel by homopolymer and bilayer coatings of polyaniline and polypyrrole. *Progr. Org. Coat.* 59: 297–303.

30. Rajagopalan, R., Iroh, J. 2001. Development of polyaniline–polypyrrole composite coatings on steel by aqueous electrochemical process. *Electroch. Acta* 46: 2443–2455.

31. Iroh, J., Zhu, Y. 2003. Electrochemical synthesis: A novel technique for processing multi-functional coatings. *Progr. Org. Coat.* 47: 365–375.

32. Seegmiller, J., Pereira da Silva, JP. 2005. Mechanism of action of corrosion protection coating for AA2024-T3 based on poly(aniline)-poly(methylmethacrylate) blend. *J. Electroch. Soc.* 152, B45–B53.

33. Breslin, C., Fenelon, A. 2005. Surface engineering: Corrosion protection using conducting polymers. *Mat. Design.* 26: 233–237.

34. Lether, FG., Wenston, PR. 1987. An algorithm for the numerical evaluation of the reversible Randles-Sevcik function. *Comp. Chem.* 11: 179–183.

11 Stimuli-Responsive Polymer Coatings

*Fabrice Ofridam, Mohamad Tarhini,
Waisudin Badri, Wei Liao, Noureddine
Lebaz, Émilie Gagnière, Denis Mangin,
Emilie Dumas, Sami Ghnimi, Abdelhamid
Errachid El Salhi, Adem Gharsallaoui,
Hatem Fessi, and Abdelhamid Elaissari*
University Claude Bernard Lyon-1

CONTENTS

11.1 INTRODUCTION

Polymers are used in most of the goods we use in our daily life, and they even form important parts of our own body, such as DNA and proteins. Since long time, polymers have been used in order to improve the quality of life. However, the real nature of the polymeric materials (natural rubber) was not known. Thanks to the economic dynamism of rubber industry in the 19th and 20th centuries, more investments were made to know about the behaviour of rubber. This event made the intense debate among researchers in this regard. Therefore, they made many efforts and developed certain polymers suitable for almost every imaginable application. This foundational study caused the development of a new category of polymers that respond to their environment by changing their physical and chemical properties. Stimuli-responsive polymers are the materials that can react by altering their properties once subjected to external stimulus. Stimuli-responsive polymers are classified into different classes based on various criteria. For instance, stimuli-responsive polymers are divided into categories of thermal-induced, light-induced, electro-responsive, pH-responsive, and magnetic-responsive polymers (Figure 11.1). Based on the nature of cross-links, stimuli-responsive polymers are classified into chemically or physically cross-linked polymers. Biodegradable stimuli-responsive polymers are also nominated as a category. According to the switching segments' nature, stimuli-responsive polymers are categorized into Tg-type stimuli-responsive polymers with amorphous switching segment and Tm-type stimuli-responsive polymers with crystalline switching segment [1]. Stimuli-responsive polymer coatings continue to be used in evermore increasingly diverse applications and sectors such as food, agriculture, drug delivery, biomedical diagnostics, sensors, textiles, and cosmetics [2].

In this chapter, we highlight some of the structures made from stimuli-responsive materials. We discuss different stimulus mechanisms, and we shed light on some important applications of that technology.

11.2 STIMULI-RESPONSIVE MATERIALS

11.2.1 STIMULI-RESPONSIVE POLYMERS (SRP)

Stimuli-responsive polymers, also known as smart or intelligent or environmentally sensitive polymers [3,4], are materials that constitute an important class of polymers today. They can react to small external changes in their immediate environment by displaying significant reversible, large, microstructural, physiological, and physicochemical changes in their properties [5,6]. One of their fundamental characteristics

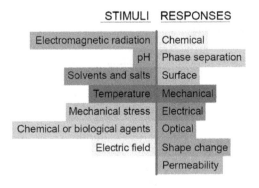

STIMULI   RESPONSES

Electromagnetic radiation	Chemical
pH	Phase separation
Solvents and salts	Surface
Temperature	Mechanical
Mechanical stress	Electrical
Chemical or biological agents	Optical
Electric field	Shape change
	Permeability

FIGURE 11.1   Synthetic polymers' possible stimuli and responses.

is that they return to their initial state as soon as the stimuli responsible for the modification in their physicochemical properties are removed [7].

Stimuli-responsive polymers can be of either natural or synthetic origin. Among natural responsive polymers, chitosan and albumin are capable of showing thermo- and pH-responsive properties [8–10]. Moreover, methylcellulose and gelatin can be mentioned with thermo-responsive properties [11]. Synthetic stimuli-responsive systems containing polymers can be designed either with a responsive polymer, or by combining a polymer with a responsive compound, the polymer serving only as a template/carrier for that compound [12]. Stimuli-responsive polymers are synthesized using polymerization techniques such as living anionic and cationic polymerization, controlled radical polymerization, atom transfer radical polymerization, and reversible addition–fragmentation chain-transfer polymerization. In the build-up of synthetic stimuli-responsive polymers, functional groups that make effective response to stimuli (charge, polarity, and solvency, for instance) are incorporated along a polymer backbone [14], resulting in abrupt and clear changes in macroscopic material properties. Response to stimuli can be defined in various ways. These responses may be associated with modifications in individual polymer chain dimensions or size, shape, surface characteristics, secondary structure, solubility, and/or intermolecular association degree. In most cases, the formation or destruction of secondary forces (hydrogen bonding, hydrophobic effects, electrostatic interactions, etc.), simple reactions (e.g., acid–base reactions) of moieties linked to the polymer backbone, and/or osmotic pressure differences are responsible for this response. Dramatic alterations in the polymeric structure, such as degradation of polymers upon the application of a specific stimulus by bond breakage in the polymer backbone or at pendant cross-linking groups, can also be mentioned as a response of a stimuli-sensitive polymer to stimuli [15].

External stimuli that provoke these changes can be divided into three main groups: physical stimuli (mechanical stress, electric, magnetic, ultrasound, light, temperature), chemical stimuli (electrochemical, pH, ionic strength), and biological stimuli (enzyme and/or biomolecules) [16,17]. Dual-stimuli-responsive polymers can be also mentioned as polymers that simultaneously respond to more than one stimulus. It should be taken into consideration that stimuli may also occur internally. The well-known examples of this case that is applied in drug delivery systems include changes in pH that occur in certain organs or diseased states, changes in temperature, the presence of specific enzymes, antigens in the human body [18], or changes in physiological conditions [19,20].

Based on their physical properties, stimuli-responsive polymers are classified into three categories:

- *Linear free polymer chains in solution*: They respond to stimuli that induce a reduction in the number of hydrogen bonds between polymer and water or the neutralization of electrical charges. Linear free soluble polymer chains precipitate with the increasing hydrophobicity. Polymers exhibiting this property have applications in bioseparation of proteins, cells, and other bioparticles.
- *Covalently cross-linked reversible and physical gels*: Hydrogels that swell in aqueous media may respond rapidly to small changes in pH, temperature, light intensity, ionic strength, magnetism, inflammation, ultrasound,

electrical fields, or biochemicals. Smart polymer hydrogels find applications in microfluidics, actuators, and sensors.

- *Smart surfaces or membranes*: Polymer chains that have the ability to undergo changes under specific stimuli conditions when adsorbed or grafted to surfaces are classified under this category. Indeed, the surfaces can be converted from hydrophilic to hydrophobic as well. Thus, surface is hydrophilic when the polymer is in its soluble state and hydrophobic when insoluble due to the action of stimuli. Separation of substances that interact differently with hydrophobic matrices can be mentioned as a potential application. Smart surfaces find applications in tissue engineering (generation/repair) and temperature-controlled or porosity-controlled separation [18–21].

The rise in interest that has characterized stimuli-responsive polymers in these last decades is related to the various possibilities brought to polymer science and the different promising applications associated with them. In fact, in addition to the applications mentioned above, stimuli-responsive polymers are used in drug delivery [22,23], smart coating, intelligent medical instruments and auxiliaries, artificial muscles and robotics [13], smart textiles and apparel [24,25], biomimetic devices, electrochemical devices, and microelectromechanical systems. Furthermore, SRP coating was used to fabricate structured particles that can range from nanoscale to microscale. By adding this responsive layer, particles can undergo modification in their structure, dimensions, and interactions, when introduced to an external stimulus. Several structured particles can be synthesized in this fashion, such as core–shell particles, micro-/nanospheres, micro-/nanogels, and nanocomposites (Figure 11.2),

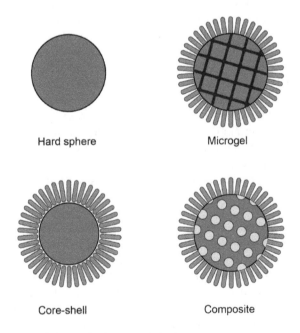

Hard sphere   Microgel

Core-shell   Composite

FIGURE 11.2 Some of the particulate morphologies.

and can be used in various technical and biomedical applications. In fact, using such a technology, mimicking a living cell can theoretically be done by combining several compartments that can communicate, exchange energy, and perform mechanical work in a single particulate structure [26].

11.2.2 Stimuli-Responsive Nanogels

Polymer nanogels are a special kind of nanomaterials since they combine the properties of nanoparticles and hydrogels. Basically, polymer nanogels are cross-linked colloidal particles formulated by swelling caused by the adsorption of large amounts of solvent. As a result, polymer nanogels are soft nanoparticles, thanks to the fluctuation of the high solvent content. However, their cross-linked three-dimensional polymeric network renders these particles resistant to dissolution [27,28]. Different materials were used to prepare polymer nanogels such as poly(N-isopropylacrylamide), polyethylene glycol, polypeptide, and chitosan [29].

The properties of stimuli-responsive nanogels (SRN) make them a good candidate for biomedical and drug delivery applications. Mainly, being sensitive to environmental stimuli allows a better control of drug release. In addition, due to their interior cross-linked network, drug loading is much higher and drug leakage is much lower in the case of nanogels compared to other nanocarriers [30]. Moreover, polymer nanogels possess a stable 3D structure promoting a stable desired size that is needed for passive drug delivery through the enhancement of permeation and retention effect. SRN were also used for active tumour targeting due to their ease of functionalization of polymer units [31]. Additionally, SRN can encapsulate multiple drugs with different solubility profiles. Jin et al. developed a triple-stimuli-responsive (protease/redox/pH) biodegradable PEGylated P(MAA/BACy) nanogel as a carrier for both hydrophilic doxorubicin and hydrophobic paclitaxel [32]. It was found that the double-drug-loaded system possesses favourable targeting ability and therapeutic effect, lower biotoxicity, longer *in vivo* circulation time, enhanced drug release profile, and a safe biodegradation [33]. These results show the validity of complex nanogel structures as drug vehicles. In fact, core–shell nanogel structures with constrained swelling to one layer were also suggested as a strategy to control multi-drug loading with desired amounts [34].

SRN are also known for their ease of synthesis. Basically, two main approaches can be used for the preparation of SRN (self-assembly of polymers, or polymerization in a heterogeneous environment). In both cases, cross-linking is an important step to ensure the colloidal stability of the nanogels. The polymerization method is usually done by emulsion polymerization, or inverse emulsion polymerization depending on the chosen continuous phase [31]. PEO-block-P(oligo(ethylene glycol) monomethyl ether methacrylate) particles were synthesized by atom transfer radical polymerization and cross-linked via inverse miniemulsion using a disulphide functionalized cross-linker. The cross-linked nanogels showed an excellent colloidal stability and controllable swelling ratio [35]. On the other hand, self-assembly of polymers is based on the formation of micelles, followed by a non-covalent interaction between the chains and cross-linking [36]. Cholesterol-bearing xyloglucan molecules were self-assembled in water forming nanogels. Cholesterol was used as a physical cross-linking point.

It was found that the obtained nanogel can successfully be coupled with hydropho-bic molecules with low molecular weight, and by incorporating galactose into their surface, they can be specifically internalized into hepatocytes, making them a good drug carrier candidate [37]. As stated before, cross-linking is a crucial step in the preparation of SRN. Different types of cross-linking can be used, such as physical cross-linking, in which the interaction between polymer moieties is done through different driving forces: hydrophobic interactions [37], electrostatic interactions, and host–guest interactions [38]. The other cross-linking strategies are disulphide-based, amine-based, click chemistry-based, and photo-induced cross-linking [31].

11.2.3 Stimuli-Responsive Membranes

Nowadays, membrane technology has widely been used in numerous applications such as drug delivery system [39], chemical separation [40], and water desalination [41], due to its remarkable properties with less energy consumption. Specifically, stimuli-responsive membranes (SRM) have gained much attention in the past decades due to the fact that they could change their physicochemical properties in response to a change in the environment stimuli [42]. SRM could be potentially designed to a sys-tem that responds to the changes in external stimuli: pH, temperature, ionic strength, light, and electric or magnetic field [40]. Functional materials (such as polymers, nanogels, and nanoparticles) with reversibly switchable physicochemical properties are generally considered as crucial resources for the preparation of SRM to provide different responsive properties. Husson et al. [43] used a two-step process to well explain the responsive mechanisms of SRM. Firstly, changes in the external environ-ment cause a specific conformation transition of a responsive functional material, and then these conformational transitions to some extent are easy to detect through naked eye or other simple technology methods. In fact, they always react to environmental stimuli to some extent by changing their conformation/polarity along the backbone, side chains, segments, or end groups. As a result, making alterations to the polymer chain length, chemical composition, architecture, and topography is the conventional method to optimize the responsive properties of SRM. There are two essential meth-ods used to prepare SRM: (1) blending of different stimuli-responsive polymers to manufacture membranes and (2) modification of existing membranes through graft-ing of stimuli-responsive polymers [44]. The former is a crucial method in the design of porous SRM. By taking this approach, the targeted mechanical properties, pore structure (porosity, pore size, and pore size distribution), and barrier structure (sym-metric and asymmetric) can be obtained. Various membrane processing techniques, such as solvent casting, phase inversion, and interpenetrating polymer networks, have been applied to develop SRM. Modification of the existing membranes is also an important method to prepare SRM, and the membrane performance is enhanced through the addition of functional materials. This method enables responsive proper-ties on the membrane surface and keeps the useful properties of the original mem-brane. Numerous polymer coating techniques, namely physical adsorption, chemical grafting, coating, layer-by-layer (LBL) assembly, chemical vapour deposition, have been used to modify the chemical surface or pores of the membranes [45].

11.3 STIMULI-RESPONSIVE POLYMERS IN COATING TECHNOLOGY

11.3.1 pH-Sensitive Polymers

pH-sensitive polymers are a group of stimuli-responsive polymers that can respond to environment pH through structural and property changes such as surface activity, chain, conformation, solubility, and configuration [46]. Materials referred to as pH-responsive polymers present the property to have potential ionizable groups in their structure. These groups are weak acidic or basic moieties sensible to pH variations and confer the pH sensitivity to the overall molecule [4]. Sensitivity to pH in response to environmental pH modification is made through the acceptance or the release of protons leading to conformational changes and changes in the colloidal behaviour of the polymer, such as flocculation, chain collapse or extension, and/or precipitation of homopolymers [46,47]. pH-responsive polymers can be produced using different polymerization techniques. Emulsion polymerization is the technique used for the preparation of most of the pH-responsive polymers due to the well-controlled size distribution. Moreover, anionic polymerization, group transfer polymerization, stable free radical polymerization, atom transfer radical polymerization, atom transfer radical polymerization, and reversible addition–fragmentation chain-transfer polymerization are other polymerization techniques used for the preparation of pH-responsive polymers [48].

According to their origin, natural pH-responsive polymers can be distinguished from synthetic ones. Among natural pH-responsive polymers, the most studied and used ones are dextran, gelatin, albumin, chitosan, hyaluronic acid, and alginic acid [11]. For synthetic pH-responsive polymers, two main categories can be identified: polymers with acidic group and polymers with basic group. Table 11.1 lists some of the basic polymers used as pH-sensitive polymers or moieties in the build-up of what is known as block pH-sensitive polymers.

Indeed, according to their structure, pH-sensitive polymers can be classified as linear homopolymer, copolymer (hydrophilic or amphiphilic block (di-, tri-, multi-block copolymers)), pH-responsive star polymers (with linear polymer chains connected to a central core), pH-responsive branched and hyper-branched polymers, pH-responsive dendrimer polymers, and pH-responsive brush and comb polymers [46].

Development of performant techniques has allowed the preparation of well-defined block copolymers that exhibit self-assemblage behaviour to external pH modification. pH-responsive block polymers can exhibit different sorts of responses following a pH modification. Thus, self-assembly such as formation of micelles, unimers, gels, vesicles, swelling, and deswelling can be noticed in the case of block copolymers [49,50].

pH-responsive polymers find application mainly in drug delivery since they are used as a coating material in controlled drug release systems or as drug carriers. Changes in pH in the gastrointestinal tract (acidic in stomach and basic in intestine) [18] and the biocompatibility of the pH-sensitive polymers make them ideal candidates for specific human drug delivery since they are able to switch from swelling to deswelling state in response to local body pH. By this process, they are capable

TABLE 11.1

Classification of Basic Acidic and Basic pH Responsive Polymers [46]

Acidic polymers	Polycarboxylic acids	Polyacrylic acid (PAAc)
		Poly(methacrylic acid) (PMMAc)
		Poly(propylacrylic acid) (PPAAc)
		Poly(4-vinylbenzoic acid) (PVBA)
		Poly(itaconic acid) (PIA)
	Polyphosphoric acids	Poly(ethylene glycol acrylate phosphate) (PEGAP)
		Poly(vinylphosphonic acid) (PVPA)
		Poly(ethylene glycol methacrylate phosphate) (PEGMP)
		Poly(4-vinyl-benzyl phosphonic acid) (PVBPA)
	Polysulphonic acids	Poly(vinylsulphonic acid) (PVSA)
		Poly(4-styrenesulphonic acid) (PSSA)
		Poly(2-acrylamido-2-methylpropane sulphonic acid) (PAMPS)
		Poly(2-acrylamido-3-sulphopropyl methacrylate potassium salt) (PKSPMA)
	Polyamino acids	Polyaspartic acid (PASA)
		Poly(L-glutamic acid) (PLGA)
		Polyhistidine (PHIS)
	Polyboronic acids	Poly(vinylphenylboronic acid) (PVPBA)
		Poly(3-acrylamidophenylboronic acid) (PAAPBA)
Basic polymers	Tertiary amine polymers	Poly(2-dimethylamino)ethyl methacrylate (PDMA)
		Poly(2-diethylamino)ethyl methacrylate (PDEA)
		Poly(2-dipropylamino)ethyl methacrylate (PDPAEMA)
		Poly(2-diisopropylamino)ethyl methacrylate (PDPA)
		Poly(N-(3-dimethylamino)-propyl) methacrylamide (PDMAPMAm)
		Poly((2-dimethylamino)ethyl acrylate) (PDMAEA)
		Poly(2-ter-butylamino)ethyl methacrylate (PtBAEMA)
		Poly(N,N-dialkylvinylbenzylamine)
		Poly(2-diethylamino)ethyl acrylamide (PDEAm)
	Morpholino, pyrrolidine, and piperazine polymers	Poly(2-N-morpholino)ethyl methacrylate (PMEMA)
		Polyacrylomorpholine (PAM)
		Poly(2-N-morpholino)ethylmethacrylamide (PMEMAm)
		Poly(N-ethylpyrrolidine methacrylate) (PEPyM)
		Poly(N-acrylo-N'-alkenyl piperazine)
	Pyridine and imidazole polymers	Poly(4-vinylpyridine) (P4VP)
		Poly(2-vinylpyridine) (P2VP)
		Poly(N-vinylimidazole) (PVI)
		Poly(6-(1-H-imidazol-1-yl)hexyl-methacrylate) (PImHeMA)
	Dendrimers	Poly(propyleneimine) dendrimer (PPI)
		Polyethylenimine dendrimer (PEI)
		Poly(amidoamine) dendrimer (PAMAM)

of retaining or release drugs in specific sites or organs according to pH. For this purpose, polymethacrylate, synthetic anionic or cationic copolymers of acrylic acid, methacrylic acid, (dimethylamino)ethyl methacrylate, methacrylic acid esters are known under the commercial name of Eudragit that are largely used as a coating material for the encapsulation of active ingredients and deliverance in specific site of human body [51]. Moreover, pH-responsive polymers find application in gene delivery and are also used in drug delivery to cancer tissues towards micelles' encapsulation using pH differences between extracellular pH of normal tissues and tumour extracellular pH [52–54].

11.3.2 Light-Sensitive Polymers

Light-responsive polymers are photodeformable in response to light stimuli. The advantage of using such materials in coating technology lies in the simplicity of operation, the avoidance of using chemical reactants and therefore limiting chemical contamination, and scalable miniaturization [55]. In addition, light sensitivity offers a remote application, ease of dosing, and accurate positioning and resolution of the response [56].

For a polymer to become light responsive, a chromophore (photoresponse functional group) should be incorporated into the polymer chain. Two categories of chromophores can be found, reversible and irreversible. Reversible chromophores, upon light excitation at a specific wavelength, undertake a reversible isomerization. This allows the alteration of the properties of the host polymer by irradiation at two different wavelengths. This reversible isomerization is important in different applications such as information storage and artificial muscles [57,58]. Spiropyran, an organic molecule that upon light excitation is engaged in a photocleavage of a spiro C-O bond, was used to prepare photochromic polymer brushes. Dense and smooth polymer films were successfully prepared and used as surfaces with switchable colour and wettability by light stimulation. A specific wavelength can convert a non-polar spiropyran to a polar one, and it can be reverted to its non-polar state using visible light [59].

On the other hand, irreversible chromophore is subject to unreversed changes upon light excitation. Therefore, equilibrium between two states does not exist. This will give the advantage of 100% photoconversion, which makes them ideal for drug delivery since this irreversibility can lead to effective drug release and better degradation when required [56]. This strategy was used in laser-induced forward transfer method (LIFT), in which biological molecules, such as protein and DNA, and even living organisms (cells) can be coated with a sacrificial polymer layer, and upon laser pulse emission, the coated molecule can be shifted from the polymer substrate to a receiver substrate in close proximity [60]. Doraiswamy et al. were able to transfer mammalian cells using polytriazines as sacrificial polymer that also acts as an irreversible chromophore [61]. Another approach was the coating of silica nanoparticles with NIR-light-responsive polymer via self-assembly. Due to the presence of coumarin moieties in the polymer, it can be disrupted by excitation by a femtosecond NIR light laser (800 nm) via a two-photon absorption process. It was found that under excitation with such light, the core–shell structure of the non-vehicle was able to release the pre-loaded drug in a controllable manner. In addition, *in vitro* results show a successful targeting of tumour cells once decorated with folic acid [62]. NIR range is considered more desirable for drug delivery than UV or visible light since it offers deep penetration without harming body tissues; however, most of the light-sensitive systems were tested *in vitro* and *in vivo* applications still require further development.

11.3.3 BIOLOGICAL ENVIRONMENT-SENSITIVE POLYMERS

Smart polymers that respond to physical and chemical stimuli such as light, pH, or temperature are desired because they are well controlled and easily measured. These polymers can be stimulated externally, offering a good control of the material properties. However, when it comes to biological systems, it is more complex and challenging. A temperature-responsive polymer, for example, cannot be targeted towards a specific receiver. Therefore, more targeted responsive systems are necessary [63].

Polymers responsive to biological environment are used for living organisms and applied in drug delivery, biomedicine, and cell engineering [64]. They can be enzyme responsive, ligand responsive, cell responsive, or generally responsive to biochemical environments. For example, responsiveness to reactive oxygen species (ROS) at the inflammation site of tumours can be exploited for targeted drug delivery by drug-loaded responsive polymer nanoparticles [65]. Doxorubicin-loaded ROS-responsive polymeric micelles were developed by Sun et al. as antitumor agents by introducing ROS-responsive thioketal as a linker between poly(ethylene glycol) and poly(ε-caprolactone). In a ROS-rich environment, the surface of these micelles is degraded and the drug was specifically released allowing a better *in vitro* anticancer activity of doxorubicin [66]. Other ROS-responsive moieties were also used for such applications, such as boronic ester and sulphide groups. However, few challenges still exist. The levels of ROS can vary from patient to patient, and therefore, these formulations must be personalized for each patient. In addition, the biocompatibility of the used linkers and their stability during circulation in normal cell environment should be addressed [67].

Enzyme-responsive polymers are another promising approach. These polymers have the advantages of being biocompatible, since enzymatic stimuli occurs naturally, and lead to more accurate targeting for the drug and therefore lead to a higher therapeutic efficiency and lower side effects [63]. Due to these advantages, enzyme-responsive polymers were applied as cell supports, injectable scaffolds, and drug delivery systems. Huang et al. used casein (a natural milk protein) as a coat for iron oxide nanoparticles. This system was loaded with doxorubicin as a model drug. It was found that these particles were stable in the acidic gastric environment in the presence of gastric protease. However, the outer casein layer was degraded in the presence of intestinal protease, allowing the maintenance of the drug's bioactivity and the enhancement of the drug delivery efficiency [68]. Synthetic polymers can also be used as enzyme-responsive drug delivery systems by introducing an enzyme-sensitive cross-linker to their matrix, rendering the polymeric structure a target to degradation by protease. Polystyrene nanoparticles were prepared by inverse miniemulsion and cross-linked with a hydrophobic peptide as a trypsin substrate. This formulation can have a promising future in nanomedicine, *in vitro* detection of specific enzymes, and cancer therapy [69]. Another example of enzyme-responsive polymers is nanoparticles sensitive to protein kinases. Nanoparticles comprising hydrophobically modified peptide substrate of a protein kinase were prepared for monitoring the activity of the latter by exploiting the enzyme to change the electrostatic interaction of the polymer aggregates in the particle [70].

Biological environment-responsive polymers were also applied as smart coatings for textiles developed as antibacterial vectors and used in several applications such as bedding and wound dressing. Kratochvil et al. developed a non-woven polymer nanofibre coating loaded with macrocyclic peptide 1, a potent non-bactericidal quorum-sensing inhibitor in *S. aureus*. The peptide was loaded to the nanofibre by electrospinning and was able to inhibit the production of bacterial hemolysins (a quorum sensing controlled virulence phenotype) and reduce the lysis of erythrocytes [71]. The release of the antimicrobial agent was in response to pathogenic bacteria specifically, and it did not respond to non-pathogenic strains. This approach can be a great alternative to antibiotic-dependent therapy, especially with the growing antibiotic resistance of *S. aureus* [72].

11.3.4 TEMPERATURE-SENSITIVE POLYMERS

Temperature-sensitive polymers (TSP) are sensitive to the temperature and alter their microstructural characteristics in response to the degree of temperature change. Temperature-sensitive polymers are well known and are the most employed polymers in drug delivery and biomaterials. Indeed, a small change in temperature is able to make new adjustments in TPS because they have a very sensitive balance between the hydrophobic and hydrophilic groups [73]. The critical solution temperature is an important property of such kind of polymers. The polymer will become insoluble after heating, if the polymer solution has a phase lower than the critical solution temperature; i.e., it has one lower critical solution temperature (LCST). The interaction strengths (hydrogen linkages) between the polymer and water molecules become inconvenient above the critical solution temperature and cause polymer swelling because of its dehydration and occurrence of hydrophobic interaction predominance [74].

The LSCT is the critical temperature at which the polymeric solution separates into two phases, going from one phase (isotropic state) to two phases (anisotropic state). LCST changes due to hydrophilic or hydrophobic incorporation existence within a polymeric system. The phase transition occurrence that causes a change in polymer is easily observable in certain temperatures.

TSP were mainly used in biomedical applications where the selected TSP should be biocompatible and non-immunogenic. To design an effective platform, it is crucial to understand the molecular mechanism behind the performance of TSP [75].

11.3.5 ELECTRICAL-SENSITIVE POLYMERS

Electrical-responsive polymers have the ability to deform (shrinking, swelling, bending) in response to an external electric field. In other words, they have the ability to transform electrical input into mechanical output. For this reason, electro-responsive polymers found their way to biomechanical applications, artificial muscles, sensing, chemical separation, and drug delivery [76].

The most researched form of electro-responsive polymers is polyelectrolyte hydrogels. In the presence of an electrical field, the ionic groups inside the gel get pushed to the anode side, which leads to swelling, or to the cathode side, which leads to shrinking and ultimately leads to a reversible deformation and volume change of the gel. Neutral polymers can also be affected by electric field; many polymer gels have been found to be subject to swelling via electrostriction (change in the length of materials in the presence of an external electric field). In addition, neutral materials can acquire a translational motion when introduced inside a non-uniform electric field via dielectrophoretic effect [77]. Both natural and synthetic polymers were used to prepare electro-responsive hydrogels, such as chitosan, chondroitin sulphate, hyaluronic acid, allylamine, and acrylonitrile.

Electrical-sensitive polymers have been studied in the development of artificial muscles. Many polymers were used in that context, but perhaps the most interesting one is polyacrylamide. This is due to several reasons: It is cheap, is easy to manipulate, is biocompatible, and have the same properties as living tissues. Moschou et al. developed an artificial muscle material based on an acrylic acid/acrylamide hydrogel blended with a conductive polypyrrole/carbon black composite. It was found that under electroactuation with a 1-volt voltage, the material showed a relatively fast response of 5 seconds and reversible bending in a neutral pH. In addition, the artificial muscle was incorporated into a bio-microelectromechanical system by immersing it in a 0.15 M NaCl solution and placing it above the polydimethylsiloxane membrane-covered microreservoir. It was found that under electroactuation, the bent hydrogel was able to push the membrane surface and pump the fluid from the reservoir into a calibrated microchannel with a low power consumption [78].

Electro-responsive polymers were also used in drug delivery. Kagatani et al. were able to produce poly(dimethylaminopropylacrylamide) gel loaded with insulin. This gel was administered subcutaneously in rats, and by stimulation with an external constant current of 1 mA, a pulsatile decrease in plasma glucose level was observed. The release is caused by the electrokinetic flow of dissolved insulin with the water from the gel, explained by electrophoresis of the counterions and electro-osmosis of water molecules in the cross-linked polyelectrolyte gel network [79].

11.4 APPLICATIONS OF STIMULI-RESPONSIVE MATERIALS

Due to their reversible responsiveness to environmental stimulus, stimuli-responsive polymer coatings were used in several applications. In this part, we will highlight some of these applications, the concept of their use, and recent advancements in their respective field. In addition, some of the recent applications are summarized in Table 11.2.

11.4.1 Food

Stimuli-responsive polymer systems have proven to be an efficient tool to use in food-packing systems with a great potential to improve the safety, quality, and traceability of food products, as well as its convenience for consumers [100]. In addition,

TABLE 11.2

Applications of Stimuli-Responsive Polymers

Application	Polymer	Responsiveness	Formulation	References
Food	Chitosan	pH and temperature	Colorimetric temperature indicator for monitoring food quality	[80]
	Chitosan	pH	Chitosan-based emulsion system	[81]
	Polyaniline	Microbial breakdown products	Films	[82]
	Polymers	Oxygen	Oxygen-scavenging film	[83]
	Phosphorescent Pt-porphyrin	CO_2	Sensors	[84]
Agriculture	Carboxymethyl chitosan	Glutathione		[85]
	Attapulgite-NH4HCO3-glyphosate/amino silicone oil-poly(vinyl alcohol)	Temperature	Core–shell nanocomposites	[86]
	Hydroxypropyl methylcellulose	Environment	Hydrogels	[87]
In vitro biomedical diagnostics	Antibody-bound poly(N-isopropylacrylamide)	Biomarkers	Polymer-coated gold nanoparticles	[88]
Sensors	Poly(N-isopropylacrylamide)	Temperature and salt	Coated nanoparticles	[89]
	Poly(N-(3-amidino)-aniline)	CO_2	Coated nanoparticles	[90]
	Chitosan (CHI)/poly(acrylic acid)	Ionic strength	Coated long-period gratings	[91]
	Poly(vinyl alcohol)	Glucose	Coated platinum electrode	[92]
Drug delivery	Ethylcellulose	pH	Polymer-coated insulin-loaded intestinal patches	[93]
	Eudragit® S100 and Eudragit® L100-55	pH	Polymer-coated insulin-loaded nanoparticles	[94]
	Ethylcellulose	Light	Polymer-coated insulin-loaded gold nanoparticles	[95]
	Poly(N-isopropylacrylamide-co-N,N-dimethylacrylamide)	Temperature	Core–shell nanoparticles	[96]
	Poly-N-isopropylacrylamide-co-itaconic acid	Temperature and pH	Doxorubicin-loaded core–shell nanoparticles	[97]
Cosmetics	Polyoxyethylene and polyoxypropylene	Temperature	Aqueous composition	[98]
Textile	Polyurethanes	Temperature	Polymer-coated shape-memory fabric	[99]

stimuli-responsive polymers were used in some functional food as coating materials for particle [101] or emulsion [102].

It is a well-established food quality control that is an essential research target to protect consumers against food-borne infection and to maximize the benefits of the food industries. Recent developments in smart food packaging to monitor the changes in the condition of food have attracted much attention. This technology may be incorporated in packaging materials or attached to the inside or outside of a package and aims to sense and share more information about food quality to consumer or producer during transport and storage [103]. Stimuli-responsive polymers have been regarded as a useful method to acquire packaging with responsive properties. For instance, a smart packaging system can be used to observe the distinct pH of foods with visual colorimetric changes of pH-responsive polymers inside the smart package [104]. In addition, packaging with responsive properties includes the sensors, which would detect the temperature, gas formation, microbial growth, freshness, and so on [100]. Principally, for food applications, the material used for packaging should have the 'Generally Regarded As Safe (GRAS)' status. Thus, most of the stimuli-responsive materials that can be used in actual food packaging have very limited implementation. Rather, natural polymers are promising materials in the design of smart packaging due to their compatibility and non-toxic nature on administration. Exploiting the pH-tuneable ability of chitosan, Maciel, Yoshida, and Franco [80] established a prototype of a colorimetric temperature indicator for monitoring food quality. Liu et al. [81] developed a simple yet reversible pH-responsive chitosan-based emulsion system by exploiting the pH-tuneable ability of chitosan. Based on polyaniline film, Bambang et al. [82] developed a novel colorimetric food package label as a chemical sensor for real-time monitoring of the microbial breakdown products in the headspace of packaged fish. Another packaging system from Cryovac Div., Sealed Air Corporation, used Cryovac® 0S2000™ polymer-based oxygen-scavenging film, which has applications in different food products such as smoked meat products and processed meats [83]. On the other hand, polymer sensors combining oxygen and carbon dioxide measurements in a single sensor [105] have been employed to smart packaging systems in many recently developed packaging systems. There are several accepted applications of polymer-based sensors: The ripeness of the pear is determined by detecting volatile aromatic compounds, or the freshness of the fish is determined by monitoring the concentration of volatile amines [106]. Nicolas et al. [84] developed a polymeric solid-state CO_2 sensor that uses phosphorescent Pt-porphyrin for food packaging applications.

The use of stimuli-responsive materials in the food packaging industry is a new strategy to get markedly improved packaging properties. However, the topic of using stimuli-responsive polymers in many food processing or packaging industries is still a challenge.

11.4.2 AGRICULTURE

Over the past few decades, using more advanced agricultural methods (high-tech machinery, resources, and chemicals) for analyzing or monitoring soils and crops has been a research trend in agricultural planting to get more information and evaluate their maturity and health status. Among these methods, incorporation of stimuli-responsive polymers as sensors into fertilizers and pesticides to respond to the external environment is a promising field for residue analysis of soils and crops [107,108].

The application of fertilizers and pesticides has a significantly great effect on crop productivity, and the major drawback of conventional applications, in general, is the loss depending on the mode of application and weather conditions. Instead, the applications of these chemical products have been expected to transform into smart agrarian practices with more functional properties. As such, it is very interesting to combine fertilizers and pesticides with stimuli-sensitive polymers [109]. The products used in agriculture may get the properties sensitive to pH, light, temperature, biological environment from stimuli-responsive materials, which can be used to formulate respond to surrounding microenvironment intelligently and smartly eliciting a better and controlled release. For instance, Zhiyan et al. successfully developed glutathione-responsive carboxymethyl chitosan systems for controlled release of Diuron herbicide with the pre-emergence treatment of target species (*Echinochloa crus-galli*), and their results showed that it will be a promising candidate for controlled release of pesticides in agriculture [85]. Chi et al. [86] developed core–shell attapulgite-NH_4HCO_3-glyphosate/amino silicone oil-poly(vinyl alcohol) (PVA) nanocomposites that offered temperature-responsive controlled release of glyphosate herbicide. Another work done by Liu et al. [110] used modified soybean protein isolate and carboxymethyl cellulose by the hydrazone formation to synthesize pH-responsive avermectin nanoparticles. On the other hand, hydrogels are cross-linked networks of polymers having high water retention capacity and can be used for the application of fertilizer with responsiveness towards temperature, pH, enzymes, etc. [87]. Chen et al. [87] fabricated an environmentally responsive cellulose-based hydrogel using hydroxypropyl methylcellulose for the controlled release of water-soluble fertilizers, and this study showed that it has the potential to be used as controlled release devices in horticulture and agriculture. Another study reported an anion-responsive carbon nanosystem prepared using polyethylenimine-modified hollow/mesoporous carbon nanoparticles to improve selenium utilization efficiency in vegetables [110].

The presence of stimuli-responsive polymers in fertilizers and pesticides open more chances for many possibilities in the development of modern agriculture. However, despite many positive aspects of blending fertilizers and pesticides with responsive materials, careful evaluation is still required before using these technologies for crops in the future.

11.4.3 *In Vitro* Biomedical Diagnostics

Hoffman et al. designed a smart polymer–biomolecule conjugate for immunoassay that was based on the conjugation of an antibody to poly(*N*-isopropylacrylamide) (PNIPAAm). In order to capture an antigen such as biomarker of hepatitis or AIDS, blood test sample including smart bioconjugate. Consecutively, second labelled antibody was added and that detection antibody was designed to affinity link to the same antigen. In the end, to separate the phase-labelled immune complex sandwich, the solution was warmed. This assay was similar to the enzyme-linked immunosorbent assay (ELISA), except that this was performed in solution with a last step named phase

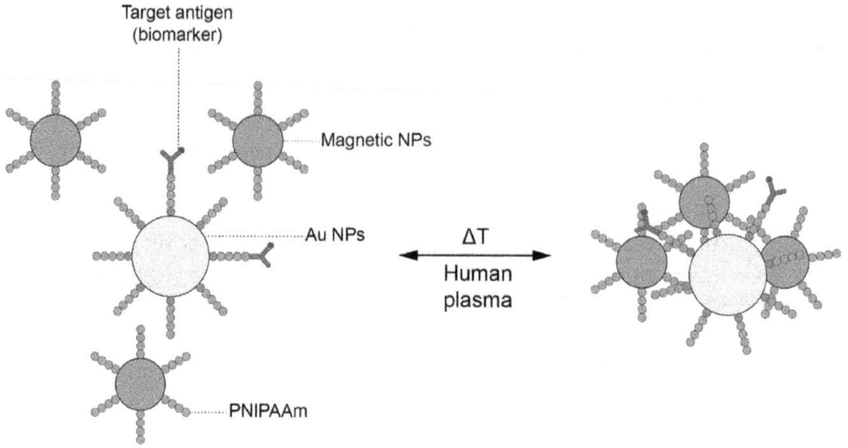

FIGURE 11.3 When PNIPAAm is coated on magnetic nanoparticles (mNP) and gold nanoparticles (Au-NP), it acts to aggregate or 'glue' the NPs together when the temperature is increased above the LCST of PNIPAAm.

separation. This method, in comparison with the typical multi-well plate ELISA, was much faster and just as accurate (Figure 11.3).

Stayton & Hoffman et al. have recently applied smart PNIPAAm technology to several novel surfaces and nanoparticle-based diagnostic systems, which use PNIPAAm coatings on microfluidic channels, on gold nanoparticles, and on magnetic nanoparticles. These smart nanoscale systems are designed and developed for clinical immunoassays [88] (Figure 11.4).

11.4.4 Drug Delivery

Due to their properties and flexibility, stimuli-responsive polymers found their way into drug delivery. They were used nanocarriers, hydrogels, micelles, and complexes that can safely carry and protect the drug through *in vivo* circulation and release it in the desired site due to environment change. In addition, SRP were used in drug delivery as coating materials for drug tablets and nanocarriers [111].

Enteric coating or enteric capsule is a polymer layer added to drug tablets to increase the efficiency of the drug by protecting the tablet from early degradation [112]. Two types of enteric coatings are popular. One is based on copolymer of methacrylic monomers, and the other is based on ethylcellulose. Both types have a pH-sensitive profile. They are hydrophobic at gastric pH that restricts the release of the drug, and they become hydrophilic at intestinal pH allowing the drug to be released in the intestines, thereby protecting it from gastric degradation. This will improve the efficiency of oral drug delivery [111]. This characteristic was exploited to deliver peptides and protein through oral route. Mucoadhesive intestinal patches loaded with insulin were coated with ethylcellulose to secure a safe passage for a therapeutic dose of insulin through the digestive system to be released near intestinal mucosa. *In vivo* results showed that these patches induced dose-dependent hypoglycaemia in normal rats with

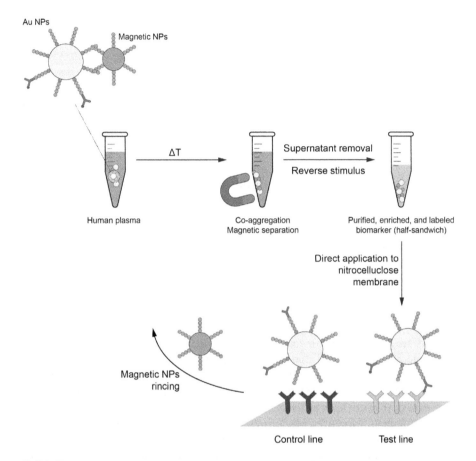

FIGURE 11.4 Biomarkers in blood plasma test samples at RT are captured by antibodies bound to PNIPAAm-coated gold nanoparticles (NPs) that are then thermally aggregated with PNIPAAm-coated magnetic NPs and isolated and concentrated by a magnetic field. After washing the aggregates and then lowering the temperature below the LCST, the NPs are dispersed and flowed onto a lateral flow strip for biomarker assay.

a maximum drop in blood glucose levels of 75% [93]. In another study, pH-sensitive chitosan and poly(ɣ-glutamic acid) nanoparticles were loaded with insulin, freeze-dried, and coated with Eudragit® S100 and Eudragit®L100-55 capsules. After *in vivo* oral administration in diabetic rats, it was found that intestinal absorption of insulin was enhanced and the relative bioavailability was about 20% [94]. Light-responsive polymer coating was also used for insulin delivery. Gold nanoparticles loaded with insulin and coated with hydrophobic ethylcellulose were developed. By applying near-infrared irradiation, gold nanoparticles are heated, which led to reversible collapse of the polymer network, resulting in the formation of porous structure, and therefore, a rapid diffusion of insulin was allowed. It was also found that by controlling the irradiation, insulin dosing in diabetic rats can be controlled [95].

Stimuli-responsive polymers were also used as coating materials in the development of core–shell organic–inorganic drug nanovehicles. Magnetite nanoparticles were coated with the temperature-responsive biopolymer poly(N-isopropylacryl-amide-co-N,N-dimethylacrylamide) to form a core–shell structure with a size of 8 nm. Physicochemical characterization demonstrated the suitability of such system in magnetic drug delivery and in magnetic resonance imaging by combining the advantages of magnetic nanoparticles and stimuli-responsive polymers [96]. Core–shell nanocarriers were also used for the delivery of anticancer agents. Magnetic core–shell nanoparticles were synthesized using poly-N-isopropylacrylamide-co-itaconic acid containing thiol side groups as a thermo- and pH-responsive polymer. Particles were used as a vehicle for doxorubicin with a size of 50 nm and a high loading capacity (55%). The pH and temperature triggered the release of doxorubicin at tumour tissue environment, which was 39% higher than in the physiological conditions. In addition, this carrier shows no cytotoxicity while inducing an efficient anticancer effect *in vitro*, suggesting its suitability as a drug carrier in general and as an anticancer agent in specific [97].

SRP coatings are, in conclusion, a very promising agent for drug delivery, especially when it comes to oral administration of therapeutic peptides. Until now, this topic remains a challenge in pharmaceutical research.

11.4.5 SENSORS

Sensors are devices that can convert an input received from a surrounding environment into an output signal that can be transformed into readable data. The ability of stimuli-responsive polymers to interact with the surrounding environment makes them a good candidate for the fabrication of sensors. Different stimuli-responsive sensors were developed, such as those responsive to temperature, pH, CO_2, and biological environment. Moreover, stimuli-responsive polymer-coated nanoparticles have gained attention in sensing research due to the optical properties of nanoparticles.

Gold nanoparticles coated with poly(N-isopropylacrylamide) were developed as colorimetric temperature and salt sensors. Polymer was synthesized by RAFT polymerization and grafted into nanoparticles by the ligand exchange process. It was found that with increasing temperature, these particles can change their colour from red to purple to blue due to the electronic coupling of the surface plasma on resonance caused by the aggregation of gold nanoparticles. In addition, it was found that the range of temperature sensitivity increases in the presence of salt due to charge screening [89]. In another study, gold nanoparticles were used to develop CO_2 sensors. Poly(N-(3-amidino)-aniline), a CO_2-responsive polymer, was grafted onto the surface of the particles, and in the presence of dissolved CO_2, it can swell and detach from the surface of the particles due to the protonation of its amidine group and the formation of hydrophilic amidinium. This change can lead to the aggregation of the nanoparticles and a change in colour. The limit of detection of this sensor was identified at 0.0024 hPa [90]. SRP were also used to produce highly sensitive fibre optic marine salinity sensors. Long-period gratings were coated with ionic-strength-responsive chitosan (CHI)/poly(acrylic acid) polyelectrolyte multi-layers by layer-by-layer assembly method. In the presence of NaCl, a shrinkage in the coating layer and salt ion doping occur, leading to an increase in water content and resulting in a light shifting from red to blue. The blue-shift increases with the increase in salt concentration. It was also noticed that polymer-coated systems have a higher sensitivity than the non-coated ones [91].

Biosensing was also a target for SRP. Platinum electrodes coated with a polymer–polymer complex of phenyl boronic acid with tertiary amine moieties and poly(vinyl alcohol) (PVA) were developed as glucose level sensors. The system was tested in a glucose-rich Dulbecco's phosphate-buffered saline solution at pH 7.4. It was found that in the presence of glucose, a change in the swelling of the polymer layer is induced, which leads to an increase in the diffusion of ion species and thus increases the measurable current change. This glucose sensor was proven to be of high selectivity and suitable for measuring the physiological concentration of glucose since current changes were proportional to glucose concentration in the range 0–300 mg/dL [92]. Another glucose sensing strategy was to develop an edible electrochemical biosensor with an important resistance to extreme acidic conditions. A direct and stable glucose monitoring was achieved using edible materials such as olive oil and activated charcoal as carbon paste biosensors. These materials were able to protect the embedded glucose oxidase enzyme from strong acidic conditions. Moreover, the electrode was coated with pH-responsive enteric coatings (Eudragit® L100 and Eudragit® E PO), which allow the control of the activation at different values of pH corresponding to gastric and intestinal conditions [113].

The presence of a high range of biocompatible SRP opens up the opportunity for many possibilities in the development of biosensors. However, an improvement in selectivity, stability, sensitivity, and output readability is always required [114].

11.4.6 COSMETICS

In general, polymers are the second-largest category of ingredients in the composition of cosmetics and personal care products. Stimuli-responsive polymers are interesting in cosmetics because they help the formulation of cosmetics to exhibit desired characteristics in the storage container. Upon the heating (above the threshold temperature), the thermally reversible viscosifying polymers show a dramatic increase in the solution viscosity. In the aqueous compositions consisting of polymers possessing a poly(2-acrylamido-2-methylpropanesulphonic acid) backbone and grafted chains of polyoxyethylene and polyoxypropylene or polyoxybutylene, thermal gelling was observed.

It is possible to use gelation in response to the stimuli such as temperature as a processing aid, for instance, in the case of multiple emulsions preparation. To obtain water/oil/water multiple emulsions normally, it is needed, for the first time, to prepare the internal emulsion in a separate way from the external base and afterwards to mix slightly the phases together to prevent the internal water phase from escaping into the external aqueous phase. Processing complications are raised when the viscosity of the oil phase is considerably higher than that of either of the aqueous phases. According to a research performed by Mercier, usage of amphipathic reverse thermal gelling polymers can increase the viscosity of both the internal and external aqueous phase, allowing a one pot mixing of multiple phase emulsions without breaking the W/O while avoiding the mixing of the internal aqueous phase and the external aqueous phase [98].

11.4.7 TEXTILES

In order to bring an added value to functional textiles, coating and lamination can be the interesting strategies. Due to the wider applications of the functional textiles, the use of these techniques is boosting since coating and lamination extend the range of smart performance properties of textiles. To design fabrics with functional properties, coating and lamination techniques are employed. Furthermore, thanks to the widespread application of coating and lamination over the range of technical textiles sectors, improve functionality and durability. Coating and lamination are including: WVP, waterproofness, increased abrasion, flame, and UV resistance or phase change materials. Usually, coatings (often in liquid form) are applied to fabrics during the preparation stage. The preparation of laminate membrane is needed to be applied on the textile consecutively. Coated and laminated textiles widespread application across the different types of technical textile sectors are including antibacterial coatings, waterproof breathable, water vapour and air permeable, UV resistance, phase change materials, etc. [25,115]. The way in which the coatings and lamination stick to the fabric surface determines the method of their communication with the fabric. Since coating is applied in liquid form on the fabric, it covers the surface of the fabric and penetrates the fabric structure, filling the air pockets and bridging the interstices. Occasionally, the end product is defined as a composite that is attributed to the combination of textile and non-textile. The tear strength and thermomechanical properties determine the mechanical properties of the fabric [99,116,117].

The textile institutes describe a coated textile as a material composed of two or more layers, at least one of which is a textile fabric and at least one of which is a substantially continuous polymeric layer. This polymeric layer is administered in liquid form with a solvent or water base that evaporates leaving the polymer behind and is applied to one or both the sides. Laminated textiles are formed from one or more layers of textile and non-textile components. Based on the Textile Institute laminated or combined fabric is a material composed of two or more layers, at least one of which is a textile fabric, bonded closely together by means of an added adhesive, or by the adhesive properties of one or more of the component layers. Smart clothing is employed to keep the users warm by default, for example, in winter and then to help the body spontaneously cool down when the wearer sweats [118].

11.5 CONCLUSION

Stimuli-responsive polymers that can be of either natural or synthetic origin react to changes or stimuli such as pH, temperature, and light. Indeed, stimuli are divided into three main categories: physical stimuli, chemical stimuli, and biological stimuli. Thanks to this characteristic, stimuli-responsive polymers offer a vast range of functionalities in food, agriculture, drug delivery, biomedical diagnostics, sensors, textiles, and cosmetics and continue to show a great promise. The stimuli-responsive materials are divided into four classes: stimuli-responsive polymers, stimuli-responsive nanogels, stimuli-responsive structured particles, and stimuli-responsive membranes. The presence of a wide range of polymers with a responsive ability to external stimuli, opens the possibility to a vast array of applications. While natural polymers were used in food, agriculture, and drug delivery for human administration, packaging, and microbial and oxidation protection, synthetic polymers are also used as sensors, scavenging agents, and many more. Finally, despite the vast knowledge we already have in the topic, more promising findings are being uncovered on yearly basis that can open the doors to a larger niche of applications.

REFERENCES

1. M.A. Jahid, J. Hu, and H. Zhuo, 'Stimuli-responsive polymers in coating and laminating for functional textile," In *Smart Textile Coatings and Laminates*, W. C. Smith, Ed. Elsevier, 2019, pp. 155–173.
2. A. Zhang, K. Jung, A. Li, J. Liu, and C. Boyer, 'Recent advances in stimuli-responsive polymer systems for remotely controlled drug release', *Prog. Polym. Sci.*, vol. 99, p. 101164, Dec. 2019.
3. I. Galaev, '"Smart" polymers and what they could do in biotechnology and medicine', *Trends Biotechnol.*, vol. 17, no. 8, pp. 335–340, Aug. 1999.
4. Y. Qiu and K. Park, 'Environment-sensitive hydrogels for drug delivery', *Adv. Drug Deliv. Rev.*, vol. 53, no. 3, pp. 321–339, Dec. 2001.
5. M. R. Aguilar, C. Elvira, A. Gallardo, B. Vázquez, and J. S. Román, 'Smart polymers and their applications as biomaterials', *Smart Polym.*, vol. 3, p. 27, Jan. 2007.
6. S. J. Grainger, and M. E. H. El-Sayed, 'Stimuli-sensitive particles for drug delivery,' In *Biologically-Responsive Hybrid Biomaterials*, E. Jabbari and A. Khademhosseini, Eds. World Scientific, 2010, pp. 171–190.
7. M. A. C. Stuart et al., 'Emerging applications of stimuli-responsive polymer materials', *Nat. Mater.*, vol. 9, no. 2, pp. 101–113, Feb. 2010.
8. L. Klouda and A. G. Mikos, 'Thermoresponsive hydrogels in biomedical applications', *Eur. J. Pharm. Biopharm.*, vol. 68, no. 1, pp. 34–45, Jan. 2008.
9. C.-Y. Chuang, t.-m. Don, and W.-Y. Chiu, 'Synthesis of chitosan-based thermo- and pH-responsive porous nanoparticles by temperature-dependent self-assembly method and their application in drug release', *J. Polym. Sci. Part Polym. Chem.*, vol. 47, pp. 5126–5136, Oct. 2009.
10. M. Fathi, P. SahandiZangabad, S. Majidi, J. Barar, H. Erfan-Niya, and Y. Omidi, 'Stimuli-responsive chitosan-based nanocarriers for cancer therapy', *BioImpacts BI*, vol. 7, no. 4, pp. 269–277, 2017.
11. S. Chatterjee and P. Chi-Leung Hui, 'Review of stimuli-responsive polymers in drug delivery and textile application', *Molecules*, vol. 24, no. 14, Jul. 2019.
12. E. Cabane, X. Zhang, K. Langowska, C. G. Palivan, and W. Meier, 'Stimuli-responsive polymers and their applications in nanomedicine', *Biointerphases*, vol. 7, no. 1, p. 9, Feb. 2012.
13. 'Stimuli-responsive polymers: Fundamental considerations and applications | SpringerLink'. [Online]. Available: https://link.springer.com/article/10.1007/s13233-017-5088-7. [Accessed: 06-Dec–2019].
14. A. Lendlein and V. P. Shastri, 'Stimuli-sensitive polymers', *Adv. Mater.*, vol. 22, no. 31, pp. 3344–3347, 2010.
15. D. Roy, J. Cambre, and B. Sumerlin, 'Future perspectives and recent advances in stimuli-responsive materials', *Prog. Polym. Sci.*, vol. 35, pp. 278–301, Feb. 2010.
16. H. Almeida, M. H. Amaral, and P. Lobão, 'Temperature and pH stimuli-responsive polymers and their applications in controlled and self- regulated drug delivery', *J. Appl. Pharm. Sci.*, vol. 2, p. 10.
17. F. Liu and M. Urban, 'Recent advances and challenges in designing stimuli-responsive polymers', *Prog. Polym. Sci.*, vol. 35, pp. 3–23, Feb. 2010.

18. P. Bawa, V. Pillay, Y. E. Choonara, and L. C. du Toit, 'Stimuli-responsive polymers and their applications in drug delivery', *Biomed. Mater. Bristol Engl.*, vol. 4, no. 2, p. 022001, Apr. 2009.
19. J. Kopeček, 'Hydrogel biomaterials: A smart future?', *Biomaterials*, vol. 28, no. 34, pp. 5185–5192, Dec. 2007.
20. S. Kim, J.-H. Kim, O. Jeon, I. C. Kwon, and K. Park, 'Engineered Polymers for Advanced Drug Delivery', *Eur. J. Pharm. Biopharm. Off. J. Arbeitsgemeinschaft Pharm. Verfahrenstechnik EV*, vol. 71, no. 3, pp. 420–430, Mar. 2009.
21. A. Kumar, A. Srivastava, I. Galaev, and B. Mattiasson, 'Smart polymers: Physical forms and bioengineering applications', *Prog. Polym. Sci.*, vol. 32, pp. 1205–1237, Oct. 2007.
22. D. Kuckling, 'Stimuli-Responsive Gels', *Gels*, vol. 4, no. 3, p. 60, Sep. 2018.
23. A. Raza, T. Rasheed, F. Nabeel, U. Hayat, M. Bilal, and H. M. N. Iqbal, 'Endogenous and exogenous stimuli-responsive drug delivery systems for programmed site-specific release', *Molecules*, vol. 24, no. 6, Mar. 2019.
24. 'Shape memory fibers prepared via wet, reaction, dry, melt, and electro spinning - diagram, schematic, and image 01'. [Online]. Available: http://www.patentsencyclopedia.com/imgfull/20090093606_01. [Accessed: 06-Dec-2019].
25. M. A. Jahid, J. Hu, and H. Zhuo, 'Stimuli-responsive polymers in coating and laminating for functional textile,' In *Smart Textile Coatings and Laminates*, W. C. Smith, Ed. Elsevier, 2019, pp. 155–173.
26. M. Motornov, Y. Roiter, I. Tokarev, and S. Minko, 'Stimuli-responsive nanoparticles, nanogels and capsules for integrated multifunctional intelligent systems', *Prog. Polym. Sci.*, vol. 35, no. 1–2, pp. 174–211, Jan. 2010.
27. Z. Jiang, J. Chen, L. Cui, X. Zhuang, J. Ding, and X. Chen, 'Advances in stimuli-responsive polypeptide nanogels', *Small Methods*, vol. 2, no. 3, p. 1700307, Mar. 2018.
28. M. Vicario-de-la-Torre and J. Forcada, 'The potential of stimuli-responsive nanogels in drug and active molecule delivery for targeted therapy', *Gels*, vol. 3, no. 2, p. 16, 2017.
29. K. Madhusudana Rao, K. S. V. Krishna Rao, and C.-S. Ha, 'Functional stimuli-responsive polymeric network nanogels as cargo systems for targeted drug delivery and gene delivery in cancer cells', in *Design of Nanostructures for Theranostics Applications*, vol. 64, no. 9, Elsevier, pp. 243–275, 2018.
30. J. Ryu, R. T. Chacko, S. Jiwpanich, S. Bickerton, R. P. Babu, and S. Thayumanavan, 'Self-cross-linked polymer nanogels: A versatile nanoscopic drug delivery platform', *J. Am. Chem. Soc.*, vol. 132, no. 48, pp. 17227–17235, Dec. 2010.
31. K. Madhusudana Rao, K. S. V. Krishna Rao, and C.-S. Ha, 'Functional stimuli-responsive polymeric network nanogels as cargo systems for targeted drug delivery and gene delivery in cancer cells,' In *Design of Nanostructures for Theranostics Applications*, vol. 64, no. 9, A. M. Grumezescu, Ed. Elsevier, 2018, pp. 243–275.
32. S. Jin, D. Li, P. Yang, J. Guo, J. Q. Lu, and C. Wang, 'Redox/pH stimuli-responsive biodegradable PEGylated P(MAA/BACy) nanohydrogels for controlled releasing of anticancer drugs', *Colloids Surf. Physicochem. Eng. Asp.*, vol. 484, pp. 47–55, 2015.
33. S. Jin et al., 'Biodegradation and toxicity of protease/redox/pH stimuli-responsive PEGlated PMAA nanohydrogels for targeting drug delivery', *ACS Appl. Mater. Interfaces*, vol. 7, no. 35, pp. 19843–19852, 2015.
34. W. Richtering and A. Pich, 'The special behaviours of responsive core-shell nanogels', *Soft Matter*, vol. 8, no. 45, pp. 11423–11430, 2012.
35. J. K. Oh, C. Tang, H. Gao, N. V. Tsarevsky, and K. Matyjaszewski, 'Inverse miniemulsion ATRP: A new method for synthesis and functionalization of well-defined water-soluble/cross-linked polymeric particles', *J. Am. Chem. Soc.*, vol. 128, no. 16, pp. 5578–5584, Apr. 2006.
36. S. Hajebi et al., 'Stimulus-responsive polymeric nanogels as smart drug delivery systems', *Acta Biomater.*, vol. 92, pp. 1–18, 2019.

37. S. I. Sawada, H. Yukawa, S. Takeda, Y. Sasaki, and K. Akiyoshi, 'Self-assembled nano-gel of cholesterol-bearing xyloglucan as a drug delivery nanocarrier', *J. Biomater. Sci. Polym. Ed.*, vol. 28, no. 10–12, pp. 1183–1198, 2017.

38. Y. Sasaki and K. Akiyoshi, 'Nanogel engineering for new nanobiomaterials: From chaperoning engineering to biomedical applications', *Chem. Rec.*, p. n/a-n/a, Sep. 2010.

39. W.-C. Yang, R. Xie, X.-Q. Pang, X.-J. Ju, and L.-Y. Chu, 'Preparation and characteriza-tion of dual stimuli-responsive microcapsules with a superparamagnetic porous mem-brane and thermo-responsive gates', *J. Membr. Sci.*, vol. 321, no. 2, pp. 324–330, Aug. 2008.

40. A. J. Brown et al., 'Interfacial microfluidic processing of metal-organic framework hol-low fiber membranes', *Science*, vol. 345, no. 6192, pp. 72–75, Jul. 2014.

41. S.-C. Low and Q.-H. Ng, 'Progress of stimuli responsive membranes in water treat-ment,' In *Advanced Nanomaterials for Membrane Synthesis and its Applications*, L. Woei-Jye, I. Ahmad Fauzi, I. Arun, and A.-A. Amir, Eds. Elsevier, 2019, pp. 69–99.

42. D. Wandera, S. R. Wickramasinghe, and S. M. Husson, 'Stimuli-responsive mem-branes', *J. Membr. Sci.*, vol. 357, no. 1–2, pp. 6–35, Jul. 2010.

43. S. M. Husson, 'Synthesis aspects in the design of responsive membranes', in *Responsive Membranes and Materials*, D. Bhattacharyya, T. Schäfer, S. R. Wickramasinghe, and S. Daunert, Eds. West Sussex: John Wiley & Sons, Ltd, 2012, pp. 73–96.

44. B. Ma et al., 'Facile fabrication of composite membranes with dual thermo- and pH-responsive characteristics', *ACS Appl. Mater. Interfaces*, vol. 9, no. 16, pp. 14409–14421, Apr. 2017.

45. H. Nicholas and S. Rajindar, Eds., *Emerging Membrane Technology for Sustainable Water Treatment.* Elsevier, 2016.

46. G. Kocak, C. Tuncer, and V. Bütün, 'pH-Responsive polymers', *Polym. Chem.*, vol. 8, no. 1, pp. 144–176, Dec. 2016.

47. D. Schmaljohann, 'Thermo- and pH-responsive polymers in drug delivery', *Adv. Drug Deliv. Rev.*, vol. 58, no. 15, pp. 1655–1670, Dec. 2006.

48. S. Dai, P. Ravi, and K. C. Tam, 'pH-Responsive polymers: Synthesis, properties and applications', *Soft Matter*, vol. 4, no. 3, pp. 435–449, Feb. 2008.

49. J. Hu, G. Zhang, Z. Ge, and S. Liu, 'Stimuli-responsive tertiary amine methacrylate-based block copolymers: Synthesis, supramolecular self-assembly and functional appli-cations', *Prog. Polym. Sci.*, vol. 39, no. 6, pp. 1096–1143, Jun. 2014.

50. A. Moreno, J. C. Ronda, V. Cádiz, M. Galià, G. Lligadas, and V. Percec, 'pH-responsive micellar nanoassemblies from water-soluble telechelic homopolymers endcoding acid-labile middle-chain groups in their hydrophobic sequence-defined initiator residue', *ACS Macro Lett.*, vol. 8, no. 9, pp. 1200–1208, Sep. 2019.

51. C. N. Patra, R. Priya, S. Swain, G. Kumar Jena, K. C. Panigrahi, and D. Ghose, 'Pharmaceutical significance of Eudragit: A review,' *Futur. J. Pharm. Sci.*, vol. 3, no. 1, pp. 33–45, 2017.

52. T. Sim et al., 'A pH-sensitive polymer for cancer targeting prepared by one-step modulation of functional side groups', *Macromol. Res.*, vol. 27, no. 8, pp. 795–802, Aug. 2019.

53. D. J. Peeler, S. N. Thai, Y. Cheng, P. J. Horner, D. L. Sellers, and S. H. Pun, 'pH-sensi-tive polymer micelles provide selective and potentiated lytic capacity to venom peptides for effective intracellular delivery', *Biomaterials*, vol. 192, pp. 235–244, 2019.

54. J. M. Benns, J. S. Choi, R. I. Mahato, J. S. Park, and S. W. Kim, 'pH-sensitive cat-ionic polymer gene delivery vehicle: N-Ac-poly(L-histidine)-graft-poly(L-lysine) comb shaped polymer', *Bioconjug. Chem.*, vol. 11, no. 5, pp. 637–645, Oct. 2000.

55. H. S. Lim, D. Kwak, D. Y. Lee, S. G. Lee, and K. Cho, 'UV-driven reversible switching of a roselike vanadium oxide film between superhydrophobicity and superhydrophilic-ity', *J. Am. Chem. Soc.*, vol. 129, no. 14, pp. 4128–4129, Apr. 2007.

56. J. Cui and A. Del Campo, 'Photo-responsive polymers: Properties, synthesis and applications', In *Smart Polymers and their Applications*, M. R. Aguilar and J. San Román, Eds. Elsevier, 2014, pp. 93–133.

57. B. L. Feringa, R. a. Van Delden, N. Koumura, and E. M. Geertsema, 'Chiroptical molecular switches chiroptical molecular switches', *Chem. Rev.*, vol. 3, pp. 123–163, 2000.

58. T. Ikeda, J. I. Mamiya, and Y. Yu, 'Photomechanics of liquid-crystalline elastomers and other polymers', *Angew. Chem. - Int. Ed.*, vol. 46, no. 4, pp. 506–528, 2007.

59. S. Samanta and J. Locklin, 'Formation of photochromic spiropyran polymer brushes via surface-initiated, ring-opening metathesis polymerization: Reversible photocontrol of wetting behavior and solvent dependent morphology changes', *Langmuir*, vol. 24, no. 17, pp. 9558–9565, 2008.

60. M. Nagel et al., 'Aryltriazene photopolymer thin films as sacrificial release layers for laser-assisted forward transfer systems: Study of photoablative decomposition and transfer behavior', *Appl. Phys. Mater. Sci. Process.*, vol. 92, no. 4, pp. 781–789, 2008.

61. A. Doraiswamy et al., 'Excimer laser forward transfer of mammalian cells using a novel triazene absorbing layer', *Appl. Surf. Sci.*, vol. 252, no. 13, pp. 4743–4747, Apr. 2006.

62. W. Ji et al., 'Coumarin-containing photo-responsive nanocomposites for NIR light-triggered controlled drug release via a two-photon process', *J. Mater. Chem. B*, vol. 1, no. 43, pp. 5942–5949, 2013.

63. M. Zelzer and R. V. Ulijn, 'Enzyme-responsive polymers: Properties, synthesis and applications', in *Smart Polymers and their Applications*, M.R. Aguilar and J. San Román, Eds. Elsevier, 2014, pp. 166–203.

64. A. Nelson, 'Engineering interactions', *Nat. Mater.*, vol. 7, no. 7, pp. 523–525, Jul. 2008.

65. C. Alexander and K. M. Shakesheff, 'Responsive polymers at the biology/materials science interface', *Adv. Mater.*, vol. 18, no. 24, pp. 3321–3328, 2006.

66. C. Sun et al., 'A ROS-responsive polymeric micelle with a π-conjugated thioketal moiety for enhanced drug loading and efficient drug delivery', *Org. Biomol. Chem.*, vol. 15, no. 43, pp. 9176–9185, 2017.

67. S. Uthaman, K. M. Huh, and I.-K. Park, 'Tumor microenvironment-responsive nanoparticles for cancer theragnostic applications', *Biomater. Res.*, vol. 22, no. 1, p. 22, Dec. 2018.

68. J. Huang, Q. Shu, L. Wang, H. Wu, A. Y. Wang, and H. Mao, 'Layer-by-layer assembled milk protein coated magnetic nanoparticle enabled oral drug delivery with high stability in stomach and enzyme-responsive release in small intestine', *Biomaterials*, vol. 39, pp. 105–113, 2015.

69. M. Maier et al., 'Highly site specific, protease cleavable, hydrophobic peptide-polymer nanoparticles', *Macromolecules*, vol. 44, no. 16, pp. 6258–6267, 2011.

70. H. Koga et al., 'Fluorescent nanoparticles consisting of lipopeptides and fluorescein-modified polyanions for monitoring of protein kinase activity', *Bioconjug. Chem.*, vol. 22, no. 8, pp. 1526–1534, 2011.

71. M. J. Kratochvil, T. Yang, H. E. Blackwell, and D. M. Lynn, 'Nonwoven polymer nanofiber coatings that inhibit quorum sensing in Staphylococcus aureus: Toward new non-bactericidal approaches to infection control', *ACS Infect. Dis.*, vol. 3, no. 4, pp. 271–280, 2017.

72. J. Zhou, A. L. Loftus, G. Mulley, and A. T. A. Jenkins, 'A thin film detection/response system for pathogenic bacteria', *J. Am. Chem. Soc.*, vol. 132, no. 18, pp. 6566–6570, May 2010.

73. A. K. Bajpai, S. K. Shukla, S. Bhanu, and S. Kankane, 'Responsive polymers in controlled drug delivery', *Prog. Polym. Sci.*, vol. 33, no. 11, pp. 1088–1118, Nov. 2008.

74. S. R. MacEwan, D. J. Callahan, and A. Chilkoti, 'Stimulus-responsive macromolecules and nanoparticles for cancer drug delivery', *Nanomed.*, vol. 5, no. 5, pp. 793–806, Jul. 2010.

75. P. Zarrintaj et al., 'Thermo-sensitive polymers in medicine: A review', *Eur. Polym. J.*, vol. 117, pp. 402–423, Aug. 2019.
76. D. Roy, J. N. Cambre, and B. S. Sumerlin, 'Future perspectives and recent advances in stimuli-responsive materials', *Prog. Polym. Sci. Oxf.*, vol. 35, no. 1–2, pp. 278–301, 2010.
77. G. Filipcsei, J. Fehér, and M. Zrínyi, 'Electric field sensitive neutral polymer gels', *J. Mol. Struct.*, vol. 554, no. 1, pp. 109–117, 2000.
78. E. A. Moschou, S. F. Peteu, L. G. Bachas, M. J. Madou, and S. Daunert, 'Artificial muscle material with fast electroactuation under neutral pH conditions', *Chem. Mater.*, vol. 16, no. 12, pp. 2499–2502, 2004.
79. S. Kagatani, T. Shinoda, Y. Konno, M. Fukui, T. Ohmura, and Y. Osada, 'Electroresponsive pulsatile depot delivery of insulin from poly(dimethylammopropyla crylamide) gel in rats', *J. Pharm. Sci.*, vol. 86, no. 11, pp. 1273–1277, 1997.
80. H. J. Joung, M.-J. Choi, J. T. Kim, S. H. Park, H. J. Park, and G. H. Shin, 'Development of food-grade curcumin nanoemulsion and its potential application to food beverage system: Antioxidant property and *in vitro* digestion', *J. Food Sci.*, vol. 81, no. 3, pp. N745–753, Mar. 2016.
81. H. Liu, C. Wang, S. Zou, Z. Wei, and Z. Tong, 'Simple, reversible emulsion system switched by pH on the basis of chitosan without any hydrophobic modification', *Langmuir*, vol. 28, no. 30, pp. 11017–11024, Jul. 2012.
82. B. Kuswandi, Jayus, A. Restyana, A. Abdullah, L. Y. Heng, and M. Ahmad, 'A novel colorimetric food package label for fish spoilage based on polyaniline film', *Food Control*, vol. 25, no. 1, pp. 184–189, May 2012.
83. V. R. Preedy, Ed., *Essential Oils in Food Preservation, Flavor and Safety*. Elsevier, 2016.
84. N. B. Borchert, J. P. Kerry, and D. B. Papkovsky, 'A CO_2 sensor based on Pt-porphyrin dye and FRET scheme for food packaging applications', *Sens. Actuators B Chem.*, vol. 176, pp. 157–165, Jan. 2013.
85. Z. Yu et al., 'Glutathione-responsive carboxymethyl chitosan nanoparticles for controlled release of herbicides', *Mater. Sci. Appl.*, vol. 06, no. 06, pp. 591–604, 2015.
86. Y. Chi, G. Zhang, Y. Xiang, D. Cai, and Z. Wu, 'Fabrication of a temperature-controlled-release herbicide using a nanocomposite', *ACS Sustain. Chem. Eng.*, vol. 5, no. 6, pp. 4969–4975, Jun. 2017.
87. Y.-C. Chen and Y.-H. Chen, 'Thermo and pH-responsive methylcellulose and hydroxypropyl methylcellulose hydrogels containing K_2SO_4 for water retention and a controlled-release water-soluble fertilizer', *Sci. Total Environ.*, vol. 655, pp. 958–967, Mar. 2019.
88. A. S. Hoffman, 'Stimuli-responsive polymers: Biomedical applications and challenges for clinical translation', *Adv. Drug Deliv. Rev.*, vol. 65, no. 1, pp. 10–16, Jan. 2013.
89. S. Maji, B. Cesur, Z. Zhang, B. G. De Geest, and R. Hoogenboom, 'Poly(N-isopropylacrylamide) coated gold nanoparticles as colourimetric temperature and salt sensors', *Polym. Chem.*, vol. 7, no. 9, pp. 1705–1710, 2016.
90. Y. Ma, K. Promthaveepong, and N. Li, 'CO_2-responsive polymer-functionalized au nanoparticles for CO_2 sensor', *Anal. Chem.*, vol. 88, no. 16, pp. 8289–8293, Aug. 2016.
91. F. Yang, S. Sukhishvili, H. Du, and F. Tian, 'Marine salinity sensing using long-period fiber gratings enabled by stimuli-responsive polyelectrolyte multilayers', *Sens. Actuators B Chem.*, vol. 253, pp. 745–751, 2017.
92. A. Kikuchi et al., 'Glucose-sensing electrode coated with polymer complex gel containing phenylboronic acid', *Anal. Chem.*, vol. 68, no. 5, pp. 823–828, Mar. 1996.
93. K. Whitehead, Z. Shen, and S. Mitragotri, 'Oral delivery of macromolecules using intestinal patches: Applications for insulin delivery', *J. Controlled Release*, vol. 98, no. 1, pp. 37–45, 2004.
94. K. Sonaje et al., 'Enteric-coated capsules filled with freeze-dried chitosan/poly(γ-glutamic acid) nanoparticles for oral insulin delivery', *Biomaterials*, vol. 31, no. 12, pp. 3384–3394, 2010.

95. B. P. Timko et al., 'Near-infrared-actuated devices for remotely controlled drug delivery', *Proc. Natl. Acad. Sci. U. S. A.*, vol. 111, no. 4, pp. 1349–1354, 2014.

96. J. L. Zhang, R. S. Srivastava, and R. D. K. Misra, 'Core-shell magnetite nanoparticles surface encapsulated with smart stimuli-responsive polymer: Synthesis, characterization, and LCST of viable drug-targeting delivery system', *Langmuir*, vol. 23, no. 11, pp. 6342–6351, 2007.

97. M. Ghorbani, H. Hamishehkar, N. Arsalani, and A. A. Entezami, 'Preparation of thermo and pH-responsive polymer@Au/Fe$_3$O$_4$ core/shell nanoparticles as a carrier for delivery of anticancer agent', *J. Nanoparticle Res.*, vol. 17, no. 7, p. 305, 2015.

98. R. Y. Lochhead, 'The role of polymers in cosmetics: Recent trends', in *Cosmetic Nanotechnology*, vol. 961, S. E. Morgan, K. O. Havelka, and R. Y. Lochhead, Eds. Washington, DC: American Chemical Society, 2007, pp. 3–56.

99. S. Mondal and J. L. Hu, 'Water vapor permeability of cotton fabrics coated with shape memory polyurethane', *Carbohydr. Polym.*, vol. 67, no. 3, pp. 282–287, Feb. 2007.

100. S. Talegaonkar, H. Sharma, S. Pandey, P. K. Mishra, and R. Wimmer, 'Bionanocomposites: smart biodegradable packaging material for food preservation', In *Food Packaging*, A. M. Grumezescu, Ed. Elsevier, 2017, pp. 79–110.

101. J. Guo and G. Kaletunç, 'Dissolution kinetics of pH responsive alginate-pectin hydrogel particles', *Food Res. Int. Ott. Ont*, vol. 88, no. Pt A, pp. 129–139, Oct. 2016.

102. Y. Ma et al., 'Biobased polymeric surfactant: Natural glycyrrhizic acid-appended homopolymer with multiple pH-responsiveness', *J. Colloid Interface Sci.*, vol. 541, pp. 93–100, Apr. 2019.

103. D. Restuccia et al., 'New EU regulation aspects and global market of active and intelligent packaging for food industry applications', *Food Control*, vol. 21, no. 11, pp. 1425–1435, Nov. 2010.

104. M. Ghaani, C. A. Cozzolino, G. Castelli, and S. Farris, 'An overview of the intelligent packaging technologies in the food sector', *Trends Food Sci. Technol.*, vol. 51, pp. 1–11, May 2016.

105. J. Kerry and D. Papkovsky, 'Development and use of non-destructive, continuous assessment, chemical oxygen sensors in packs containing sensitive foodstuffs', *Res. Adv. Food Sci.*, vol. 3, pp. 121–140, Jan. 2002.

106. S. Cichosz, A. Masek, and M. Zaborski, 'Polymer-based sensors: A review', *Polym. Test.*, vol. 67, pp. 342–348, May 2018.

107. A. Antonacci, F. Arduini, D. Moscone, G. Palleschi, and V. Scognamiglio, 'Nanostructured (Bio)sensors for smart agriculture', *TrAC Trends Anal. Chem.*, vol. 98, pp. 95–103, Jan. 2018.

108. H. Guo, J. C. White, Z. Wang, and B. Xing, 'Nano-enabled fertilizers to control the release and use efficiency of nutrients', *Curr. Opin. Environ. Sci. Health*, vol. 6, pp. 77–83, Dec. 2018.

109. A. Singh et al., 'Advances in controlled release pesticide formulations: Prospects to safer integrated pest management and sustainable agriculture', *J. Hazard. Mater.*, p. 121525, Nov. 2019.

110. G. Liu et al., 'Hydrazone-linked soybean protein isolate-carboxymethyl cellulose conjugates for pH-responsive controlled release of pesticides', *Polym. J.*, vol. 51, no. 11, pp. 1211–1222, Nov. 2019.

111. A. S. Hoffman, 'Stimuli-responsive polymers: Biomedical applications and challenges for clinical translation', *Adv. Drug Deliv. Rev.*, vol. 65, no. 1, pp. 10–16, 2013.

112. J. Yu, Y. Zhang, H. Bomba, and Z. Gu, 'Stimuli-responsive delivery of therapeutics for diabetes treatment', *Bioeng. Transl. Med.*, vol. 1, no. 3, pp. 323–337, 2016.

113. V. Ruiz-Valdepeñas Montiel, J. R. Sempionatto, S. Campuzano, J. M. Pingarrón, B. Esteban Fernández de Ávila, and J. Wang, 'Direct electrochemical biosensing in gastrointestinal fluids', *Anal. Bioanal. Chem.*, vol. 411, no. 19, pp. 4597–4604, Jul. 2019.

114. M. Wei, Y. Gao, X. Li, and M. J. Serpe, 'Stimuli-responsive polymers and their applications', *Polym. Chem.*, vol. 8, no. 1, pp. 127–143, Dec. 2016.
115. K. Singha, 'A review on coating & lamination in textiles: Processes and applications', *Am. J. Polym. Sci.*, vol. 2, no. 3, pp. 39–49, 2012.
116. X. Ding and J. Hu, 'Temperature-sensitive polyurethane properties', 2003.
117. S. Mondal, J. L. Hu, and Z. Yong, 'Free volume and water vapor permeability of dense segmented polyurethane membrane', *J. Membr. Sci.*, vol. 280, no. 1–2, pp. 427–432, Sep. 2006.
118. P.-C. Hsu et al., 'Radiative human body cooling by nanoporous polyethylene textile', *Science*, vol. 353, no. 6303, pp. 1019–1023, Sep. 2016.

12 Self-Healing Polymer Coatings

P. Poornima Vijayan
Sree Narayana College for Women

Sharika T. Nair
Mahatma Gandhi University
St. Xavier's College Vaikom

CONTENTS

12.1 INTRODUCTION

Self-healing is defined as "the property that enables a material to repair themselves and to recover their functionality using the resources inherently available to them" (Blaiszik et al. 2010). Nature has its own self-healing policies to protect living organisms such as cell wall repair in bacteria, and wound and bone repair in human body. Inspired from nature, scientists integrate self-healing ability in materials. This approach is to find a new route toward safe and long-lasting engineering materials. The self-healing materials emerge under the category of smart or intelligent material, as they are stimulus-responsive materials where they respond over mechanical stimulus to restore the damage. Self-healing enables the damaged materials to recover their original performance within a short span of time. Refilling of damage volume and reforming bonds across the damage surfaces can restore mechanical, barrier, anti-corrosion, electrical, and aesthetic properties. These materials have a huge demand in modern society as they can find application in aircrafts, mobile phones, wearable electronics, car paints, dental fillings, tissue engineering, soft robotics, and many more. Self-healing ability can be integrated in a variety of materials including polymers (Blaiszik et al. 2010), polymer composites (Blaiszik et al. 2010), hydrogels (Taylor and

Panhuis 2016), concretes (Jonkers 2007; Hilloulin et al. 2015), and ceramics (Yang et al. 2012). The global self-healing materials market size is expected to grow at a compound annual growth rate (CAGR) of 46.1% from 2019 to 2025 ("Self-Healing Materials Market Size, Share | Global Industry Report, 2025" 2020).

Polymers and their composites are the most studied materials for their self-healing ability. Breakthrough research in self-healing polymers was reported in 2001 with the discovery of a structural epoxy matrix with the ability to autonomically heal cracks (White et al. 2001). In their work, the healing agent that has been encapsulated inside microcapsules would be released upon crack intrusion, followed by polymerization of the healing agent triggered by contact with an embedded catalyst to recover the damage. They claimed 75% recovery in their performance after healing process. Since then, numerous research studies have been carried out to discover fast self-healing polymeric materials with high healing efficiency.

With the advancement in self-healing technology, the coating of several materials including glasses (Orain et al. 1980; Choung et al. 2018), metals (Sauvant-Moynot, Gonzalez, and Kittel 2008), fabrics (Qiang et al. 2017), concretes ("Sunlight-Induced Self-Healing of a Microcapsule-Type Protective Coating | ACS Applied Materials & Interfaces" 2020), and mortars ("Sunlight-Induced Self-Healing of a Microcapsule-Type Protective Coating | ACS Applied Materials & Interfaces" 2020) using self-healing polymers has been emerged as a long-lasting surface protection strategy. Self-healing polymer coatings have the ability to repair microscale damage in the coating and restore the functionality of the substrate. Self-healing coatings enhance the service life of the substrate while reducing the maintenance cost. They offer more reliable and safer protection for structural components. An ideal self-healing coating is capable of continuously sensing and responding to damage over the entire lifetime, and a polymer coating with multiple healing cycle would fulfill this need ("Repeated Self-Healing of Microvascular Carbon Fibre Reinforced Polymer Composites - IOPscience" 2020). Apart from mechanical damage, light ("Sunlight-Induced Self-Healing of a Microcapsule-Type Protective Coating | ACS Applied Materials & Interfaces" 2020) (Odarczenko et al. 2020) and pH (Zhang et al. 2018) have also been investigated as important healing stimuli for specific applications.

This chapter begins with basic concepts of self-healing coatings and the strategies adopting for the development of self-healing coating. Subsequently, a discussion on the use of self-healing coating as an effective tool to control corrosion of metal substrates has been included by instancing some important coating systems. The chapter is then highlighting some achievements in designing multifunctional self-healing coating for specific applications which would be the part of future mobile phones, car paints, aircrafts, and skyscrapers. Last but not least, the chapter quotes the green practices in developing self-healing coatings. The chapter concludes with the enormous potential of self-healing coating to extend the safety and service life of engineering materials. Finally, the chapter forecasts the possible research in self-healing coatings.

12.2 SELF-HEALING STRATEGIES

Two major self-healing strategies have been adopted to repair the cracks and damages in protective coatings. They are intrinsic and extrinsic healing methods and are differentiated by the polymer response to the damage (Yuan et al. 2008; Vijayan and Al-Maadeed 2019). In the extrinsic method, the damage is recovered by a suitably encapsulated healing agent that is embedded in the polymer-based coating matrix, while in the intrinsic method, the polymers are able to heal the cracks by themselves owing to their latent self-healing functionality.

12.2.1 EXTRINSIC SELF-HEALING

In the case of extrinsic self-healing, the polymer materials themselves do not have self-healing capability. In order to induce self-healing capability, a healing-agent-encapsulated container has to be embedded into the polymer matrix in advance. When the fragile containers are opened by any external stimuli, the healing agent would be released into the crack planes, heals the cracks, and prevents the damage propagation. There are different types of containers to load the healing agent ("EBSCOhost I 114268242 I 'Containers' for Self-Healing Epoxy Composites and Coating: Trends and Advances" 2020). Among them, micro- or nanopolymer capsules are the commonly employed ones. In capsule-based self-healing materials, the healing agents are stored in capsules until the damage triggers the healing process. When the damage occurs, it raptures the capsule and the self-healing mechanism is triggered through the release of the healing agent into the damage region. Then, the healing agent undergoes polymerization in the presence of a suitable catalyst, leading to a healing event. Figure 12.1 schematically shows the entire healing process in a capsule-based self-healing coating. Since the fabrication is straightforward by dispersing capsules inside a matrix, the capsule-based healing concept has been commonly employed. Some of the major capsule-based self-healing coatings will be discussed in Section 14.3.

Another important extrinsic self-healing strategy is the use of vascular network, where the healing agent is sequestered in a network in the form of capillaries, hollow fibers, or hollow channels, which may be interconnected one-dimensionally (1D), two-dimensionally (2D), or three-dimensionally (3D) (Blaiszik et al. 2010). A coating consisting of a vascular network can be used to provide a uniform healing facility throughout the surface (Toohey, Sottos, and White 2009). When there is damage or crack formation in the polymer coating, the healing agent is released into the crack plane, and the network may be refilled by an external source or from an undamaged but connected region of the vasculature. For vascular healing systems, access to a large reservoir of healing agent and the ability to replenish the network enable repeated healing of multiple damage events (Hamilton, Sottos, and White 2012).

12.2.2 INTRINSIC SELF-HEALING

The intrinsic self-healing process is based on inherent reversibility of various types of molecular interactions exhibited by polymers that enable crack healing under certain external stimulation. These materials possess a latent functionality that initiates self-healing of damage via thermally reversible reactions, hydrogen bonding, ionomeric arrangements, melting of thermoplastic materials, or molecular diffusion and entanglement. Such capability is generally derived from the specific molecular structures of the polymers that allow an effective inter-chain mobility upon the supply of

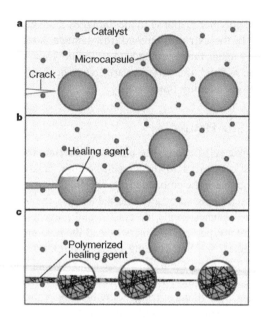

FIGURE 12.1   The schematic diagram showing the self-healing process. A microencapsulated healing agent is embedded in a structural composite matrix containing a catalyst capable of polymerizing the healing agent. (a) Cracks form in the matrix wherever damage occurs; (b) the crack ruptures the microcapsules, releasing the healing agent into the crack plane through capillary action; (c) the healing agent contacts the catalyst, triggering polymerization that bonds the crack faces closed (White et al. 2001).

energy (e.g., temperature, static load, UV light) followed by a process of restoration of the chemical or physical bond strength upon stimulus removal. The intrinsic self-healing mechanisms are classified into physical interactions and chemical interactions. Compared to extrinsic self-healing materials, the intrinsic materials provide the possibility of fully or partially repairing the initial properties multiple times.

Intrinsic self-healing polymers can recover their damage by specific reversible chemical bonds and are beneficial due to the possibility of multiple healing steps at the same location. Intrinsic self-healing materials based on chemical interactions are generally based on either non-covalent chemistries or dynamic covalent chemistries (Yang and Urban 2013). Self-healing coating developed through covalent approaches utilizes dynamic covalent bonds from reactions such as Diels–Alder reaction (Kötteritzsch et al. 2013), disulfide bond formation (Wang et al. 2020), and dynamic urea bond ("Dynamic Urea Bond for the Design of Reversible and Self-Healing Polymers I Nature Communications" 2020), while self-healing through non-covalent approaches encompasses intermolecular interactions such as hydrogen bonds, metal-to-ligand interactions (Bode et al. 2013), and host–guest interactions (Nakahata, Takashima, and Harada 2016; Xuan et al. 2017). For instance, Kötteritzsch et al. (2013) utilized polymers based on a methacrylate backbone containing furan as well as maleimide units in the side chains to achieve self-healing via reversible cross-linking by DA reaction.

12.3 SELF-HEALING COATING FOR ANTI-CORROSION

Metallic corrosion is a severe industrial issue, which causes huge economic loss. Polymer coatings have been used to protect metal parts from the corrosive environment. However, the mechanical durability of polymer coatings is a serious question, which limits the potential of these coatings. During the service period, the polymer coatings are subjected to microcracks, which would propagate to create macrocracks, and finally to ultimate failure of the coating and exposure of metal substrate to the corrosive environment. Conventionally, corrosion inhibitors have been extensively used for corrosion control in metals. This approach is inadequate to control long-term corrosion. Moreover, the direct incorporation of inhibitor molecules has weakening/plasticizing effects toward polymer coating (Alrashed, Jana, and Soucek 2019). However, polymer coatings designed to supply the corrosion inhibitors on sites of corrosive attack and locally inhibit corrosion, the so-called "self-healing anti-corrosion coatings," could effectively solve these drawbacks. The difference in response toward the corrosive environment of a regular coating and a self-healing coating is schematically shown in Figure 12.2.

In self-healing anti-corrosive coatings, the controlled release of the corrosion inhibitor facilitates a long-term supply of inhibitors which relief the metallic structures from severe corrosion. Hence, the concept of self-healing has recently been adopted as an alternative method for anti-corrosion protection. The introduction of self-healing ability into polymer coatings has been invented as a strategy to extend the service life of polymer coatings and hence the underneath metallic products. Researchers have made serious attempts to tackle the corrosion of metallic substrates using self-healing polymer coatings. Epoxy-, polyurethane-, and alkyd-based self-healing coatings were widely studied. In such extrinsic self-healing coatings, either a polymerizable healing agent or a corrosion inhibitor has been loaded inside suitable containers. Coating

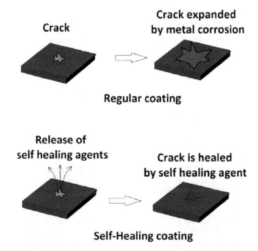

FIGURE 12.2 The behaviors of the regular and self-healing coatings in the corrosion process (Abdullayev et al. 2013).

formulations of primers containing corrosion inhibitor-loaded micro- and nanocontainers have been proved as efficient anti-corrosion coatings. Those inhibitor carriers include polymer capsules (Huang and Yang 2014), nano- and mesoporous inorganic materials (Chen et al. 2015; Keyvani, Yeganeh, and Rezaeyan 2017), layered double hydroxide (LDH) ("Unraveling EDTA Corrosion Inhibition When Interleaved into Layered Double Hydroxide Epoxy Filler System Coated onto Aluminum AA 2024-ScienceDirect" 2020; "Zn–Al Layered Double Hydroxides as Chloride Nanotraps in Active Protective Coatings - ScienceDirect" 2020; Hayatdavoudi and Rahsepar 2017), inorganic nanotubes ("TiO$_2$ Nanotubes and Mesoporous Silica as Containers in Self-Healing Epoxy Coatings | Scientific Reports" 2020; Abdullayev et al. 2013), and ceramic nanocontainers (Kartsonakis et al. 2014). Self-healing coatings have been used to protect a variety of metal substrates such as stainless steel, carbon steel, galvanized steel, aluminum alloy, and copper. The corrosion resistance of the self-healing coatings is generally evaluated with electrochemical impedance spectroscopy (EIS) test, salt spray, scanning vibrating electrode technique (SVET), and accelerated salt immersion test done over coating surface with artificial defects.

The monomer/corrosion inhibitor-encapsulated polymer capsules are a major component of the widely studied anti-corrosive polymer coatings. The microcapsules have been commonly prepared by in situ polymerization (Suryanarayana, Rao, and Kumar 2008; Zhao et al. 2012) or by interfacial polymerization (Koh et al. 2014) to form polymer shell over the healing agent droplets. Important polymer shells investigated are poly(urea-formaldehyde) (PUF) shell (Zhao et al. 2012), epoxy shell (Liu et al. 2012), polyurethane shell (Koh et al. 2014), etc. Samadzadeh et al. (2010) contributed a detailed review on micro-/nanocapsules used to prepare self-healing coatings. These encapsulated microcapsules are homogeneously dispersed into coating resin and coated over the metal substrate. Once this coating gets damaged, the capsules get ruptured and the healing agent gets leaked into the damaged site. For instance, alkyd varnish coatings (AVCs) incorporated with isophorone diisocyanate (IPDI)-encapsulated polymer capsules effectively resisted long-term corrosion of Q235 steel (Wang et al. 2014). Figure 12.3 shows SEM images of IPDI-loaded polymer capsules that were utilized to supply IPDI into the damaged coating surface to heal and stop further corrosion by the formation of water-insoluble polyurethanes as the self-healing product.

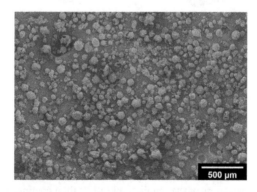

FIGURE 12.3   FE-SEM overview image of IPDI microcapsules (Wang et al. 2014).

Hexamethylene diisocyanate (HDI)-encapsulated polyurethane microcapsule-based self-healing epoxy coating showed superior anti-corrosion protection for steel substrates (Huang and Yang 2014). An accelerated salt immersion test, where the scratched coated samples were immersed in 10 wt.% NaCl solution, showed that encapsulated microcapsule-loaded coating was free from rust and severe corrosion was observed on the control coating (Figure 12.4).

At the same time, the healing agents/corrosion inhibitors are loaded in the pores or lumen of porous or tubelike inorganic nanomaterials via vacuum cycling (Vijayan, Hany El-Gawady, and Al-Maadeed 2016) ("TiO$_2$ Nanotubes and Mesoporous Silica as Containers in Self-Healing Epoxy Coatings | Scientific Reports" 2020; Chen et al. 2015). Halloysite nanotubes are naturally occurring clay minerals (Figure 12.5(I)) whose nanosized lumens can store a sufficient amount of anti-corrosive agents

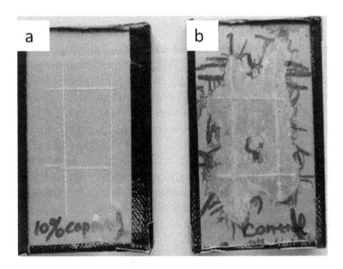

FIGURE 12.4 (a) Steel panel coated with 10 wt.% HDI microcapsule-incorporated self-healing coating and (b) steel panel coated with pure coating after immersion in 10 wt.% NaCl solution for 48 h and storage in open air for 6 months at room temperature (Huang and Yang 2014).

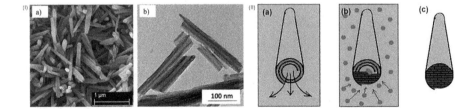

FIGURE 12.5 (I) (a) SEM and (b) TEM images of halloysite nanotubes (Abdullayev et al. 2013). (II) (a) Scheme of benzotriazole leakage from a halloysite nanotube in pure water; (b) initiation of complex formation in Cu(II) solution; and (c) nanotube after completion of the stopper (Abdullayev and Lvov 2010).

(Abdullayev and Lvov 2010). There are mainly two reported release mechanisms. In the first case, the loaded nanotubes stay sealed in dry paint, and once the coating gets damaged, the tube ends are exposed to the environmental moisture and the healing agent/corrosion inhibitor gets released (Abdullayev et al. 2013). Alternatively, in order to have a more controlled release of corrosion inhibitor, open ends of the tubes are chemically capped to seal the inhibitor inside the lumen. For example, Abdullayev and Lvov (Abdullayev and Lvov 2010) capped benzotriazole (a corrosion inhibitor)-loaded halloysite tube ends with Cu–benzotriazole complex by the reaction of benzotriazole with bulk copper ion solutions (Figure 12.5II). Saturated solution of ammonia in water dissolves the Cu–benzotriazole complex, and hence, treatment with ammonia solution acts as a trigger for the release of the sealed corrosion inhibitor. The capped halloysite nanotubes have been effectively used in acrylic latex paint for the corrosion protection of copper via the controlled release of the benzotriazole inhibitor.

Corrosion inhibitor anions are entrapped in LDH by making use of its anion exchange capacity ("Unraveling EDTA Corrosion Inhibition When Interleaved into Layered Double Hydroxide Epoxy Filler System Coated onto Aluminum AA 2024- ScienceDirect" 2020; Hang et al. 2012). Epoxy coating incorporated with 2-benzothiazolylthio-succinic acid (BTSA) (a corrosion inhibitor)-loaded LDH has been studied for its corrosion protection ability (Hang et al. 2012). The inhibitor release from BTSA-loaded LDH has been determined at different concentrations of NaCl solutions, and the release curves are shown in Figure 12.6a. The release of BTSA is based on an exchange reaction between BTSA and chloride ions, and the rate of exchange increases with the NaCl concentration. To evaluate the corrosion protection ability of epoxy coating containing BTSA-loaded LDH, electrochemical impedance spectroscopy (EIS) was used. The variation in impedance modulus at 10 mHz ($|Z|10mHz$) with immersion time in 0.5 M NaCl solution for carbon steel coated

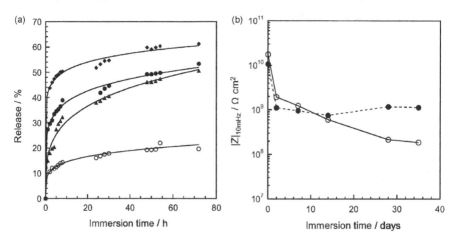

FIGURE 12.6 (a). Release curves of BTSA from LDH–BTSA NaCl solution at different concentrations: (○) 0%; (▲) 0.5%; (●) 1%; and (♦) 3%. **b.** $|Z|10mHz$ versus immersion time in 0.5 M NaCl solution for the carbon steel covered by (○) pure epoxy and (●) epoxy containing 3 wt.% LDH–BTSA (Hang et al. 2012).

with pure epoxy and epoxy containing BTSA-loaded LDH is shown in Figure 12.6b. After two days of initial exposure time, the impedance modulus for the pure epoxy coating decreased continuously with time. While it remains relatively stable for the epoxy coating containing 3% of LDH–BTSA, a self-healing anti-corrosion is clearly evident. Also, the impedance modulus of the LDH–BTSA-containing coating was higher than that of the pure epoxy coating.

Hot-dip-galvanized steel (HDG) was reported to be well protected using a self-healing coating where corrosion inhibitor-loaded ceramic nanocontainers acted as the major component (Kartsonakis et al. 2012). The loading of the corrosion inhibitor in ceramic nanocontainers has been done using the vacuum process (Kartsonakis and Kordas 2010). Cerium molybdate nanocontainers loaded with 2-mercaptobenzothiazole (MBT)-incorporated epoxy coating over HDG were proved to have superior anti-corrosion performance over a long period. The controlled release of corrosion inhibitor into required site helps to limit further corrosion. Modification of the epoxy matrix with cerium molybdate nanoparticles induces some losses in terms of barrier properties. The addition of a mixture of cerium molybdate nanoparticles and layered double hydroxides, both loaded with the corrosion inhibitor, produced a synergistic inhibition effect on epoxy coating that combines early corrosion protection and longer term inhibition (Montemor et al. 2012).

12.4 SELF-HEALING POLYMER COATINGS FOR SPECIFIC APPLICATIONS

Due to the transparency and aesthetics, glass is a prime component in building exteriors, mobile phones, and other portable electronic devices. The demand for self-healing protective coating for glass is high due to its inherent brittleness. Researchers are in constant efforts to develop transparent self-healing coatings with high healing efficiency for glass substrates. Among them, few polymer-based self-healing coatings for glass substrates were reported. Xuan et al. (2017) designed a transparent multilayer film for glass from beta-cyclodextrin (β-CD) bound with poly(ethylenimine) (PEI) and adamantane (AD) linked with poly(acrylic acid) (PAA) via the layer-by-layer (LbL) self-assembly technique. This film repeatedly heals external mechanical damage via host–guest interactions. In addition to self-healing, the film possesses high flexibility and acid resistance due to its chemical nature. Figure 12.7 shows the digital photograph, SEM image, and mechanism of self-healing of glass coated with the PAA-AD/PEI-β-CD film.

Choung et al. (2018) developed a transparent self-healing coating based on the polydimethylsiloxane (PDMS) matrix for glass. A commercially available photoinitiator (Irgacure 2959) was encapsulated inside the polymer capsule and coated over plasma-treated glass. It is essential to control the concentration of the photoinitiator-loaded capsules in the coating to minimize the reflection, diffraction, refraction, and scattering of light of the coating. PDMS was poured into the capsule-coated glass, and chambers were created in PDMS using a 3D printed stamp. Photopolymerizable hydrogel based on poly(ethylene)glycol diacrylate (PEGDA) as monomer and deionized water as solvent were confined in to the chambers. The mechanism of repair of damage on the glass is the released hydrogel precursor from the ruptured chambers

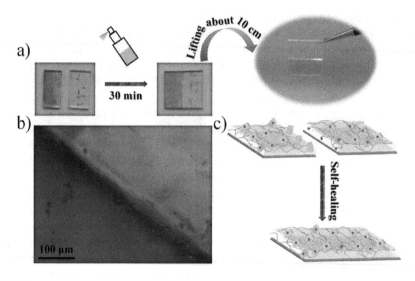

FIGURE 12.7 (a) Digital photograph of the healing process after cutting off the glass with the PAA-AD/PEI-β-CD film; (b) optical microscope photograph of healing the PAA-AD/PEI-β-CD film on the glass; (c) scheme of the healing mechanism after cutting off the glass with the PAA-AD/PEI-β-CD film (Xuan et al. 2017).

mixed with the photoinitiator from the ruptured capsules at the crack site and polymerized using UV light. Based on the intrinsic self-healing method, Wang et al. (2020) designed a transparent self-healing polyurethane (TPU) coating with polytetramethylene ether glycol (PTMEG) as the soft segment and aliphatic isocyanate and aliphatic disulfide as the hard segments. In this coating system, a self-healing efficiency of about 39% was achieved with the cooperative action of dynamic disulfide bonds and hydrogen bonding. Along with self-healing ability, this coating has high transparency, and the long-term optical stability makes it suitable to coat glass substrates.

The harsh environment in space shortens the life span of aerospace coatings. Guo et al. (2016) developed UV-triggered self-healing coatings as spacecraft coatings. They have put forward an approach where sufficiently available UV radiation in space acts as a stimulus for self-healing coating. Therein, different from usual practice, epoxy resin monomer (as the healing agent) and cationic photoinitiator are successfully encapsulated into a single SiO_2 microcapsule via a combined interfacial/ in situ polymerization technique and incorporated into epoxy coating. The release of epoxy resin and cationic photoinitiator into the damaged site upon the fracture of SiO_2 capsule is followed by UV-triggered cationic polymerization of epoxy to heal the crack with high healing efficiency. The two important advantages of this coating system for application in spacecraft are the capacity of the single SiO_2 microcapsule to withstand thermal cycling in a simulated space environment and the lack of oxygen sensitivity of the UV-triggered cationic polymerization of epoxy. Figure 12.8 shows the SEM images of SiO_2 capsules before and after thermal cycling and the recovery of the scratch in microcapsule-loaded epoxy coating after UV irradiation for 30 min.

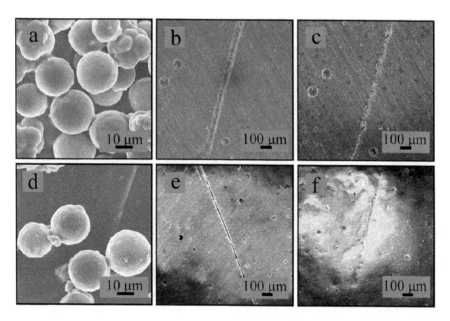

FIGURE 12.8 SEM images of the pristine microcapsules (a), the microcapsules after thermal cycling in the temperature range from −50°C to 110°C for 5 days (d), the scribe region of the self-healing coating containing the pristine microcapsules before (b) and after (c) UV irradiation for 30 min, and the scribe region of the self-healing coating containing the thermally cycled microcapsules before (e) and after (f) UV irradiation for 30 min (Guo et al. 2016).

Another crucial area where the contribution of self-healing coating is expected is medical implants. Different from coatings used for other applications, the coating components (i.e., skeleton coatings and corrosion inhibitors) for medical implants should satisfy the criteria such as biosafety and biocompatibility. Magnesium (Mg) and its alloy are suitable to fabricate cardiovascular and orthopedic implants (Chen et al. 2014). Due to their rapid degradation in physiological ambient, there is a need for a biocompatible self-healing coating for magnesium (Mg) and its alloy. For the first time, Xiong et al. (2019) developed a biocompatible, pH-responsive self-healing coating for biomedical-grade Mg alloy by compositing silk fibroin and K_3PO_4 (where silk fibroin acts as the skeleton coating and PO_4^{3-} ions act as the corrosion inhibitor). The self-healing ability of the coating was studied using artificially scratched coating immersed in Hanks' solution for 24 h. The scanning vibrating electrode technique (SVET), a nondestructive tool to visualize current densities and accurately express the localized cathodic and anodic activities, was employed to follow the self-healing process in Mg alloy coated with silk fibroin–K_3PO_4 composite. Figure 12.9 shows the SVET 3D maps of electric current density measured above the defected surface of Mg alloy coated with silk fibroin alone and Mg alloy coated with silk fibroin–K_3PO_4 composite.

After immersion of Mg alloy coated with silk fibroin and K_3PO_4 for 120 min in Hanks' solution, both the anodic and cathodic current density verged to zero plane (Figure 12.8(B3)) due to the self-healing action of the coating to prevent corrosion of the underlying metal. However, Mg alloy coated with silk fibroin sustained a negative

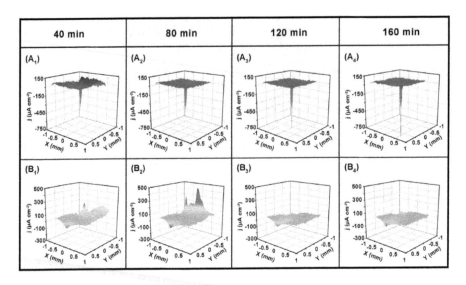

FIGURE 12.9 VET3D maps of the electric current density measured above the defected surface of the Mg alloy coated with silk fibroin alone (A1-A4) and Mg alloy coated with silk fibroin– K_3PO_4 composite (B1-B4) (Xiong et al. 2019).

peak evolution during the entire immersion period, due to significant cathodic current density as the material chose anodic sacrifice protection. The mechanism behind the pH-responsive healing in silk fibroin–K_3PO_4 composite coating is the release of PO_4^{3-} rapidly in alkaline pH and its reaction with Mg^{2+} ion (released as a result of corrosion of Mg) to get $Mg_3(PO_4)_2$. Thus, the formed $Mg_3(PO_4)_2$ salt deposits a passive film that heals the defect and stops further corrosion.

The multifunctionality of smart materials is a much-discussed topic due to their ability to give best performance by exactly mimicking the biological system ("Biomimetic Multifunctional Materials: A Review | SpringerLink" 2020). Among such attempts, multifunctionality was also introduced into polymer coatings. Promising self-healing coatings on glass and metal surface for outdoor applications were reported to be achieved by the integration of self-cleaning ability (hydrophobicity) with self-healing (Wu et al. 2014; Wang et al. 2019; Sahoo et al. 2017). Wang et al. (2019) prepared a self-healing coating with water-resistant hydrophobic surface. In this coating, the multifunctionality was introduced by the micro-/nanostructures of hydrophobic microspheres made from poly(acrylic acid-*co*-trifluoroethyl methacrylate). While the carboxylic acid groups on the surface of silica@polymer core–shell nanocontainers (nanorod) facilitates pH-responsive release of the encapsulated healing inhibitor, the fluorinated groups of the polymer nanocontainers impart hydrophobicity to the epoxy coating (Figure 12.10).

12.5   GREEN SELF-HEALING COATINGS

Green approaches in developing self-sealing polymer coatings mainly focus on providing adequate corrosion inhibition or quick healing via the use of natural materials (Riaz, Nwaoha, and Ashraf 2014). It is challenging to develop fully green self-healing

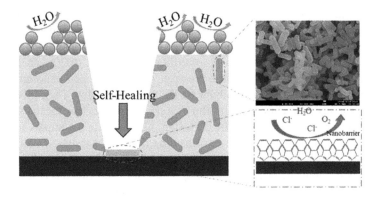

FIGURE 12.10 Mechanism of anti-corrosion performance for the hydrophobic self-healing coating (Wang et al. 2019).

coatings having the same protection performance as that of traditional counterparts (Zheludkevich et al. 2011). The self-healing coatings developed via the green strategy have potential application in packing, fashion, and biomedical industries.

Encapsulating natural materials as healing agents or corrosion inhibitors in a microcapsule is considered as a good green processing approach. Zulkifli et al. (2017) used an eco-friendly corrosion inhibitor that was extracted from henna leaves (HLE) and was incorporated in an acrylic resin coating to protect the aluminum alloy. HLE can effectively inhibit corrosion for a variety of metals exposed to a wide range of electrolytes.

Naturally available drying oils, such as tung oil, linseed oil, and sunflower oil, are proved as efficient self-healing agents. These natural oils undergo oxidation in the presence of atmospheric oxygen and form a film of polymerized dried oils to protect the coating surface. For instance, Samadzadeh et al. (2011) investigated the capability of the encapsulated tung oil as a scratch-healing agent for self-healing coatings. Tung oil-encapsulated urea-formaldehyde (UF) shell was dispersed in epoxy resin. Upon the damage of the coating, tung oil was released from ruptured microcapsules and it healed the scratch in the coating matrix. Abdipour, Rezaei, and Abbasi (2018) synthesized UF capsules containing linseed oil by in situ polymerization and used them to develop self-healing epoxy coatings for anti-corrosion protection.

A few researchers have reported the preparation of self-healing coatings containing microcapsules with ethyl cellulose as the shell material and plant oil as the core material. For example, Mirabedini, Dutil, and Farnood (2012) developed a water-based latex coating containing ethyl cellulose microcapsule that is encapsulated with rapeseed oil as the healing agent. The deformation and disruption of latex films containing microcapsules filled with colored rapeseed oil were examined under applied tensile load using color coordinate measurements. These color measurements confirmed the release of oil, and the released oil could act as a healing agent and restore mechanical properties of the latex film by filling the microcracks. Es-haghi et al. (2014) prepared microcapsules using ethyl cellulose and filled them with linseed oil as the healing agent via the solvent evaporation method. They reported that these microcapsules offer an attractive prospect for water-based self-healing coatings with enhanced mechanical properties.

In a greener approach, natural polymers can be used as pre-layer reservoirs for environment-friendly corrosion inhibitors. Among the natural polymers, both chitosan and cellulose nano-/microfibers have been reported as efficient green reservoirs for corrosion inhibitors (Carneiro et al. 2012; Nawaz et al. 2020). Zheludkevich et al. (2011) successfully developed a pre-layer of cerium-doped chitosan that acts as a reservoir for the corrosion inhibitor for the aluminum alloy. Their investigation described the use of a chitosan thin film as a pre-layer reservoir for the immobilization of cerium(III) cations (as the corrosion inhibitor) via a complex formation. Ghosh and Urban (2009) developed self-healing polyurethane-based coatings containing an oxetane-substituted chitosan as the reservoir. The proposed mechanism of action is the four-member oxetane rings open to create two reactive ends upon mechanical damage of the polymer network. When exposed to ultraviolet light, chitosan chain breaks, and forms cross-links with the reactive oxetane end to repair the network.

Yabuki, Kawashima, and Fathona (2014) reported that cellulose nanofibers act as pathways to release corrosion inhibitors into damaged areas on epoxy coatings over carbon steel substrates. They pointed out that cellulose nanofibers deformed and shortened in the presence of water, thereby facilitating the release of corrosion inhibitors into the scratched area. Vijayan et al. (2017) immobilized both the epoxy monomer and amine curing agent on cellulose nanofibers and incorporated them to an epoxy coating to protect carbon steel for seawater applications. While the amine curing agent was chemically bonded with fibers, epoxy monomer was physically adhered on the surface of cellulose nanofibers. Here, the self-healing was triggered by water. Upon surface damage, the water enters into the crack and the nanofibers get deformed and hence release the epoxy monomer. The released epoxy monomer subsequently reacts with the active functional group in curing agents to recover the damage.

Further, the use of bio-based polymers as a matrix for self-healing coating emerged as a green strategy in this area. Ryu, Oh, and Kim (2016) prepared a bio-based self-healing polymer using a cross-linking mechanism between polybutylene furanoate (PBF) and bismaleimide (BM) by Diels–Alder reactions, and it was blended with bio-based polyurethane (BPU) to improve the liquidity, elasticity, and mechanical properties. BPU is a good reinforcer of the furan-based self-healing polymer with improved self-healing ability and eco-friendly performance.

12.6 CONCLUSION AND FUTURE PERSPECTIVES

Self-healing polymer coatings have been proved as the future protection strategy for various substrates. They are excellent smart materials and effectively protect and extend the life span of metal, glass, concrete, etc. Self-healing anti-corrosion coatings are the best choice toward the protections of metals and alloys in the oil and gas industry, aircrafts, buildings, automobiles, medical implants, etc. Highly transparent self-healing coatings offer long-lasting protective coatings for automotive laminated glass, flexible displays, mobile phone screens, etc. The scope of integration of multifunctionality in self-healing coatings is enormous and needs to explore greener strategies in developing them.

REFERENCES

Abdipour, H., M. Rezaei, and F. Abbasi. 2018. "Synthesis and characterization of high durable linseed oil-urea formaldehyde micro/nanocapsules and their self-healing behaviour in epoxy coating." *Progress in Organic Coatings* 124 (November): 200–212. doi: 10.1016/j.porgcoat.2018.08.019.

Abdullayev, E., V. Abbasov, A. Tursunbayeva, V. Portnov, H. Ibrahimov, G. Mukhtarova, and Y. Lvov. 2013. "Self-healing coatings based on halloysite clay polymer composites for protection of copper alloys." *ACS Applied Materials & Interfaces* 5 (10): 4464–71. doi: 10.1021/am400936m.

Abdullayev, E., and Y. Lvov. 2010. "Clay nanotubes for corrosion inhibitor encapsulation: Release control with end stoppers." *Journal of Materials Chemistry* 20 (32): 6681–87. doi: 10.1039/C0JM00810A.

Alrashed, M. M., Sadhan Jana, and M. D. Soucek. 2019. "Corrosion performance of polyurethane hybrid coatings with encapsulated inhibitor." *Progress in Organic Coatings* 130 (May): 235–43. doi: 10.1016/j.porgcoat.2019.02.005.

"Biomimetic Multifunctional Materials: A Review | SpringerLink." 2020. Accessed April 2. https://link.springer.com/article/10.1007/s42247-019-00051-7.

Blaiszik, B. J., S. L. B. Kramer, S. C. Olugebefola, J. S. Moore, N. R. Sottos, and S. R. White. 2010. "Self-healing polymers and composites." *Annual Review of Materials Research* 40 (1): 179–211. doi: 10.1146/annurev-matsci–070909-104532.

Bode, S., L. Zedler, F. H. Schacher, B. Dietzek, M. Schmitt, J. Popp, M. D. Hager, and U. S. Schubert. 2013. "Self-healing polymer coatings based on crosslinked metallosupramolecular copolymers." *Advanced Materials* 25 (11): 1634–38. doi: 10.1002/adma.201203865.

Carneiro, J., J. Tedim, S. C. M. Fernandes, C. S. R. Freire, A. J. D. Silvestre, A. Gandini, M. G. S. Ferreira, and M. L. Zheludkevich. 2012. "Chitosan-based self-healing protective coatings doped with cerium nitrate for corrosion protection of aluminum alloy 2024." *Progress in Organic Coatings* 75 (1): 8–13. doi: 10.1016/j.porgcoat.2012.02.012.

Chen, T., R. Chen, Z. Jin, and J. Liu. 2015. "Engineering hollow mesoporous silica nanocontainers with molecular switches for continuous self-healing anticorrosion coating." *Journal of Materials Chemistry A* 3 (18): 9510–16. doi:10.1039/C5TA01188D.

Chen, Y., Z. Xu, C. Smith, and J. Sankar. 2014. "Recent advances on the development of magnesium alloys for biodegradable implants." *Acta Biomaterialia* 10 (11): 4561–73. doi: 10.1016/j.actbio.2014.07.005.

Choung, T., J. Lim, D.-J. Won, and J. Kim. 2018. "Chamber/capsule-integrated self-healing coating on glass for preventing crack propagation." *Macromolecular Materials and Engineering* 303 (4): 1800041. doi: 10.1002/mame.201800041.

"Dynamic Urea Bond for the Design of Reversible and Self-Healing Polymers | Nature Communications." 2020. Accessed April 3. https://www.nature.com/articles/ncomms4218?page=1.

"EBSCOhost | 114268242 | 'Containers' for Self-Healing Epoxy Composites and Coating: Trends and Advances." 2020. Accessed April 3. https://web.a.ebscohost.com/abstract-?direct=true&profile=ehost&scope=site&authtype=crawler&jrnl=1788618X&AN=114268242&h=9GuxrK0rvD%2f5Ekb1bdeuFyXvuLwnqqR2yhGnBgzMJza4Bv5o1ubDzaEyXw5ilJ2m%2fzT%2bzD%2fIZO8FQKDdikA9fw%3d%3d&crl=c&resultNs=AdminWebAuth&resultLocal=ErrCrlNotAuth&crlhashurl=login.aspx%3fdirect%3dtrue%26profile%3dehost%26scope%3dsite%26authtype%3dcrawler%26jrnl%3d1788618X%26AN%3d114268242.

Es-haghi, H., S. M. Mirabedini, M. Imani, and R. R. Farnood. 2014. "Preparation and characterization of pre-silane modified ethyl cellulose-based microcapsules containing linseed oil." *Colloids and Surfaces A: Physicochemical and Engineering Aspects* 447 (April): 71–80. doi: 10.1016/j.colsurfa.2014.01.021.

Ghosh, B., and M. W. Urban. 2009. "Self-repairing oxetane-substituted chitosan polyurethane networks." *Science* 323 (5920): 1458–60. doi: 10.1126/science.1167391.

Guo, W., Y. Jia, K. Tian, Z. Xu, J. Jiao, R. Li, Y. Wu, L. Cao, and H. Wang. 2016. "UV-triggered self-healing of a single robust SiO_2 microcapsule based on cationic polymerization for potential application in aerospace coatings." *ACS Applied Materials & Interfaces* 8 (32): 21046–54. doi: 10.1021/acsami.6b06091.

Hamilton, A. R., N. R. Sottos, and S. R. White. 2012. "Pressurized vascular systems for self-healing materials." *Journal of The Royal Society Interface* 9 (70): 1020–28. doi: 10.1098/rsif.2011.0508.

Hang, T. T. X., T. A. Truc, N. T. Duong, N. Pébère, and M.-G. Olivier. 2012. "Layered double hydroxides as containers of inhibitors in organic coatings for corrosion protection of carbon steel." *Progress in Organic Coatings, Application of Electrochemical Techniques to Organic Coatings*, 74 (2): 343–48. doi: 10.1016/j.porgcoat.2011.10.020.

Hayatdavoudi, H., and M. Rahsepar. 2017. "Smart inhibition action of layered double hydroxide nanocontainers in zinc-rich epoxy coating for active corrosion protection of carbon steel substrate." *Journal of Alloys and Compounds* 711 (July): 560–67. doi: 10.1016/j.jallcom.2017.04.044.

Hilloulin, B., K. Van Tittelboom, E. Gruyaert, N. De Belie, and A. Loukili. 2015. "Design of polymeric capsules for self-healing concrete." *Cement and Concrete Composites* 55 (January): 298–307. doi: 10.1016/j.cemconcomp.2014.09.022.

Huang, M., and J. Yang. 2014. "Salt spray and EIS studies on HDI microcapsule-based self-healing anticorrosive coatings." *Progress in Organic Coatings* 77 (1): 168–75. doi: 10.1016/j.porgcoat.2013.09.002.

Jonkers, H. M. 2007. "Self healing concrete: A biological approach." In *Self Healing Materials: An Alternative Approach to 20 Centuries of Materials Science*, edited by S. van der Zwaag, 195–204. Springer Series in Materials Science. Dordrecht: Springer Netherlands. doi: 10.1007/978-1-4020-6250-6_9.

Kartsonakis, I. A., E. Athanasopoulou, D. Snihirova, B. Martins, M. A. Koklioti, M. F. Montemor, G. Kordas, and C. A. Charitidis. 2014. "Multifunctional epoxy coatings combining a mixture of traps and inhibitor loaded nanocontainers for corrosion protection of AA2024-T3." *Corrosion Science* 85 (August): 147–59. doi: 10.1016/j.corsci.2014.04.009.

Kartsonakis, I. A., A. C. Balaskas, E. P. Koumoulos, C. A. Charitidis, and G. C. Kordas. 2012. "Incorporation of ceramic nanocontainers into epoxy coatings for the corrosion protection of hot dip galvanized steel." *Corrosion Science* 57 (April): 30–41. doi: 10.1016/j.corsci.2011.12.037.

Kartsonakis, I. A., and G. Kordas. 2010. "Synthesis and characterization of cerium molybdate nanocontainers and their inhibitor complexes." *Journal of the American Ceramic Society* 93 (1): 65–73. doi: 10.1111/j.1551-2916.2009.03310.x.

Keyvani, A., M. Yeganeh, and H. Rezaeyan. 2017. "Application of mesoporous silica nanocontainers as an intelligent host of molybdate corrosion inhibitor embedded in the epoxy coated steel." *Progress in Natural Science: Materials International* 27 (2): 261–67. doi: 10.1016/j.pnsc.2017.02.005.

Koh, E., N.-K. Kim, J. Shin, and Y.-W. Kim. 2014. "Polyurethane microcapsules for self-healing paint coatings." *RSC Advances* 4 (31): 16214–23. doi: 10.1039/C4RA00213J.

Kötteritzsch, J., S. Stumpf, S. Hoeppener, J. Vitz, M. D. Hager, and U. S. Schubert. 2013. "One-component intrinsic self-healing coatings based on reversible crosslinking by Diels–Alder cycloadditions." *Macromolecular Chemistry and Physics* 214 (14): 1636–49. doi: 10.1002/macp.201200712.

Liu, X., H. Zhang, J. Wang, Z. Wang, and S. Wang. 2012. "Preparation of epoxy microcapsule based self-healing coatings and their behavior." *Surface and Coatings Technology* 206 (23): 4976–80. doi: 10.1016/j.surfcoat.2012.05.133.

Mirabedini, S. M., I. Dutil, and R. R. Farnood. 2012. "Preparation and characterization of ethyl cellulose-based core–shell microcapsules containing plant oils." *Colloids and Surfaces A: Physicochemical and Engineering Aspects* 394 (January): 74–84. doi: 10.1016/j.colsurfa.2011.11.028.

Montemor, M. F., D. V. Snihirova, M. G. Taryba, S. V. Lamaka, I. A. Kartsonakis, A. C. Balaskas, G. C. Kordas, et al. 2012. "Evaluation of self-healing ability in protective coatings modified with combinations of layered double hydroxides and cerium molibdate nanocontainers filled with corrosion inhibitors." *Electrochimica Acta* 60 (January): 31–40. doi: 10.1016/j.electacta.2011.10.078.

Nakahata, M., Y. Takashima, and A. Harada. 2016. "Highly flexible, tough, and self-healing supramolecular polymeric materials using host–guest interaction." *Macromolecular Rapid Communications* 37 (1): 86–92. doi: 10.1002/marc.201500473.

Nawaz, M., S. Habib, A. Khan, R. A. Shakoor, and R. Kahraman. 2020. "Cellulose microfibers (CMFs) as a smart carrier for autonomous self-healing in epoxy coatings." *New Journal of Chemistry*, March. doi: 10.1039/C9NJ06436B.

Sahoo, B. N., S. Nanda, J. A. Kozinski, and S. K. Mitra. 2017. "PDMS/camphor soot composite coating: Towards a self-healing and a self-cleaning superhydrophobic surface." *RSC Advances* 7 (25): 15027–40. doi: 10.1039/C6RA28581C.

Odarczenko, M., D. Thakare, W. Li, S. P. Venkateswaran, N. R. Sottos, and S. R. White. 2020. "Sunlight-activated self-healing polymer coatings." *Advanced Engineering Materials* 22 (3): 1901223. doi: 10.1002/adem.201901223.

Orain, R., H. Heuser, H. Ohlenforst, and R. Pelzer. 1980. Safety Window Comprising Self-Healing Polymeric Layer. United States US4232080A, filed March 10, 1977, and issued November 4, 1980. https://patents.google.com/patent/US4232080A/en.

Qiang, S., K. Chen, Y. Yin, and C. Wang. 2017. "Robust UV-cured superhydrophobic cotton fabric surfaces with self-healing ability." *Materials & Design* 116 (February): 395–402. doi: 10.1016/j.matdes.2016.11.099.

"Repeated Self-Healing of Microvascular Carbon Fibre Reinforced Polymer Composites - IOPscience." 2020. Accessed March 29. https://iopscience.iop.org/article/10.1088/0964-1726/23/11/115002/meta.

Riaz, U., C. Nwaoha, and S. M. Ashraf. 2014. "Recent advances in corrosion protective composite coatings based on conducting polymers and natural resource derived polymers." *Progress in Organic Coatings* 77 (4): 743–56. doi: 10.1016/j.porgcoat.2014.01.004.

Ryu, Y. S., K. W. Oh, and S. H. Kim. 2016. "Synthesis and characterization of a furan-based self-healing polymer." *Macromolecular Research* 24 (10): 874–80. doi: 10.1007/s13233-016-4122-5.

Samadzadeh, M., S. Hatami Boura, M. Peikari, A. Ashrafi, and M. Kasiriha. 2011. "Tung oil: An autonomous repairing agent for self-healing epoxy coatings." *Progress in Organic Coatings, Coatings Science International 2010*, 70 (4): 383–87. doi: 10.1016/j.porgcoat.2010.08.017.

Samadzadeh, M., S. Hatami Boura, M. Peikari, S. M. Kasiriha, and A. Ashrafi. 2010. "A review on self-healing coatings based on micro/nanocapsules." *Progress in Organic Coatings* 68 (3): 159–64. doi: 10.1016/j.porgcoat.2010.01.006.

Sauvant-Moynot, V., S. Gonzalez, and J. Kittel. 2008. "Self-healing coatings: An alternative route for anticorrosion protection." *Progress in Organic Coatings, Workshop on Application of Electrochemical Techniques to Organic Coatings*, 63 (3): 307–15. doi: 10.1016/j.porgcoat.2008.03.004.

"Self-Healing Materials Market Size, Share | Global Industry Report, 2025." 2020. Accessed April 3. https://www.grandviewresearch.com/industry-analysis/self-healing-materials.

"Sunlight-Induced Self-Healing of a Microcapsule-Type Protective Coating | ACS Applied Materials & Interfaces." 2020. Accessed March 29. https://pubs.acs.org/doi/10.1021/am302728m.

Suryanarayana, C., K. C. Rao, and D. Kumar. 2008. "Preparation and characterization of microcapsules containing linseed oil and its use in self-healing coatings." *Progress in Organic Coatings* 63 (1): 72–78. doi: 10.1016/j.porgcoat.2008.04.008.

Taylor, D. L., and M. i. h. Panhuis. 2016. "Self-healing hydrogels." *Advanced Materials* 28 (41): 9060–93. doi: 10.1002/adma.201601613.

"TiO$_2$ Nanotubes and Mesoporous Silica as Containers in Self-Healing Epoxy Coatings | Scientific Reports." 2020. Accessed March 26. https://www.nature.com/articles/srep38812.

Toohey, K. S., N. R. Sottos, and S. R. White. 2009. "Characterization of microvascular-based self-healing coatings." *Experimental Mechanics* 49 (5): 707–17. doi: 10.1007/s11340-008-9176-7.

"Unraveling EDTA Corrosion Inhibition When Interleaved into Layered Double Hydroxide Epoxy Filler System Coated onto Aluminum AA 2024- ScienceDirect." 2020. Accessed March 26. https://www.sciencedirect.com/science/article/abs/pii/S0169131713002329.

Vijayan, P. P., and M. Al-Maadeed. 2019. "Self-repairing composites for corrosion protection: a review on recent strategies and evaluation methods." *Materials* 12 (17): 2754. doi: 10.3390/ma12172754.

Vijayan, P. P., Y. M. H. El-Gawady, and M. A. S. A. Al-Maadeed. 2016. "Halloysite nanotube as multifunctional component in epoxy protective coating." *Industrial & Engineering Chemistry Research* 55 (42): 11186–92. doi: 10.1021/acs.iecr.6b02736.

Vijayan, P. P., A. Tanvir, Y. H. El-Gawady, and M. Al-Maadeed. 2017. "Cellulose nanofibers to assist the release of healing agents in epoxy coatings." *Progress in Organic Coatings* 112 (November): 127–32. doi: 10.1016/j.porgcoat.2017.07.008.

Wang, X., H. Zhang, B. Yang, L. Wang, and H. Sun. 2020. "A colorless, transparent and self-healing polyurethane elastomer modulated by dynamic disulfide and hydrogen bonds." *New Journal of Chemistry*, March. doi: 10.1039/C9NJ06457E.

Wang, J.-K., Q. Zhou, J.-P. Wang, S. Yang, and G. L. Li. 2019. "Hydrophobic self-healing polymer coatings from carboxylic acid- and fluorine-containing polymer nanocontainers." *Colloids and Surfaces A: Physicochemical and Engineering Aspects* 569 (May): 52–58. doi: 10.1016/j.colsurfa.2019.02.050.

Wang, W., L. Xu, X. Li, Y. Yang, and E. An. 2014. "Self-healing properties of protective coatings containing isophorone diisocyanate microcapsules on carbon steel surfaces." *Corrosion Science* 80 (March): 528–35. doi: 10.1016/j.corsci.2013.11.050.

White, S. R., N. R. Sottos, P. H. Geubelle, J. S. Moore, M. R. Kessler, S. R. Sriram, E. N. Brown, and S. Viswanathan. 2001. "Autonomic healing of polymer composites." *Nature* 409 (6822): 794–97. doi: 10.1038/35057232.

Wu, G., J. An, X.-Z. Tang, Y. Xiang, and J. Yang. 2014. "A versatile approach towards multifunctional robust microcapsules with tunable, restorable, and solvent-proof superhydrophobicity for self-healing and self-cleaning coatings." *Advanced Functional Materials* 24 (43): 6751–61. doi: 10.1002/adfm.201401473.

Xiong, P., J. L. Yan, P. Wang, Z. Jia, W. Zhou, W. Yuan, Y. Li, et al. 2019. "A PH-sensitive self-healing coating for biodegradable magnesium implants." *Acta Biomaterialia, 10th BIOMETAL2018- International Symposium on Biodegradable Metals*, 98 (October): 160–73. doi: 10.1016/j.actbio.2019.04.045.

Xuan, H., J. Ren, J. Zhang, and L. Ge. 2017. "Novel highly-flexible, acid-resistant and self-healing host-guest transparent multilayer films." *Applied Surface Science* 411 (July): 303–14. doi: 10.1016/j.apsusc.2017.03.129.

Yabuki, A., A. Kawashima, and I. W. Fathona. 2014. "Self-healing polymer coatings with cellulose nanofibers served as pathways for the release of a corrosion inhibitor." *Corrosion Science* 85 (August): 141–46. doi: 10.1016/j.corsci.2014.04.010.

Yang, H. J., Y. T. Pei, J. C. Rao, and J. T. M. De Hosson. 2012. "Self-healing performance of Ti$_2$AlC ceramic." *Journal of Materials Chemistry* 22 (17): 8304–13. doi: 10.1039/C2JM16123K.

Yang, Y., and M. W. Urban. 2013. "Self-healing polymeric materials." *Chemical Society Reviews* 42 (17): 7446–67. doi: 10.1039/C3CS60109A.

Yuan, Y. C., T. Yin, M. Z. Rong, and M. Q. Zhang. 2008. "Self healing in polymers and polymer composites. concepts, realization and outlook: A review." *Express Polymer Letters* 2 (4): 238–50. doi: 10.3144/expresspolymlett.2008.29.

Zhang, F., P. Ju, M. Pan, D. Zhang, Y. Huang, G. Li, and X. Li. 2018. "Self-healing mechanisms in smart protective coatings: A review." *Corrosion Science* 144 (November): 74–88. doi: 10.1016/j.corsci.2018.08.005.

Zhao, Y., W. Zhang, L.-p. Liao, S.-j. Wang, and W.-j. Li. 2012. "Self-healing coatings containing microcapsule." *Applied Surface Science, International Vacuum Congress (IVC-18)*, 258 (6): 1915–18. doi: 10.1016/j.apsusc.2011.06.154.

Zheludkevich, M. L., J. Tedim, C. S. R. Freire, S. C. M. Fernandes, S. Kallip, A. Lisenkov, A. Gandini, and M. G. S. Ferreira. 2011. "Self-healing protective coatings with 'green' chitosan based pre-layer reservoir of corrosion inhibitor." *Journal of Materials Chemistry* 21 (13): 4805–12. doi: 10.1039/C1JM10304K.

"Zn–Al Layered Double Hydroxides as Chloride Nanotraps in Active Protective Coatings - ScienceDirect." 2020. Accessed March 26. https://www.sciencedirect.com/science/article/pii/S0010938X11005312.

Zulkifli, F., N. Ali, M. S. M. Yusof, M. I. N. Isa, A. Yabuki, and W. B. Wan Nik. 2017. "Henna leaves extract as a corrosion inhibitor in acrylic resin coating." *Progress in Organic Coatings* 105 (April): 310–19. doi: 10.1016/j.porgcoat.2017.01.017.

13 Hydrophobic and Hydrophilic Polymer Coatings

Sanjay Remanan
Indian Institute of Technology

Harikrishnan Pulikkalparambil, Sanjay Mavinkere Rangappa, Suchart Siengchin, and Jyotishkumar Parameswaranpillai
King Mongkut's University of Technology North Bangkok

Narayan Chandra Das
Indian Institute of Technology

CONTENTS

13.1 INTRODUCTION

Understanding the wetting behavior of solid surfaces by liquids plays a pivotal role in material surface engineering. Ideally, when a liquid drop is placed over a solid surface, it may either undergo complete spreading into the bulk, or form a spherical droplet on the surface. The former is called superwetting, and the latter is called anti-wetting surfaces. In general, material surfaces showing high affinity for water spread it evenly, while those with less affinity repel it. Hence, the point of contact between a liquid and a solid has a significant influence in describing the wetting property of the surface of the solid. The wetting behavior of solid surfaces may precisely be analyzed by contact angle measurements. The angle measured between the liquid, solid, and solid–air interface is called the *contact angle* (Gesser and Paul 2000). In other words, it may be defined as the angle between the liquid–solid interface and the tangent point of the curve at the contact point of solid, liquid, and air (Otitoju, Ahmad, and Ooi 2017).

Figure 13.1 shows the schematic representation of the motion of liquid droplets on a solid substrate. It is evident from Figure 13.1a that when the contact area between a solid substrate and liquid droplets is large and the angle formed is small (<90°), this type of surface is called a *hydrophilic surface*, while on the other hand, when liquid droplets make a small contact area and a very high contact angle (>90°) as seen in Figure 13.1b, these kind of surfaces are termed as a *hydrophobic surface* (Genzer and Marmur 2011). Furthermore, it is observed that the affinity of dirt particles varies and behaves differently on different surfaces. For instance, when the adhesion between the dirt particles and the substrate is high, the droplet just passes through. However, on a substrate that is topographically decorated, the dirt particles have difficulty in adhering to it. As the liquid droplet rolls off the substrate, it picks up the dirt particles and hence cleans up the substrate.

Young studied the static contact angle on the smooth surfaces and reported the mechanism behind the wetting of a surface (Bonn et al. 2009). The surface energy

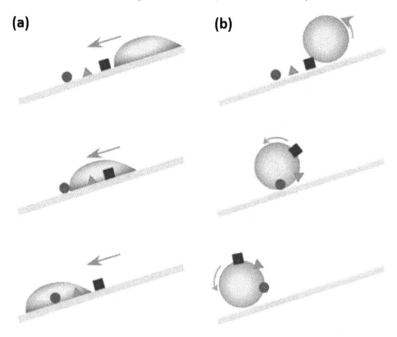

FIGURE 13.1 Schematic representation depicting the motion of a liquid droplet on a substrate covered with dirt: (a) hydrophilic surface and (b) hydrophobic surface. (Adopted from Genzer and Marmur 2011, with permission from Materials Research Society.)

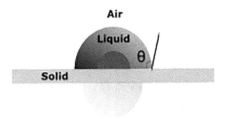

FIGURE 13.2 Depiction of the contact angle of a surface measured considering the solid–liquid, liquid–air, and solid–air interfaces.

between the solid–liquid, liquid–air, and solid–air boundaries plays a significant role in the wetting of the surface (Figure 13.2 and Figure 13.5a). If the wetted surface is more energetically favorable than the non-wetting surface, the static contact angle will be less than 90° and forms a hydrophilic surface. On the other hand, if the surface is more energetically favorable toward the non-wetting surface, it forms a hydrophobic surface with a water contact angle greater than 90°. The term *energetically favorable* is associated with the *spreading parameter* (S), which is the difference between the surface energy of the solid in wet and dry conditions. A wet surface is said to be energetically favorable when its spreading parameter is greater than zero, and a non-wetting surface is said to be energetically favorable when its spreading parameter is less than zero. When the S is greater than zero, liquids spread all over the high-energy solid surface (hydrophilic surface), and when the S is less than zero, droplets form an equilibrium liquid droplet cap over the low-energy solid surface (hydrophobic surface). A surface with extremely high water contact angle (>150°) is called a *superhydrophobic or ultraphobic* surface, and a surface having a very low water contact angle (<10°) is called a *superhydrophilic* surface.

Neinhuis and Barthlott identified about 200 different plant species having superhydrophobic properties (Neinhuis and Barthlott 1997) that help to clean the plant leaves in such a way that dirt particles are carried away along with the rolling water. This property is popularly known as *lotus effect*, which finally results in a cleaned surface (Figure 13.3a–d). The ultraphobic property is a result of the combined surface micropattern of the leaf and the presence of epitaxial wax. The lotus leaf roughness is due to the presence of papillose epidermal cells that contain epitaxial wax. The microscale roughness along with the tubular wax structure of the papillose imparts the superhydrophobicity and is termed as *lotus effect* (Blossey 2003). The ultrahydrophobicity also imparts high anti-adhesive property, which results in high anti-wetting property between the liquid and solid surfaces. Superhydrophobic property of the plant leaves is deteriorated by its aging and decay; in nature, plants either follow a repairing process, or give up the damaged leaf to sustain the property. Similarly, self-healable and sustainable polymer coatings can be designed for long service life in outdoor applications. Hence, a biomimetic approach to designing a superhydrophobic self-cleaning surface is found to be one of the most important industrially adaptable techniques. In the near future, few application areas such as dust-free products including apparels and paints may show high potential for self-cleaning materials.

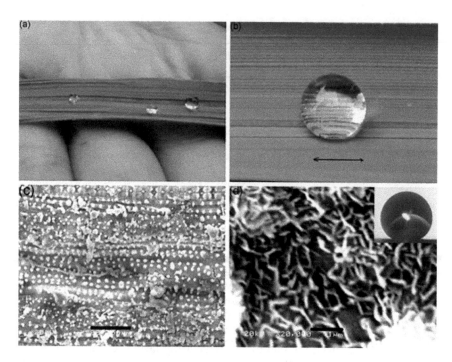

FIGURE 13.3 (a) and (b) An almost ball-shaped water droplet on a non-wettable rice leaf, (c) and (d) SEM images of a rice leaf with different magnifications and the contact angle ($157° \pm 2°$) observed is shown in the inset (d) (Adopted from Guo and Liu 2007, with permission from Elsevier.)

13.2 THEORY AND MECHANISM OF HYDROPHOBICITY AND HYDROPHILICITY

Self-cleaning surfaces are the new era in material science that has seen considerable interests in industrial and research fields. Basically, self-cleaning surfaces are possible in two ways, viz. hydrophilic and hydrophobic materials; coatings based on these categories are widely being used in the automobile and construction industries (https://www.aculon.com/repellency-treatments/hydrophobic-coatings/, https://www.neverwet.com/). Hydrophilic coating and hydrophobic coating have the tendency to clean the dirt present on the surfaces and are accomplished using two different mechanisms.

Hydrophilic coating helps the sliding motion of the water droplets on its surface and allows the dirt to interact with the coated surface chemically. Superhydrophilic surfaces can expel the dirt by a spreading mechanism. The dirt spreads over the superhydrophilic surface and forms a thin layer. The chemical interactions with the coated surface lead to the degradation of the dirt particles (Syafiq et al. 2018). This kind of decomposition is visualized in photoinduced superhydrophilic coatings. One such example is the TiO_2-assisted photocatalyst-coated surface that decomposes the organic dirt molecules (Adachi et al. 2018). Superhydrophilic surface can wash away the dirt through the sliding motion of water molecules. Water molecules spread beneath the dirt and wash them away from the surface (Nishimoto and Bhushan 2013). On the other hand, the hydrophobic coating allows the rolling action of the water droplets, which is found less likely to leave any dirt particles on the surface compared to the sliding motion of the water droplets. Hence, the movement of water, especially the rolling of water droplets, is significant in the case of self-cleaning coating. The rolling action of the water droplets depends upon hysteresis (it is the difference between the advancing and receding angles, Figure 13.4, (the average of these two angles is called the contact angle)) in the contact angles during the advancement or receding of a droplet (Eral and Oh 2013). This is influenced by the interaction of droplets with different interfaces such as surface–water, water–air, and surface–air interfaces. The contact angle hysteresis property is vastly influenced by the chemical composition and surface texture of the material, and for an ideal case, the hysteresis must be zero to achieve least resistance. But in practice, when the hysteresis is less than 10°, the surface tends to show self-cleaning and extreme water-repellent properties. Thus, it is clear that the static contact angle over the coating surface is ideal for the measurement of hydrophobicity or hydrophilicity. To achieve the least hysteresis (~0°), the contact angle of the surface must be high enough over which water droplets rolls smoothly. For a surface with given surface roughness, Wenzel and Cassie droplets have different hysteresis values. Cassie droplets show much less hysteresis than Wenzel droplets. This is due to the difference in the wetting of the surface. In Wenzel model, liquid droplets wet the rough surface, while in Cassie model, liquid droplets tend to sit on the peaks of the rough surface (He, Lee, and Patankar 2004). So, dirt removal can be efficiently accomplished with a hydrophobic surface with least rolling resistance compared to hydrophilic surface (low roll-off angle; the roll-off angle is defined as the minimum angle required for the droplet to roll over the coating surface, Figure 13.5b) (Marmur 2004). For instance, a lower wet area of a solid surface has a lower roll-off angle because the force required to keep the liquid in contact with the solid is lower. This can be visibly observed from the high-contact-angle surfaces in which the wetting area regime is considerably small. A surface with a high contact

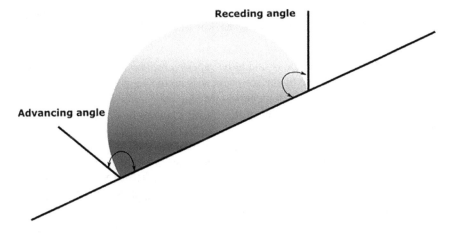

FIGURE 13.4   A droplet on a surface showing advancing and receding angles.

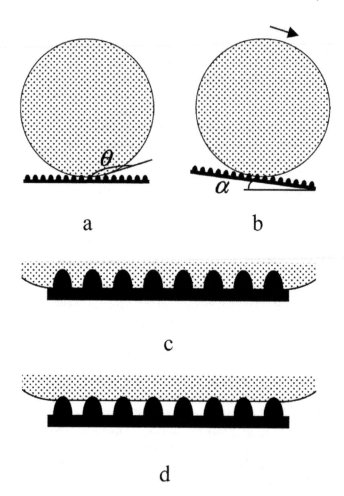

FIGURE 13.5 A droplet on a rough surface: (a) the contact angle, (b) the roll-off angle, (c) the homogeneous wetting regime, and (d) the heterogeneous wetting regime. (Adopted from Marmur 2004, with permission from American Chemical Society.)

angle and a least roll-off angle is said to be *superhydrophobic or ultraphobic*. Yang et al. reported the influence of surface roughness on the superhydrophobicity of the material. An increase in the root-mean-square roughness value increases the contact angle, and it is nearly independent of the fractal dimensions of the surface (Yang, Tartaglino, and Persson 2006).

Designing an ultraphobic surface requires knowledge on the topography of the material. Some polymeric materials, such as fluorocarbon, silicone, and polyurethane, are typical examples of having a very high contact angle (~110°–120°). These polymer materials are often used for "easily cleanable surface" applications. However, self-cleaning may not be achieved in this due to the high roll-off angle of water droplets.

As discussed, the superhydrophobicity of the lotus (*Nelumbo nucifera*) leaves arises from the micropatterns present on the leaf surface. The quest for achieving a very high contact angle led to the exploration of an ultraphobic plant leaf surface having rough patterned structures seen under microscopy. Wenzel developed a model for the determination of the contact angle of a rough surface, which is different from that of the smooth surface consideration in Young's model (Wang et al. 2012). Wenzel model addresses this major disadvantage of the Young's equation as the latter only considers the atomically smooth surfaces. But in practice, atomically flat surfaces are rare and they mostly exhibit at least some degree of surface roughness. Wenzel suggested that if a sample surface favors wetting, its wettability will be enhanced on the rougher surface. Similarly, for a non-wetting surface, wettability reduces with higher surface roughness.

In the Wenzel model, the macro-sized liquid droplet is placed over a surface which has microlevel roughness, and the angle formed between the droplet and solid surface is called θ. It was then compared with the angle (θ_0) formed when the water droplet is placed over a smooth surface. In this model, Wenzel considered that water molecules can penetrate into the air pockets present on the rough surface (homogeneous surface, Figure 13.5c) and the water contact angle is given by the relation:

$$\cos\theta = R_f \cos\theta_0 \tag{13.1}$$

where R_f is the roughness factor and is defined as the ratio of the solid–liquid area (A_{SL}) to its projection on a flat plane (A_F) (Koch, Amirfazli, and Elliott 2014). Mathematically, it is represented as follows:

$$R_f = \frac{A_{SL}}{A_F} \tag{13.2}$$

The Wenzel model finds that with an increase in surface roughness, the hydrophilic surface becomes more hydrophilic, while the hydrophobic surface becomes more hydrophobic. When Young's contact angle on the smooth surface is less than 90°, roughness can reduce the apparent contact angle leading to superwetting or the formation of superhydrophilic surfaces (Otitoju, Ahmad, and Ooi 2017).

Cassie and Baxter mathematically analyzed the water contact angle based on the surface (composite interface, Figure 13.5d) roughness. In this model, air pockets were present on the rough surface, in between the water droplet and the solid surface, which was taken into account. Hence, the contact angle formation depends upon the area fraction of the solid and area fraction of the air.

The Cassie–Baxter model considers that a non-wetting liquid might not penetrate into the cavities, which form an air gap between solid and liquid surfaces. This can be called interface and, respectively, termed as solid–liquid, liquid–gas, and solid–gas interfaces. Hence, this model is an extension of the Wenzel work in which composite interface contribution is considered. By considering the fractional area contribution of wet surfaces and air pockets, the Cassie–Baxter equation for apparent contact angle can be written as follows:

$$\cos\theta_{CB} = f_s \cos\theta_c + f_v \cos\theta_v \tag{13.3}$$

where θ_c is the intrinsic contact angle on the original smooth surface and f_s and f_v are the area contributions of the solid and air, respectively (Sigmund and Hsu 2016).

Bhushan et al. reported the Cassie–Baxter equation as an extension of Wenzel model and mathematically wrote it as follows:

$$\cos\theta = R_f \cos\theta_0 - f_{LA}\left(R_f \cos\theta_0 + 1\right) \tag{13.4}$$

where f_{LA} is the fractional flat geometric area of the liquid–air interfaces under the droplet. For a hydrophobic surface, the contact angle increases with the increase in f_{LA} for both smooth and rough surfaces. For a hydrophilic surface, the contact angle increases with f_{LA}, but at higher f_{LA} values, the surface can become hydrophobic. This may not be achieved due to the instability of the formed air pockets (Bhushan, Jung, and Koch 2009).

Water droplets placed on the micropatterned surface have a different approach toward hydrophobic and hydrophilic surfaces. In the case of hydrophilic surface, the water droplet favors the texture and flows to the valley of pattern and spreads out, and this leads to a decrease in the water contact angle. Hence, for a hydrophilic surface, an increase in surface roughness may decrease the contact angle. On the other hand, a water droplet present on the hydrophobic surface requires a very high energy to flow through the patterns; hence, water droplet recedes on this textured surface. This increases the water contact angle (>150°) to an ultra-high value compared to the smooth surface.

A biomimetic approach for the preparation of hydrophobic surfaces is already being explored, and several different observations are made. The typical examples are the surface texture on gecko feet, nanopillars present on a lotus leaf, and impressions present on moth wings (Figure 13.6). A water strider moves at a very high speed over the water surface. This is also due to the presence of a superhydrophobic surface on its feet. Hence, observations from nature help in designing a surface with patterns that have more tendencies to show a higher contact angle compared to the smooth surface.

Different approaches can be used to fabricate the rough surface to achieve the high static contact angle. Techniques such as nanoimprint lithography (NIL), soft lithography, electrodeposition, and micromolding are typically used for surface preparation. Sol–gel, electrospinning, layer-by-layer deposition, and plasma technique are the other methods used (Zhang et al. 2008). Molding process is comparatively cheap and easy to replicate from the plant leaf texture and from the silicon microstructures. Figure 13.7a shows that the presence of microimpressions over the lotus leaf imparts a very high contact angle and hence superhydrophobicity (Barati Darband et al. 2018). Figure 13.7b shows the positive replica of the leaf made on a PDMS substrate, which has more or less the same morphological features. Hence, nature-inspired patterned surfaces form hierarchical structures that can increase the water contact angle, which is advantageous for developing superhydrophobic surfaces. Guo et al. investigated the influence of nanopillars on the water contact angle of the polymer surfaces (Guo et al. 2004). With an increase in nanopillar diameter, a decrease in the water contact angle was observed. This is due to the increase in the air surface fraction present among the cavities, which increases the contact angle. Similarly,

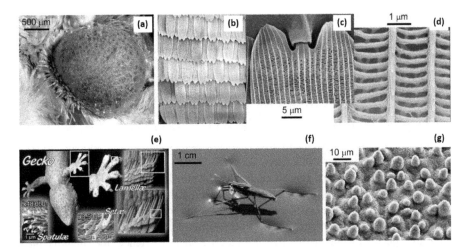

FIGURE 13.6 Images showing how biological functionality, ranging from reflective characteristics to hydrophobicity in insect feet and plant leaves, arises from natural surfaces. (a) Scanning electron microscopy image of a moth's eye. (b–d) Details of butterfly wings at different levels of magnification, showing that they are composed of hundreds of thousands of scales with complex hierarchical structures. (e) Details of gecko feet. (f) Water strider (*Gerris remigis*) walking on the surface of a lake. (g) Close-up of a lotus leaf (*Nelumbo nucifera*), an example of a superhydrophobic plant surface. (Adopted from Genzer and Marmur 2011, with permission from Materials Research Society.)

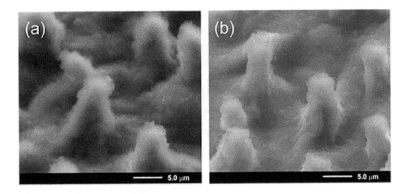

FIGURE 13.7 Higher-magnification SEM images of (a) the natural lotus leaf and (b) its positive replica. They almost have the same surface morphology on both the microscale and nanoscale. (Adopted from Sun et al. 2005, with permission from American Chemical Society.)

Parameswaranpillai et al. reported that the nanostructuring of poly(ethylene oxide)–poly(propylene oxide)–poly(ethylene oxide) (PEO–PPO–PEO) triblock copolymer in epoxy thermosets improves the water contact angle values. This is due to two reasons: (1) the protected triblock copolymer domains in the epoxy matrix and (2) the interaction of PEO–PPO–PEO with free hydroxyl groups in epoxy networks that are responsible for lowering the contact angles (Parameswaranpillai et al. 2017).

Superhydrophilic or superwetting surfaces have the tendency to spread out water (liquid) present on the surface, and this is mostly in the form of films than droplets. For such surfaces, the roughness factor in the Wenzel equation is always greater than 1 ($r > 1$) and the water spread over the surfaces helps in the application of removal of dirt and staining materials, and oil-in-water separation, biomedical, anti-fogging, anti-corrosive, and pervaporation applications. Tuning the surface chemistry to increase the surface energy and making the pattern to increase the surface roughness is the key feature to achieve the desired hydrophilicity. There are different preparation methods to fabricate superhydrophilic coatings, such as plasma treatment, ultraviolet and laser irradiation, sol–gel, phase inversion, anodization, and etching, the details of which can be found in a recent review article (Otitoju, Ahmad, and Ooi 2017).

Surface energy is a scalar quantity, and it is the property of an area of the surface. The term surface energy and surface tension are two distinct quantities, but sometimes used interchangeably. However, *surface tension* is a vector quantity and measured perpendicular to the line drawn on a surface. Hence, the quantity measured during the contact angle determination is surface tension. Surface tension is the quantity involved in contact angle equilibrium (Law and Zhao 2016). For a superhydrophobic coating material, a low surface energy with desired surface texture is the key parameter to achieve the property, while for a superhydrophilic material, the surface energy is higher.

13.3 HYDROPHOBIC AND HYDROPHILIC POLYMER COATINGS, METHODS, AND SPECIFIC APPLICATIONS

Different additives are present in the coatings, such as a base polymer, cross-linking agents, pigments, reaction inhibitors, accelerators, and weather-resistant materials. The automotive applications currently demand scratch-resistant, self-healable, and dust-resistant coatings. The specific property requirements depend upon the polymer/cross-linking system used as the coating material, and the three-dimensional network formation of this coating facilitates the attainment of the desired property. This polymeric coating over the substrate is applied either via a batch or via a continuous process.

13.3.1 SURFACE COATING METHODS

Different surface coating methods can be adopted for the preparation of hydrophobic or hydrophilic surfaces. Surface modification by coating is a simple and quickly adopted method. Essential elements of coating units comprise unwinding, applying, drying, cooling, and winding-up processes. The commonly used surface coating methods are knife coating, dip coating, spray coating, and spin coating.

 i. *Knife coating*:
    Knife coating is a simple coating method in which excess coating material is applied onto the substrate and spread out using a metered knife with a controlled speed. An industrial knife coating head may be used to achieve the controlled coating over the substantial length of substrates. The knife

used for coating the substrate is called the doctor blade, and depending upon its configuration, it is classified as a floating knife, knife over roll, knife over the fixed table, and knife over conveyor (Shim 2013). The commonly used coating method is the knife over roll in which coating thickness is controlled by the gap between the roller and the knife used. This technique is used to get the precise coating thickness. The knife coating method is widely used in the textile industry where the doctor blade helps to penetrate the coating medium into the fabric, and it is usually used when relatively less coating thickness is required.

ii. *Dip coating*:

The dip coating method comprises of simple immersion and withdrawal steps. In this method, the substrate is immersed into the coating medium and, after a specified time, withdrawn from the medium at a predefined speed and the coated substrate is allowed to dry. It is a low-cost process in which the thickness can be easily adjusted. Coating medium, coating speed, viscosity, lift-off angle, surface tension, gravitational force, dwell time, and viscous drag can significantly affect the quality of the dip coating process. This process is widely used for the modification of fibers and objects with intricate structures, where a high production output is generally needed to regulate the uniform composition of the liquid mixture after a specified time (Gutoff and Cohen 2010; ten Elshof 2015; Wypych 2016).

iii. *Spray coating*:

In this technique, the coating medium is directly applied over the substrate through spraying. Different types of techniques exist, of which pressed air vaporization, airless pressure spray, hot-flame spray, electrostatic spray are some examples. Pressed air spray is the commonly used method in which air and coating materials are pressed out through a nozzle (Shim 2013). Spray coating may generate porosity that is undesirable when concerned with the anti-corrosive-type applications where oxidative stress affects the exposed area.

iv. *Spin coating*:

It is the simplest coating method in which a uniform polymer deposition is achieved on the substrate. The coating material is diluted in a suitable solvent placed at the center of a substrate, and it is rotated at a high speed leading to the deposition of the material. The principle behind spin coating is the centrifugal force which causes the liquid to spread over the substrate and form the film (Smith, Inomata, and Peters 2013). In this process, the substrate spins around an axis (spin-up), which is perpendicular to the coating area (Makhlouf 2011), followed by spin-off and evaporation. During spin-up (Figure 13.8a), the coating medium spreads all over the substrate due to the centrifugal force and the spin-off process initiates when the coating is completed (Figure 13.8b).

Spin coating thickness depends upon the rheological properties of the liquid as well as the rotating speed of the substrate. A high speed results in thinning of the coating medium and uniform evaporation of the solvent (Yilbas, Al-Sharafi, and Ali 2019).

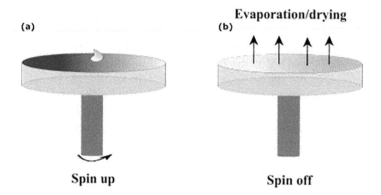

FIGURE 13.8 Spin coating process; (a) spin-up of coating medium using centrifugal force and (b) spin-off and evaporation of the solvent.

The solvent used for spin coating is removed partly during the coating process and subsequently dried at a higher temperature (Nguyen 2012). A major limitation of the method is in the size of the coating substrate. As the size increases, the high-speed rotation of the substrate becomes difficult. Another drawback is the loss of materials during the coating process, which reduces the coating efficiency.

13.3.2 Applications

13.3.2.1 Anti-Fogging

Appearance of fog on the glass surface makes the vision unclear and hazy. An increase in the hydrophilic character increases the surface wettability by improving the surface energy and roughness. The hydrophilic surfaces exhibit an excellent resistance to fog formation on the glass surface. The presence of droplets over a non-treated surface allows scattering of the incident light, resulting in the hazy or foggy appearance over the glass surface. When the contact angle decreases (an increase in hydrophilic character), a sudden spreading out of the droplet and the formation of a thin film were observed. This decrease in water contact angle is attributed to the high surface energy of the hydrophilic surfaces, which contributes to the liquid film formation. Hence, the adhered fog spreads like a film and imparts a clear vision by allowing less scattering (Syafiq et al. 2018).

13.3.2.2 Anti-Corrosive Coatings

Superhydrophobic surfaces can be employed for the application of anti-corrosive coating attributed to the formation of low corrosion current and high corrosion potential after coating on a substrate (Zhang et al. 2016). These two parameters can be measured from the Tafel extrapolation method and can be used in predicting the protection efficiency of the coating.

$$\text{Protection efficiency} = \frac{I_{corr} - I_{corr(c)}}{I_{corr}} \times 100 \qquad (13.5)$$

where I_{corr} and $I_{corr(c)}$ are the corrosive currents measured without coating and with coating, respectively. This corrosion protection efficiency depends on the following three aspects:

1. Water sorption of the coating.
2. Transport of water droplets in coating.
3. Accessibility of water droplets to interact with the coating/substrate interface.

Superhydrophobic polymer materials have a very high water contact angle (>150°), which can fulfill the above requirements. Hence, compared to conventional coating, superhydrophobic materials are ideal for achieving better protection efficiency. A superhydrophobic coating material for anti-corrosive applications such as UV-curable epoxy acrylate coating was developed by Peng et al., and it exhibited an increased water contact angle of 153±1° and consistent contact angle values for more than one month (Peng et al. 2013). The developed coating showed a protection efficiency of 87% against the stainless steel coating, which has a protection efficiency of 75%. In another study, Weng et al. developed a nature-inspired electroactive anti-corrosive epoxy coating (Figure 13.9a–d) which showed a protection efficiency of 87% with a water contact angle of ca. 155° (Weng et al. 2011). Lenon et al. electro-deposited conducting polymers such as polythiophene on to the stainless steel as an

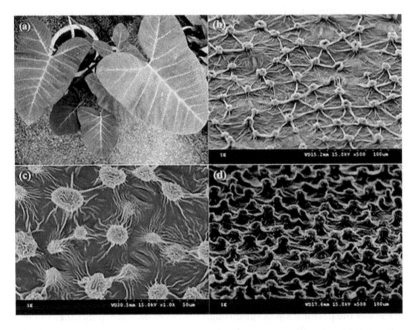

FIGURE 13.9 (a) Photograph of the *Xanthosoma sagittifolium* leaves. (b) SEM image of the leaf. (c) SEM images of the imprinted layers of superhydrophobic electroactive epoxy (SEE) showing a top view of the surface. (d) SEM images of the imprinted layers of SEE showing a cross-sectional view of the surface. (Adopted from Weng et al. 2011, with permission from American Chemical Society.)

anti-corrosive coating and obtained a protection efficiency of >95% with a contact angle of 152°±2° (de Leon, Pernites, and Advincula 2012). Other similar studies of thiophene-derived polymers that show higher anti-wetting and superhydrophobic characters are reported in the literature (Darmanin et al. 2010). Electrochemically synthesized polyaniline-based conductive superhydrophobic composite coating was developed, which showed a contact angle of about 161° (Bhandari, Singha, and Khastgir 2017). An electroactive polyimide coating shows a water contact angle of 155° and has a lower corrosion current and a higher corrosion voltage compared to the non-superhydrophobic coating material. This electroactive superhydrophobic coating exhibits an excellent barrier resistance against strong oxidative environments (Chang et al. 2013).

13.3.2.3 Specialty Coatings

Superhydrophobic materials can be employed in the specialty coatings, which have multiple properties such as self-healing and shape-memory applications. Behera et al. developed a polyurethane-based composite coating that has improved water contact angle and mechanical properties (Behera, Mondal, and Singha 2018). Polyurethane is a widely used industrial polymer and is used as an adhesive in surface sealants, sports utilities, and automotive interior applications. One of the major drawbacks of polyurethane is its low stability at high temperatures, which can be improved by the incorporation of various fillers. The addition of fillers can influence the surface energy of the coating; for instance, the addition of polyhedral oligomeric silsesquioxanes (POSS) increases the water contact angle to 141.3°±1.1°, which increases the material's hydrophobic behavior.

The porous polymeric coating prepared by photo- and thermal initiation results in highly superhydrophobic polymers with a very high contact angle (>170°) (Levkin, Svec, and Fréchet 2009). Porous coating of poly(butyl methacrylate-co-ethylene dimethacrylate) and poly(styrene-co-divinylbenzene) is two such examples and shows ultraphobicity. The ultraphobic nature arises from the presence of porogens in the polymerization system. Porogens are the solvents added to the polymerization reaction, which leads to phase separation when the growing cross-linked structure gains a critical size. This phase separation results in the formation of a highly porous structure from globules that have a hierarchical surface, which helps in achieving the superhydrophobicity. Genzer and Efimenko developed mechanically assembled monolayers from an elastomeric material, which shows a very high water contact angle and superhydrophobicity. The modified surface shows excellent non-wetting and non-permeability properties (Genzer and Efimenko 2000). A PDMS-based aerosol-assisted chemical vapor deposition (AACVD) coating was achieved on copper meshes for oil/water separation, and it showed a maximum water contact angle of 167° with an oil separation efficiency >99% (Figure 13.10). Superhydrophobic-coated meshes allow the permeation of oil through the pores as water molecules get repelled from the membrane surface (Crick, Gibbins, and Parkin 2013). Yilgor et al. reported that the coating of hydrophobic silica over various substrates shows a water contact angle of about 170°±1°. The hydrophobic coating on various polymeric surfaces such as polystyrene, epoxy, poly(methyl methacrylate), and polycarbonate was studied (Yilgor et al. 2012). A hydrophobic patterned surface is an approach created by

FIGURE 13.10 SEM images showing (a) plain copper mesh with a pore diameter of 170 μm and (b) copper mesh coated with Sylgard polymer by aerosol-assisted chemical vapor deposition. (Adopted from Crick et al. 2013, with permission from the Royal Society of Chemistry.)

a simple template method that leaves a 3D micropatterned surface over the polyethylene film. This hydrophobic pattern has an excellent stability in terms of abrasion resistance, which persists even after more than 5000 abrasion cycles (Xu, Mondal, and Lyons 2011; Asthana et al. 2014).

13.3.2.4 Self-Healing

The superhydrophobic coating has promising applications such as anti-corrosive and self-cleaning. When exposed to the outdoor environment, the low-surface-energy coating materials often fail under different environmental stresses such as sunlight, wind, scratching by animals, and sands. The exposed area is in contact with the atmospheric moisture or undergoes oxidation reaction leading to corrosion (Pulikkalparambil, Siengchin, and Parameswaranpillai 2018, Verho et al. 2011, Pulikkalparambil et al. 2018). The durable coating can be developed by improving the surface mechanical properties and wear resistance of the coating, which improves the slow degradation of the coating, keeping the hierarchal roughness and superhydrophobicity intact.

As mentioned in Introduction, the biological hydrophobic behavior of plants and insects is renewed or regenerated from time to time. This regeneration or repairing helps organisms to sustain their non-wettability over their whole lifetime. Mimicking this self-healing property to design coatings that have timely regeneration of the surface wetting characteristics is an important aspect. For the man-made materials, this regeneration or restoration can be achieved through the self-healing mechanism of polymer materials. Self-healing is defined as the autonomous healing of the internal

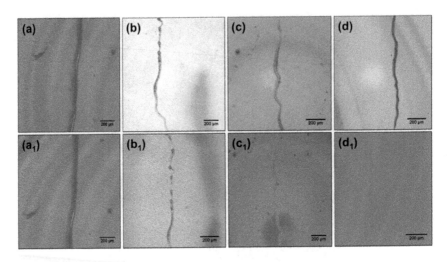

FIGURE 13.11 Optical microscope images of self-healing behavior of scratched films of furan-containing polyurethane (FPU) functionalized with different concentrations of maleimide-functionalized polyhedral oligomeric silsesquioxanes (POSS-M): (a) (FPU), (b) FPU-POSS-M6, (c) FPU-POSS-M12, and (d) FPU-POSS-M36. Their respective healed images are a_1, b_1, c_1, and d_1. (Adopted from Behera et al. 2018, with permission from American Chemical Society.)

cracks or damages generated in the polymer matrix and the restoration of the material properties of the damaged part via autonomous healing (Pulikkalparambil et al. 2020). The process is more or less analogous to the healing of human skin from any cuts (Figure 13.11).

In the literature, few works have been reported to study the regeneration of superhydrophobic behavior after the damage to the surface. Li et al. developed a fluoroalkylsilane-based superhydrophobic coating that showed a contact angle of about 157° (Li, Li, and Sun 2010). The coated surface undergoes different stresses, for example oxygen plasma etching, which degrade the surface coating and decrease the hydrophobicity. At this point, the self-healing process starts and restores the wetting property by transferring the silane-based coating from the bulk (Figure 13.12) of the coating to the damaged area. Similar to self-repairing, a lubricating surface was prepared by Wong et al. for ice repellency, pressure stability, and enhanced optical transparency (Wong et al. 2011). Durable self-healing, the composite fabric developed by Wang et al. showed an excellent restoration of the superhydrophobic property from various physical and chemical stresses such as abrasion and acid treatment (Wang et al. 2011).

13.3.2.5 Textiles

Superhydrophobic surfaces can be used for advanced applications such as smart textiles, in which the modified fabrics serve as an electromagnetic interference (EMI) shielding material, sensor, separation, and anti-bacterial material (Hu et al. 2012; Liu, Xin, and Choi 2012; Ghosh et al. 2018). Also, superhydrophilic materials help in easily washing out the stain and dirt present on the textile by allowing the complete spreading of the liquid in the fabric.

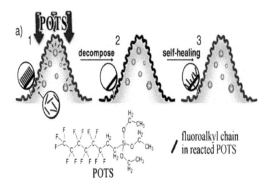

FIGURE 13.12 (a) Working principle of self-healing superhydrophobic coatings: (1) The porous polymer coating with micro- and nanoscaled hierarchical structures can preserve an abundance of healing agent units of reacted fluoroalkylsilane; (2) the top fluoroalkylsilane layer is decomposed and the coating loses its superhydrophobicity; (3) the preserved healing agents can migrate to the coating surface and heal the superhydrophobicity. (Adopted from Li et al. 2010, with permission from the WILEY-VCH Verlag GmbH & Co. KGaA, Weinheim.)

Hydrophobic textiles are developed from fabric materials having low surface energy, and because of their low tilt angle and tuned diameter, superhydrophobicity can be achieved. Another approach to designing a hydrophobic surface is the modification of low-cost textiles such as cotton with hydrophobic coating materials. For instance, the modification is achieved via sol–gel or different coating methods. Coating of conductive carbon black on to the cotton fabric increases the water contact angle from 42° to 102°, the fluorinated coating on nylon 6,6 polymer shows a water contact angle of 149°, a cotton fabric coated with silver paint shows a water contact angle of more than 140° (Ghosh et al. 2019; Satoh, Nakazumi, and Morita 2003; Ghosh et al. 2018). Figure 13.13a–c shows the developed superhydrophobic surface with a self-lubricating tendency experimented with different substances such as crude oil, blood, and ice on the coated surface (Wong et al. 2011). Figure 13.13d shows that the hierarchal surface of a coated cotton fabric decreases the surface energy and increases the surface roughness, leading to the formation of a superhydrophobic textile surface.

13.3.2.6 Separation Applications

Membrane desalination (MD) is a recent and emerging separation technology to cater the complete rejection of salt ions from the feed water. During this process, the warmer feed water stream flows on one side of the microporous membrane and a cooler distillate flows on the permeate side (Drioli, Ali, and Macedonio 2015). The temperature difference across the membrane creates a variation in the vapor pressure, which facilitates the transfer of water vapors across the membrane to the permeate side. Theoretically, the MD process has a desalination efficiency of about 100%. The major requirement for the MD is the material hydrophobicity. A hydrophobic membrane surface allows the permeation of water vapors while restricting the wetting characteristics of the surface. This increase in hydrophobicity ameliorates the diffusion of water molecules across the

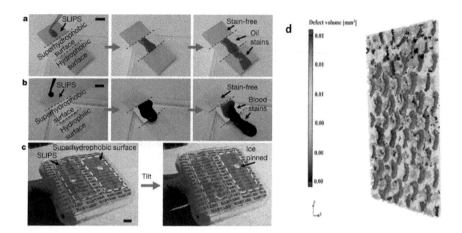

FIGURE 13.13 (a) Movement of light crude oil on a substrate composed of a slippery liquid-infused porous surface (SLIPS), a superhydrophobic Teflon porous membrane, and a flat hydrophobic surface. Note the slow movement on and staining of the latter two regions. (b) Comparison of the ability to repel blood by a SLIPS, a superhydrophobic Teflon porous membrane, and a flat hydrophilic glass surface. Note the slow movement on and staining of the latter two regions. (c) Ice mobility on a SLIPS (highlighted in green) compared to strong adhesion to an epoxy-resin-based nanostructured superhydrophobic surface. (d) μ-CT image of a superhydrophobic cotton fabric after brush coating treatment at thickness of 0.3 mm. (Figure 13.13 (a–c) adopted from Wong et al. 2011, with permission from Nature Publishing Group and Figure 13.13 (d) adopted from Ghosh et al. 2019, with permission from Wiley-VCH Verlag GmbH & Co. KGaA, Weinheim.)

membrane. Recently, Xu et al. have developed the MD using poly(ethylene chlorotri-fluoroethylene) (ECTFE) membrane that showed a water contact angle of more than 120° (Xu et al. 2019). Similarly, a poly(vinylidene fluoride) (PVDF)-based electrospun nanofiber membrane showed a contact angle of about 115° (Santoro et al. 2019), and a superhydrophobic PVDF-based composite membrane showed a contact angle of about 163°. These are the examples of the hydrophobic surfaces used in the MD process (Razmjou et al. 2012).

Other applications, such as separation of bacteria and organic foulants from the polluted water, are achieved by tuning the hydrophilic and hydrophobic characteristics of the membrane. A hydrophilic membrane is more resistant to bacterial growth and biofouling due to the presence of a hydration layer over the membrane surface. This hydration layer reduces the interaction of membrane surface with the hydrophobic bacteria and other biological proteins, which reduces the growth of biofouling on the membrane surface (Nguyen, Roddick, and Fan 2012). The hydrophilic character of the membrane increases the antifouling and anti-bacterial behaviors during the filtration. Hydrophilic polymers such as chitosan, cellulose acetate, polyethylene glycol, and polyvinyl alcohol and fillers such as ZnO, Fe_3O4, and GO are introduced to enhance the membrane hydrophilicity and thereby anti-bacterial and antifouling resistances (Mukherjee and De 2018; Remanan et al. 2018).

13.4 CONCLUSIONS AND OUTLOOK

Depending upon the surface energy and surface roughness of the materials, surfaces show different wetting behaviors to water. Water contact angle measurement is a direct and straight forward approach to understanding the wetting characteristics. An increase in surface roughness and surface energy gives a low water contact angle, which indicates the hydrophilicity of the material, while an increase in surface roughness and a decrease in surface energy increase the water contact angle and make the surface hydrophobic. Variations in these two parameters can tune the material's wetting behavior to either superhydrophilic or superhydrophobic nature. Self-healing and shape-memory polymer materials are the other emerging research areas where superhydrophobic or superhydrophilic coatings are applicable. These self-healing polymeric coatings can regenerate the surface wetting characteristics when used in the extreme stress environments, while shape-memory polymer coatings can stimulate the activities of a shape-memory solar panels used in the satellite. Superhydrophobic smart textiles for electronic and anti-staining fabric applications and superhydrophobic membrane materials for membrane distillation (MD) are the other promising research areas. Hence, surface wetting property is an essential aspect in the material science that can be used for various industrial applications.

REFERENCES

Adachi, T., S. S. Latthe, S. W. Gosavi, et al. 2018. Photocatalytic, superhydrophilic, self-cleaning TiO_2 coating on cheap, light-weight, flexible polycarbonate substrates. *Applied Surface Science* 458: 917–923.

Asthana, A., T. Maitra, R. Büchel, M. K. Tiwari, and D. Poulikakos. 2014. Multifunctional superhydrophobic polymer/carbon nanocomposites: Graphene, carbon nanotubes, or carbon black? *ACS Applied Materials & Interfaces* 6 (11): 8859–8867.

Barati Darband, G., M. Aliofkhazraei, S. Khorsand, S. Sokhanvar, and A. Kaboli. 2018. Science and engineering of superhydrophobic surfaces: Review of corrosion resistance, chemical and mechanical stability. *Arabian Journal of Chemistry* 13: 1763–1802.

Behera, P. K., P. Mondal, and N. K. Singha. 2018. Self-healable and ultrahydrophobic polyurethane-POSS hybrids by Diels–Alder "Click" reaction: A new class of coating material. *Macromolecules* 51 (13): 4770–4781.

Bhandari, S., N. K. Singha, and D. Khastgir. 2017. Synthesis of graphene-like ultrathin polyaniline and its post-polymerization coating on nanosilica leading towards superhydrophobicity of composites. *Chemical Engineering Journal* 313: 1302–1310.

Bhushan, B., Y. C. Jung, and K Koch. 2009. Micro-, nano- and hierarchical structures for superhydrophobicity, self-cleaning and low adhesion. *Philosophical Transactions of the Royal Society A: Mathematical, Physical and Engineering Sciences* 367 (1894): 1631–1672.

Blossey, R. 2003. Self-cleaning surfaces — virtual realities. *Nature Materials* 2 (5): 301–306.

Bonn, D., J. Eggers, J. Indekeu, J. Meunier, and E. Rolley. 2009. Wetting and spreading. *Reviews of Modern Physics* 81 (2): 739–805.

Chang, K.-C., H.- I. Lu, C.-W. Peng, et al. 2013. Nanocasting technique to prepare lotus-leaf-like superhydrophobic electroactive polyimide as advanced anticorrosive coatings. *ACS Applied Materials & Interfaces* 5 (4): 1460–1467.

Crick, C. R., J. A. Gibbins, and I. P. Parkin. 2013. Superhydrophobic polymer-coated copper-mesh; membranes for highly efficient oil–water separation. *Journal of Materials Chemistry A* 1 (19): 5943–5948.

Darmanin, T., E. T. d. Givenchy, S. Amigoni, and F. Guittard. 2010. Hydrocarbon versus fluorocarbon in the electrodeposition of superhydrophobic polymer films. *Langmuir* 26 (22): 17596–17602.

de Leon, A. C. C., R. B. Pernites, and R. C. Advincula. 2012. Superhydrophobic colloidally textured polythiophene film as superior anticorrosion coating. *ACS Applied Materials & Interfaces* 4 (6): 3169–3176.

Drioli, E., A. Ali, and F. Macedonio. 2015. Membrane distillation: Recent developments and perspectives. *Desalination* 356: 56–84.

Eral, H. B., and J. M. Oh. 2013. Contact angle hysteresis: A review of fundamentals and applications. *Colloid and Polymer Science* 291 (2): 247–260.

Genzer, J., and K. Efimenko. 2000. Creating long-lived superhydrophobic polymer surfaces through mechanically assembled monolayers. *Science* 290 (5499): 2130–2133.

Genzer, J., and A. Marmur. 2011. Biological and synthetic self-cleaning surfaces. *MRS Bulletin* 33 (8): 742–746.

Gesser, H. D., and K. Paul. 2000. A demonstration of surface tension and contact angle. *Journal of Chemical Education* 77 (1): 58.

Ghosh, S., P. Das, S. Ganguly, et al. 2019. 3D-enhanced, high-performing, super-hydrophobic and electromagnetic-interference shielding fabrics based on silver paint and their use in antibacterial applications. *ChemistrySelect* 4 (40): 11748–11754.

Ghosh, S., S. Remanan, S. Mondal, et al. 2018. An approach to prepare mechanically robust full IPN strengthened conductive cotton fabric for high strain tolerant electromagnetic interference shielding. *Chemical Engineering Journal* 344: 138–154.

Guo, C., L. Feng, J. Zhai, et al. 2004. Large-area fabrication of a nanostructure-induced hydrophobic surface from a hydrophilic polymer. *ChemPhysChem* 5 (5): 750–753.

Guo, Z., and W. Liu. 2007. Biomimic from the superhydrophobic plant leaves in nature: Binary structure and unitary structure. *Plant Science* 172 (6): 1103–1112.

Gutoff, E. B., and E. D. Cohen. 2010. Chapter 13- Water- and solvent-based coating technology. In *Multilayer Flexible Packaging*, edited by J. R. Wagner. Boston, MA: William Andrew Publishing.

He, B., J. Lee, and N. A. Patankar. 2004. Contact angle hysteresis on rough hydrophobic surfaces. *Colloids and Surfaces A: Physicochemical and Engineering Aspects* 248 (1): 101–104.

Hu, J., H. Meng, G. Li, and S. I. Ibekwe. 2012. A review of stimuli-responsive polymers for smart textile applications. *Smart Materials and Structures* 21 (5): 053001.

Koch, B. M. L., A. Amirfazli, and J. A. W. Elliott. 2014. Wetting of rough surfaces by a low surface tension liquid. *The Journal of Physical Chemistry C* 118 (41): 23777–23782.

Liu, Y. J., H. Xin, and C.-H. Choi. 2012. Cotton fabrics with single-faced superhydrophobicity. *Langmuir* 28 (50): 17426–17434.

Law, K.-Y., and H. Zhao. 2016. *Surface Wetting: Characterization, Contact Angle, and Fundamentals*. Switzerland: Springer.

Levkin, P. A., F. Svec, and J.M. J. Fréchet. 2009. Porous polymer coatings: A versatile approach to superhydrophobic surfaces. *Advanced Functional Materials* 19 (12): 1993–1998.

Li, Y., L. Li, and J. Sun. 2010. Bioinspired self-healing superhydrophobic coatings. *Angewandte Chemie International Edition* 49 (35): 6129–6133.

Makhlouf, A. S. H. 2011. 1- Current and advanced coating technologies for industrial applications. In *Nanocoatings and Ultra-Thin Films*, edited by A. S. H. Makhlouf, and I. Tiginyanu: UK: Woodhead Publishing.

Marmur, A. 2004. The lotus effect: Superhydrophobicity and metastability. *Langmuir* 20 (9): 3517–3519.

Mukherjee, M., and S. De. 2018. Antibacterial polymeric membranes: A short review. *Environmental Science: Water Research & Technology* 4 (8): 1078–1104.

Neinhuis, C., and W. Barthlott. 1997. Characterization and distribution of water-repellent, self-cleaning plant surfaces. *Annals of Botany* 79 (6): 667–677.

Nguyen, N.-T. 2012. Chapter 4- Fabrication technologies. In *Micromixers (Second Edition)*, edited by N.-T. Nguyen. Oxford: William Andrew Publishing.

Nguyen, T., F. A. Roddick, and L. Fan. 2012. Biofouling of water treatment membranes: A review of the underlying causes, monitoring techniques and control measures. Membranes 2 (4): 804–840.

Nishimoto, S., and B. Bhushan. 2013. Bioinspired self-cleaning surfaces with super-hydrophobicity, superoleophobicity, and superhydrophilicity. *RSC Advances* 3 (3): 671–690.

Otitoju, T. A., A. L. Ahmad, and B. S. Ooi. 2017. Superhydrophilic (superwetting) surfaces: A review on fabrication and application. *Journal of Industrial and Engineering Chemistry* 47: 19–40.

Parameswaranpillai, J., S. K. Sidhardhan, P. Harikrishnan, et al. 2017. Morphology, thermo-mechanical properties and surface hydrophobicity of nanostructured epoxy thermosets modified with PEO-PPO-PEO triblock copolymer. *Polymer Testing* 59: 168–176.

Peng, C.-W., K.-C. Chang, C.-J. Weng, et al. 2013. UV-curable nanocasting technique to pre-pare bio-mimetic super-hydrophobic non-fluorinated polymeric surfaces for advanced anticorrosive coatings. *Polymer Chemistry* 4 (4): 926–932.

Pulikkalparambil, H., M. R. Sanjay, S. Siengchin, A. Khan, M. Jawaid, and C. I. Pruncu. 2020. Chapter 17- Self-repairing hollow-fiber polymer composites. In *Self-Healing Composite Materials*, edited by A. Khan, M. Jawaid, S. N. Raveendran, and A. M. Ahmed Asiri. UK: Woodhead Publishing

Pulikkalparambil, H., S. Siengchin, and J. Parameswaranpillai. 2018. Corrosion protective self-healing epoxy resin coatings based on inhibitor and polymeric healing agents encapsulated in organic and inorganic micro and nanocontainers. *Nano-Structures & Nano-Objects* 16: 381–395.

Pulikkalparambil, H., S. A. Varghese, S. Siengchin, and J. Parameswaranpillai. 2018. Thermally mendable and improved hydrophilic bioepoxy/PEG-PPG-PEG blends for coating applica-tion. *Materials Research Express* 6 (2): 025307.

Razmjou, A., E. Arifin, G. Dong, J. Mansouri, and V. Chen. 2012. Superhydrophobic modifi-cation of TiO2 nanocomposite PVDF membranes for applications in membrane distil-lation. *Journal of Membrane Science* 415–416: 850–863.

Remanan, S., M. Sharma, S. Bose, and N. C. Das. 2018. Recent advances in preparation of porous polymeric membranes by unique techniques and mitigation of fouling through surface modification. *ChemistrySelect* 3 (2): 609–633.

Santoro, S., I. Vidorreta, I. Coelhoso, et al. 2019. Experimental evaluation of the thermal polarization in direct contact membrane distillation using electrospun nanofiber mem-branes doped with molecular probes. *Molecules* 24 (3): 638.

Satoh, K., H. Nakazumi, and M. Morita. 2003. Preparation of super-water-repellent fluori-nated inorganic-organic coating films on nylon 66 by the sol-gel method using micro-phase separation. *Journal of Sol-Gel Science and Technology* 27 (3): 327–332.

Shim, E. 2013. Chapter 10- Bonding requirements in coating and laminating of textiles. In *Joining Textiles*, edited by I. Jones, and G. K. Stylios. UK: Woodhead Publishing.

Sigmund, W. M., and S. H. Hsu. 2016. Cassie–Baxter model. *J Encyclopedia of Membranes* 310–311.

Smith, R., H. Inomata, and C. Peters. 2013. Chapter 4- historical background and applica-tions. In *Introduction to Supercritical Fluids*, edited by R. Smith, H. Inomata, and C. Peters. The Netherlands: Elsevier.

Sun, M., C. Luo, L. Xu, et al. 2005. Artificial lotus leaf by nanocasting. *Langmuir* 21 (19): 8978–8981.

Syafiq, A., B. Vengadaesvaran, A. K. Pandey, and N. A. Rahim. 2018. Superhydrophilic smart coating for self-cleaning application on glass substrate. *Journal of Nanomaterials* 2018: 10.

ten Elshof, J. E. 2015. Chapter 4- Chemical solution deposition techniques for epitaxial growth of complex oxides. In *Epitaxial Growth of Complex Metal Oxides*, edited by G. Koster, M. Huijben, and G. Rijnders. UK: Woodhead Publishing.

Verho, T., C. Bower, P. Andrew, S. Franssila, O. Ikkala, and R. H. A. Ras. 2011. Mechanically durable superhydrophobic surfaces. *Advanced Materials* 23 (5): 673–678.

Wang, B., Y. Zhang, L. Shi, J. Li, and Z. Guo. 2012. Advances in the theory of superhydrophobic surfaces. *Journal of Materials Chemistry* 22 (38): 20112–20127.

Wang, H., Y. Xue, J. Ding, L. Feng, X. Wang, and T. Lin. 2011. Durable, self-healing super-hydrophobic and superoleophobic surfaces from fluorinated-decyl polyhedral oligomeric silsesquioxane and hydrolyzed fluorinated alkyl silane. *Angewandte Chemie International Edition* 50 (48): 11433–11436.

Weng, C.-J., C.-H. Chang, C.-W. Peng, et al. 2011. Advanced anticorrosive coatings prepared from the mimicked Xanthosoma sagittifolium-leaf-like electroactive epoxy with synergistic effects of superhydrophobicity and redox catalytic capability. *Chemistry of Materials* 23 (8): 2075–2083.

Wong, T.-S., S. H. Kang, S. K. Y. Tang, et al. 2011. Bioinspired self-repairing slippery surfaces with pressure-stable omniphobicity. *Nature* 477 (7365): 443–447.

Wypych, G. 2016. Chapter 18-Fillers in different processing methods. In *Handbook of Fillers* (Fourth Edition), edited by G. Wypych. Canada: ChemTec Publishing.

Xu, K., Y. Cai, N. T. Hassankiadeh, et al. 2019. ECTFE membrane fabrication via TIPS method using ATBC diluent for vacuum membrane distillation. *Desalination* 456: 13–22.

Xu, Q. F., B. Mondal, and A. M. Lyons. 2011. Fabricating superhydrophobic polymer surfaces with excellent abrasion resistance by a simple lamination templating method. *ACS Applied Materials & Interfaces* 3 (9): 3508–3514.

Yang, C., U. Tartaglino, and B. N. J. Persson. 2006. Influence of surface roughness on super-hydrophobicity. *Physical Review Letters* 97 (11): 116103.

Yilbas, B. S., A. Al-Sharafi, and H. Ali. 2019. Chapter 3- surfaces for self-cleaning. In *Self-Cleaning of Surfaces and Water Droplet Mobility*, edited by B. S. Yilbas, A. Al-Sharafi, and H. Ali. The Netherlands: Elsevier.

Yilgor, I., S. Bilgin, M. Isik, and E. Yilgor. 2012. Facile preparation of superhydrophobic polymer surfaces. *Polymer* 53 (6): 1180–1188.

Zhang, X., F. Shi, J. Niu, Y. Jiang, and Z. Wang. 2008. Superhydrophobic surfaces: From structural control to functional application. *Journal of Materials Chemistry* 18 (6): 621–633.

Zhang, D., L. Wang, H. Qian, and X. Li. 2016. Superhydrophobic surfaces for corrosion protection: A review of recent progresses and future directions. *Journal of Coatings Technology and Research* 13 (1): 11–29.

14 Antifouling, Antibacterial, and Bioactive Polymer Coatings

Anna Sienkiewicz and Piotr Czub
Cracow University of Technology

CONTENTS

14.1 INTRODUCTION

Bacterial contamination relates to various surfaces including both medical (e.g., medical devices, implants, wound dressings) and industrial applications (e.g., food packages, industrial pipes, separation membranes, and marine). It is a globally serious problem, leading to health issues or threat to the limited efficiency and lifetime of appliances. In general, bacteria adhere on these surfaces and are able to survive by the formation of so-called biofilms. The development of these sessile-structured communities is shown in Figure 14.1. The environment of biofilms provides ideal conditions for the bacteria living, allowing for safe metabolism and increased resistance to antibiotics and host immunological defense. Because of the complex exopolysaccharides matrix protecting the attached microorganisms from the antimicrobial agents, even 1000-fold compared to bacteria growing in suspension (Campoccia, Montanaro and Arciola, 2013), the created biofilm is difficult to destroy (Davies, 2003). Mainly due to this reason, all actions leading to inhibition of microorganisms' proliferation process and the creation of the pathogens' biofilm are extremely important (Glinel et al., 2012). In the process of bacterial adhesion, two main stages can be distinguished: (1) interaction between the bacterial cell surface and the material surface (rapid and easily reversible) and (2) interplay between the so-called adhesion proteins on bacterial cell wall (fimbriae or bacterial pilli) and binding molecules of the surface material (relatively slow, reversible, and often termed irreversible process) (Lichter, Van

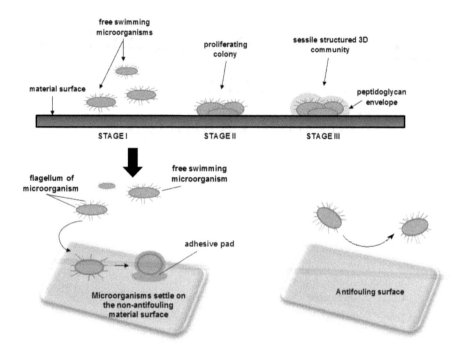

FIGURE 14.1 Illustration of bacterial adhesion during the process of biofilm formation. (Sufficiently adapted, modified, and redrawn based on Davies, 2003 and Lichter, Van Vliet and Rubner, 2009.)

Vliet and Rubner, 2009). It is worth nothing that not one specific feature, but a variety of different factors have their influence on the process of bacterial adhesion. They include both specific and nonspecific characteristics of pathogenic microorganisms, the surface properties, and environmental conditions including temperature, concentrations of glucose and oxygen, and sustained fluid shear flows.

14.2   ANTIBACTERIAL POLYMER COATINGS

Numerous medical interventions require the introduction of a medical device into the body. This procedure, on the one hand, is beneficial for the health improvement; on the other hand, it increases the risk of numerous complications, which include infection, inflammation, and initiation of a wound-healing response. Therefore, the research on antibacterial polymer coatings is focused on designing layers, which will be characterized by one of the following features: antibacterial agent release, contact killing, and antiadhesion/bacteria-repelling (Cloutier, Mantovani and Rosei, 2015).

14.2.1   ANTIBACTERIAL AGENT RELEASE COATINGS

The main role of antibacterial agent release coatings is discharging (*via* diffusion into the aqueous medium, followed by erosion/degradation or hydrolysis of covalent bonds) the antibacterial compounds over time. This way, a high antibacterial dose is delivered just in the specific areas where it is needed the most. Antibiotics or antiproliferative drugs are released from the surface in therapeutic concentrations to the certain areas, at the same time avoiding the toxic effects of drugs on the whole body system, as well as tailoring selection of antibiotics toward specific pathogens associated with exact implant infections. Antibiotic release is achieved using a wide variety of coatings, both non- and biocompatible polymer coatings. However, it is worth noting that due to the difficult diffusion through the pores of the polymer, releasing the antibiotics from nonbiocompatible materials is limited to just a certain portion of loaded amount, and additionally, after the procedure is completed, nonbiodegradable polymer matrices have to be removed. On the other hand, biodegradable polymers allow for the delivery of higher doses of antibiotics, and in most cases, their degradation products are common metabolites. Among the variety of biodegradable coatings obtained for the controlled delivery of antibiotics, the most often mentioned in the literature are poly(propylenefumarate/methylmethacrylate), collagen, polyanhydrides, polyorthoesters, and polylactide-co-glycolide.

An interesting example of polymer coatings, characterized by good durability and flexibility in terms of antibacterial agent release, is a mixture of poly(butylmethacrylate) and poly(ethylene-co-vinyl acetate) copolymers (Chudzik et al., 2005). These coatings were designed to release the antimicrobial (e.g., antibiotics such as vancomycin or norfloxacin) and/or antithrombotic agents (e.g., heparin, hirudin and coumadin) and are recommended both for self-expanding/balloon-expandable stents and urinary catheters. The authors of this invention claim the total combined concentrations of both polymers in the coating composition between 0.25% and about 70% by weight, with the bioactive agent dissolved or suspended in the coating mixture at a concentration of 0.01%–90% by weight. The medical device is covered by the polymer coating by dipping or spraying, and allowed to cure by solvent evaporation.

Another noteworthy study on the agent release polymer coatings presents biomimetic trilayer polymeric coatings that combine controlled NO release with surface-immobilized active heparin (Zhou and Meyerhoff, 2005). The coating, containing a dense polymer bottom layer, a middle layer (polymer matrix) doped with a lipophilic diazeniumdiolate-type NO [diazeniumdiolated dibutylhexanediamine (DBHD/N_2O_2)], and an aminated polymer (PVC or PU) top-coating allowing for the direct attachment of heparin, is synthesized to mimic the nonthrombogenic properties of the endothelial cell (EC) layer that lines the inner wall of healthy blood vessels. This invention combines the antiplatelet adhesion and activation, which are characteristics of NO release polymers, with the reduction of thrombosis by the immobilization of heparin. Figure 14.2 shows the synthesis of aminated polyurethane and covalently binding of heparine and aminated polymer through 1-ethyl-3-(3-dimethylaminopropyl)carbodiimidehydrochloride (EDC) and (N-hydroxysuccinimide) NHS as coupling agents.

14.2.2 CONTACT-KILLING POLYMER COATINGS

The contact-killing task of polymer coatings is performed through disrupting microorganisms' cell membranes by antimicrobial compounds, which are covalently anchored to the material surface by flexible, hydrophobic polymeric chains. This way, the antibacterial function is performed constantly without the risk of running out of

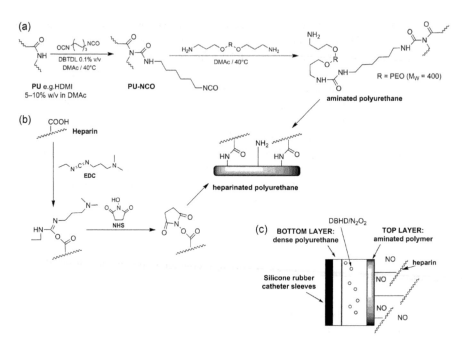

FIGURE 14.2 Schematic representation of (a) the synthesis of aminated polyurethane, (b) covalent binding of aminated polymer and heparin through EDC and NHS as coupling agents, and (c) dual acting biomimetic coating with combined NO release and surface-bound heparin (top layer: aminated polymer with surface-bound heparin; middle layer: a polymer matrix with NO donor DBHD/N$_2$O$_2$, and bottom layer: dense polymer). (Sufficiently adapted, modified, and redrawn based on Zhou and Meyerhoff, 2005.)

released antimicrobial compounds. The killing of bacteria is performed by either release-killing of antibacterial moiety from a matrix (such as antibiotics, phenols, and heavy metals using various methods such as spray or dip coating and hydrogel trapping) (Chung, Papadakis and Yam, 2003; Li et al., 2006) or contact killing of antibacterial surfaces (Wu et al., 2016). The most effective compounds for contact-killing coatings are either cationic compounds (quaternary ammonium compounds, chitosan, antimicrobial peptides, phosphonium salts, titanium oxide particles, etc.) or enzymes (Popa et al., 2003; Green, Fulghum and Nordhaus, 2011; Muñoz-Bonilla and Fernández-García, 2012).

Figure 14.3 shows 3-(trimethoxysilyl)-propyldimethyloctadecyl ammonium chloride (QAS), quaternized poly(4-vinyl-N-alkylpyridinium bromide) (PVP) and quaternized poly(2-(dimethylamino)ethyl methacrylate (PDMAEMA), examples of intensively studied synthetic biocidal quaternary ammonium compounds (QACs), which are used as contact-based bactericidal surfaces.

Although the mechanism of action of QACs is not fully discovered, they exhibit strong contact-killing activity toward both Gram-positive and Gram-negative bacteria through (1) destabilization of the intracellular matrix of a bacterium by the ion-exchange mechanism with Ca^{2+} and Mg^{2+} ions in the cytoplasmic membrane and (2) destructive influence on the cytoplasmic membrane causing the leakage of intracellular fluid

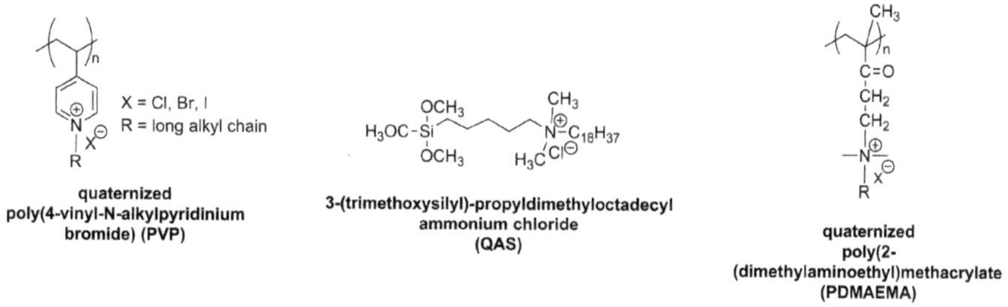

FIGURE 14.3 Chemical structures biocidal QACs used as contact-based bactericidal surfaces. (Sufficiently adapted, modified, and redrawn based on Yu, Wu, and Chen, 2015.)

(Yu and Wu, 2015). The antimicrobial effect of QACs is related to strong affinity and damaging interactions between the positively charged quaternary nitrogen of QACs and the negatively charged head groups of acidic phospholipids in microorganisms' membranes (Elena and Miri, 2018).

On the other side, it is important to highlight that the technique using leachable biocides is characterized by two main disadvantages: termed application and generating the increasing drug resistance throughout the microbial species. So, in order to achieve biocidal effect without releasing biocide into the environment, antimicrobial moieties are irreversibly (covalently) affixed to the surface of material. Polymeric antimicrobial compounds such as tertiary amine 2-(dimethylamino)ethyl methacrylate (DMAEMA) can be attached to the surfaces of common materials including glass and paper using atom transfer radical polymerization and applied as antibacterial treatment of food packaging, everyday household items, and military applications (Lee et al., 2004). Figure 14.4 presents the synthetic pathway for the atom transfer radical polymerization (ATRP) and subsequent quaternization of DMAEMA on solid surfaces. The process involves the reaction of 2-bromoisobutyryl bromide with the hydroxyl groups of the cellulose in filter paper and the free amine groups on amino glass slides *via*, respectively, esterification and amidation. Throughout this reaction, the active initiator of ATRP is obtained and subsequently used to polymerize DMAEMA to the initiated surfaces in the presence of Cu(I)Br and the ligand 2,2′-bipyridine as catalysts. Additionally, as shown in Figure 14.4b, propionyl bromide can be mixed with stoichiometrically varying amounts of 2-bromoisobutyryl to differ the density of active ATRP initiation sites on the paper.

Another important issue toward preventing microbial contamination is inhibiting the microbial colonization on surfaces. Recently, it is mostly performed by application of water-soluble antimicrobial compounds, leading unfortunately to rapid development of resistant strains and environmental problems in the short time. Therefore, the invention of the coating materials based on polyethylenimine (PEI) that can kill pathogenic microorganism and stay bound to surfaces seems very promising (Hoque et al., 2015). Figure 14.5 presents the synthesis of colorless branched *N*-alkyl-*N*-methyl PEIs by *E. Clarke* methylation of branched PEIs and subsequent quatemization with alkyl bromide.

Obtained coatings display excellent compatibility with both medically relevant polymer, such polylactic acid, and commercial paints. Most of all they are effective

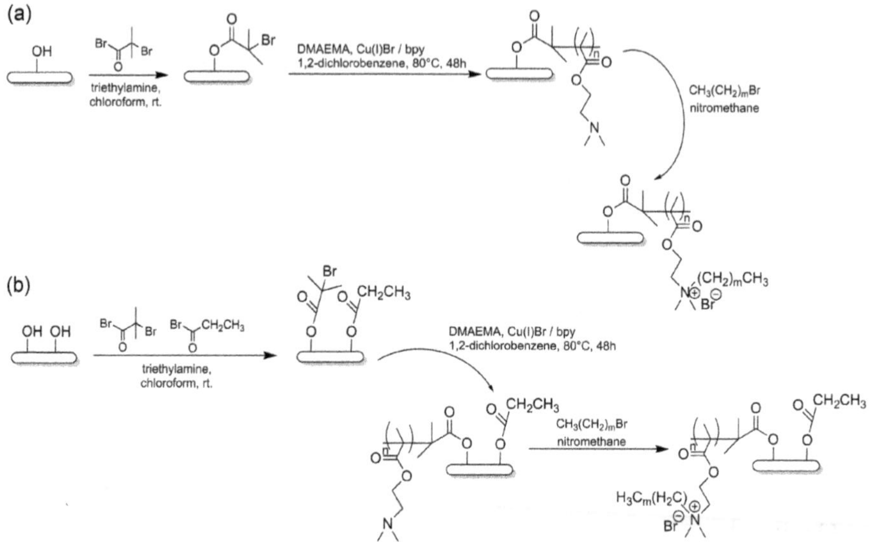

(a)

(b)

FIGURE 14.4 The atom transfer radical polymerization and subsequent quaternization of DMAEMA on solid surfaces. (Sufficiently adapted, modified, and redrawn based on Lee et al., 2004.)

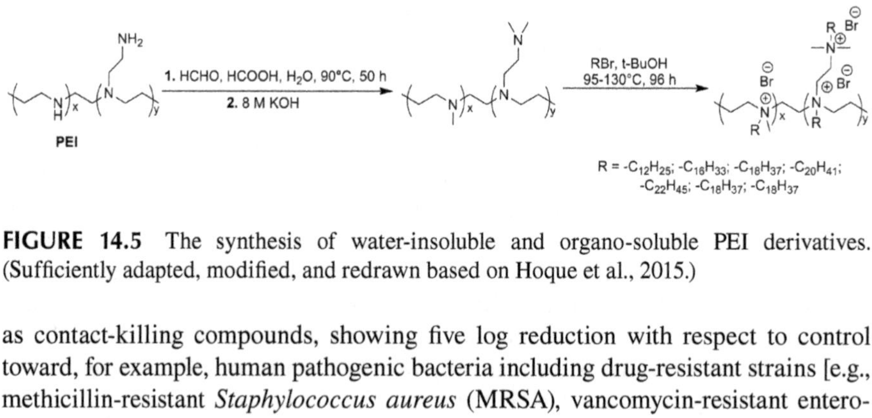

$R = -C_{12}H_{25}; -C_{16}H_{33}; -C_{18}H_{37}; -C_{20}H_{41};$
$-C_{22}H_{45}; -C_{18}H_{37}; -C_{18}H_{37}$

FIGURE 14.5 The synthesis of water-insoluble and organo-soluble PEI derivatives. (Sufficiently adapted, modified, and redrawn based on Hoque et al., 2015.)

as contact-killing compounds, showing five log reduction with respect to control toward, for example, human pathogenic bacteria including drug-resistant strains [e.g., methicillin-resistant *Staphylococcus aureus* (MRSA), vancomycin-resistant entero-cocci (VRE), and beta lactam-resistant *Klebsiella pneumoniae*] and pathogenic fungi (*Candida* spp. and *Cryptococcus* spp.). Additionally, it was found that linear polymers were more active and exhibit a higher killing rate than the branched one.

14.2.3 ANTIADHESION/BACTERIA-REPELLING POLYMER COATINGS

Antiadhesion/bacteria-repelling polymer coatings prevent formation of microbial bio-film *via* the noncytotoxic mechanisms. It might be realized by introducing, for example, highly negatively charged polymers (electrostatic repulsion), similar hydrogel forming polymers (steric repulsion) mostly based on poly(ethylene glycol), or special polymers with low surface energy (ultrahydrophobic repulsion) (Chen et al., 2016).

Based on performed studies (Venkateswaran et al., 2014), polymers with anti-adhesive characteristics against a range of clinically important Gram-negative and Gram-positive bacteria were selected from the library of 381 polyacrylates/acrylamides and polyurethanes. The tests performed against both individual microbe or microbial mixtures revealed that PA13, copolymer of methylmethacrylate and dimethylacrylamide (9:1 monomer ratio); PA515, copolymer of methoxyethylmethacrylate, diethylaminoethylacrylate, and methylmethacrylate (6:1:3 monomer ratio); and PU83, polyurethane synthesized from poly(ethyleneglycol)$_{900}$ and 4,4′-methylen ebis(cyclohexylisocyanate), with 1,4-butanediol as a chain extender (1:2:1 monomer ratio) exhibit the best antiadhesion/bacteria-repelling properties. In Figure 14.6, the chemical structures of copolymers PA13 and PA515 are presented. Additionally, it was found that coatings of a polyurethane-based and a silicone-based intravenous catheter with PA13 significantly reduce bacterial binding, which makes it potential antibiotic-free bacteria-repellent coatings for medical devices.

A recently presented, very interesting approach (Lin et al., 2018) is based on obtaining durably antibacterial and bacterially antiadhesive cotton fabrics coated with antibacterial cationic fluorine-containing polymers. The fabrics are prepared by spray coating of antibacterial quaternary ammonium monomers with different alkyl chain lengths and fluorine-containing monomers. Figure 14.7 shows the synthesis route for the antibacterial and bacterially antiadhesive cotton fabric.

Presented antibacterial monomers, polymers, and fabrics exhibit good antibacterial activities against both *S. aureus* and *Escherichia coli*, and slightly better for *S. aureus*. Moreover, alkyl chain length and contents of the antibacterial monomers, as well as the add-on percentage of polymer, has greater influence on the antibacterial properties of the fabrics. Furthermore, it was found that incorporation of fluorine component into the polymer enhances the antibacterial activity and bacterial antiadhesion of the treated fabrics due to the low surface energy-induced hydrophobicity.

Another very important group of coating materials are obtained in order to prevent microbial infection causing the implant failure. During the surgery, implants are susceptible to bacterial contamination from both skin and mucous membranes. The most critical pathogenic moment leading to development of infection on biomaterials is the formation of biofilm, starting with bacterial adhesion. There are two stages

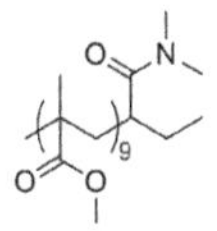

PA13
methylmethacrylate and dimethylacrylamide
co-polymer
(9:1 monomer ratio)

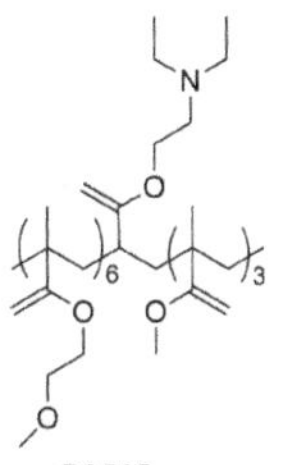

PA515
methoxyethylmethacrylate, diethylaminoethylacrylate and methylmethacrylate
co-polymer
(6:1:3 monomer ratio)

FIGURE 14.6 Examples of copolymers with antiadhesive characteristics against Gram-negative and Gram-positive bacteria. (Sufficiently adapted, modified, and redrawn based on Venkateswaran et al., 2014.)

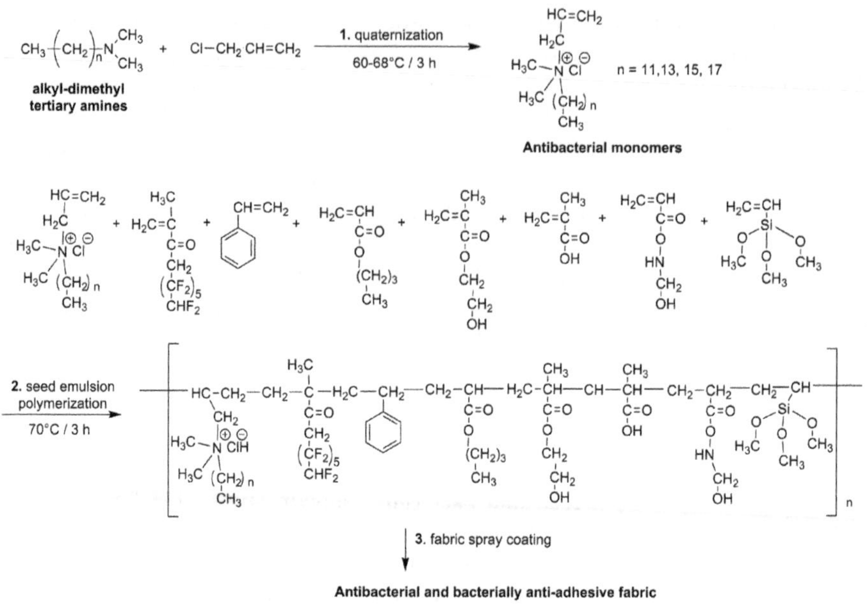

FIGURE 14.7 Synthesis route for the antibacterial and bacterially antiadhesive cotton fabric. (Sufficiently adapted, modified, and redrawn based on Lin et al., 2018.)

of bacterial adhesion: (1) the initial interaction between bacterial cell surfaces and material surfaces and (2) interactions between proteins on the bacterial surface structures and binding molecules on the material surface (Chouirfa et al., 2019). Among the tested solutions, applied mainly for titanium implants, coating with a thermoresponsive polymer *N*-isopropylacylamide (PIPAAM) is very promising. The antiadhesive titanium surface is obtained by coating with polyglycidyl methacrylate using the initiated chemical vapor deposition technique, and in the next stage, grafting with an amine group terminated PIPAAM by the ring-opening reaction (Lee et al., 2015). The in vitro tests performed on the samples of titanium coated with PIPAAM indicate that bacteria causing peri-implantitis and nosocomial infections are effectively detached by lowering the temperature.

14.3 BIOACTIVE POLYMER COATINGS

The main simultaneous tasks of bioactive coatings are (1) prevention of bacterial attachment and subsequent biofilm formation, causing device-related infections and failure of devices, and (2) promoting host cell adhesion and tissue healing (Bazaka et al., 2010). Bioactive surfaces with the optimal properties must be characterized by appropriately high and well-controlled binding capacities for biomolecules as well as properties preventing denaturation of the immobilized biomolecules (Yu et al., 2011). Bioactive surfaces might release in controlled manner antibiotics such as vancomycin, amoxicillin and gentamicin (Stigter et al., 2004) or contain alternative agents such as silver ions, nitric oxide, bioactive antibodies and naturally occurring biocidal compounds. One of the large areas where bioactive coatings are intensively studied is joint arthroplasty. These coatings are designed to supplement the function of current implants, mainly metals such as cobalt chrome alloy, stainless steel, or titanium alloy. Coatings used for such applications must be biocompatible, not generating significant immune or foreign-body response; "osteoconductive" in its promotion of osteoblasts to adhere to, proliferate, and grow on the surface of the implant to form bone-implant bonding, and "osteoinductive," being able to employ various stem cells to induce differentiation into osteogenic cells. These materials additionally must be characterized by sufficient mechanical stability and antimicrobial properties minimizing the risk of prosthetic infection (Zhang et al., 2014).

A relatively large group of materials recently intensively studied due to their superior osteoconductivity are nanoscale bioactive glasses. In general, bioactive glasses are reactive materials that are able to bond to bone in physiological environment and consist of a silicate network incorporating sodium, calcium, and phosphorus in different relative proportions (Boccaccini et al., 2010). Bioactive silicate glasses, in comparison to other, for example, sintered hydroxyapatite, display better bioactive properties, because they dissolute into the products, which rise the expression of genes controlling osteogenesis, potentially exhibit angiogenic properties and might locally deliver selected ions monitoring specific cell functions or can be considered as hemostatic agents (Xynos et al., 2000; Gorustovich, Roether and Boccaccini, 2009). Composites formed by combination of nanoscale bioactive glasses with polymer matrices exhibit enhanced performance in biomedical applications, by, for example, a faster release of ions and a higher protein adsorption and thus enhanced bioactivity. An example of such multilayer bioactive polymer coating is the material obtained by incorporation of bioactive glass nanoparticles into chitosan. Within the obtained material, polymer matrix from chitosan provides viscoelastic properties, while the nano-glass particles are responsible for bioactivity of the hybrid organic–inorganic structure. In vitro studies show that the multilayers induced the formation of apatite (Couto, Alves and Mano, 2009).

Another, very interesting and intensively developing area of bioactive polymer coatings are technologies related to so-called nanocoating technique, which are applied for textile finishing. The method is based on deposition of very thin (20–40 µm) polymeric coats, containing appropriate functional nanoparticles on the surfaces of textiles, which might be used as, e.g., protective clothing, sports clothes, or special products for medical applications. The functional particles consist of submicro-spheres of SiO_2, TiO_2, and SiO_2-TiO_2 hybrid system with "nanoislets" of metallic silver as an antibacterial agent might be incorporated into either hydrophobic (with microporous structure) or hydrophilic (with compact structure) polyurethane coats during their production (Brzeziński et al., 2007). The obtained bacteriostatic effect is resistant to repeated washings, while there is no release of the submicro-powders from the structure of the coating material and no migration to the surface of olyurethane membrane.

The other studied area of polymer coating with bioactive properties covers edible coatings, which recently gained more attention due to the positive results in food preservation. Currently applied edible coatings include polysaccharides (e.g., cellulose derivatives, starch, chitin, and gums), proteins (soy, milk, gelatin, corn zein, or wheat gluten), and lipids (oils, waxes, and resins). Edible coatings act as moisture and gas barriers; they might also preserve the color and texture of the product. Moreover, edible coatings can be utilized as encapsulating matrices to bioactive compounds that improve the quality of food products, by introducing antioxidants, antimicrobials, flavors, and probiotics. Besides obvious advantages of bioactive compounds, they may also exhibit some disadvantages such as off flavors and early loss of functionality, which usually are mitigated by encapsulation process (Ayala-Zavala et al., 2011). The encapsulation might be performed *via* incorporation of bioactive compounds into edible coatings, which could serve as a role of matrix: (1) protecting them against the environmental factors such as temperature (Kayaci and Uyar, 2012) and light, to avoid chemical degradation by UV light and oxygen (Durand et al., 2010; Ponce et al., 2008), (2) reducing the risk of pathogens' growth on food surfaces (Del Toro-Sánchez et al., 2010; Chiu and Lai, 2010) and (3) providing a functional product with health benefits to the consumer by improving solubility and controlling bioactive compound release (Ayala-Zavala et al., 2008; Fabra et al., 2012). There are different encapsulation strategies applied: (1) bioactive compounds affixed on the external surface of the film in contact with the environment, (2) at the interface between the film and the food, (3) among multilayered edible coating or (4) dispersed among different sections of the film (Quirós-Sauceda et al., 2014).

14.4   ANTIBIOFOULING POLYMER SURFACES

Biofouling is a process of colonization of all surfaces, exposed to the natural environment, by microorganisms. In biomedical contexts, fouling typically is considered as the formation of biofilms (Stoodley et al., 2002), related to biomedical implants, biosensors, and carriers for targeted drug delivery, while marine and industrial biofouling is typically related to macrofouling (Dobretsov and Qian, 2002) and inorganic fouling (Shirazi, Lin and Chen, 2010). Four recognized stages of the macrofouling process, as shown in Figure 14.8, include (1) the formation of a conditioning film, (2) the development of microbial biofilm, (3) the accumulation of unicellular eukaryotes like spores of macroalgae and (4) the attachment of multicellular foulers such as mollusks and barnacles (Chambers et al., 2006). Created biofilms interrupt the flow of ions and water between the substrate surface and surrounding environment by acting as a diffusion barrier. The settlement of microorganisms on ship hulls produces high frictional resistance and increased weight, resulting in increases in both fuel consumption and the frequency of dry-docking. On the other side, considering the medical aspect of biofouling, its reduction could significantly decrease the inflammatory responses including leukocyte activation, tissue fibrosis, thrombosis coagulation, and infection (Chen et al., 2010).

The performance of antibiofouling polymer coatings is based on the low interfacial energy with surrounding liquid, low intermolecular forces of interaction with biomolecules, or both, so that an adhered cell is easily released under low shear stresses. In general, compounds applied as antibiofouling coatings may act throughout hydrophobic or hydrophilic properties. Hydrophobic surfaces, due to high interfacial energy with water (e.g., about 52 mJ/m^2 for poly(dimethylsiloxane)), exhibit good "fouling release"

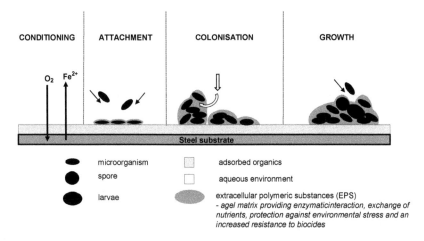

FIGURE 14.8 Stages of biofouling. (Sufficiently adapted, modified, and redrawn based on Chambers et al., 2006.)

properties, whereas the hydrophilic one, which are characterized by relatively low values of polymer-water interfacial energy (e.g., below 5 mJ/m^2 for poly(2-hydroxyethyl methacrylate), show resistance to protein adsorption and cell adhesion (Krishnan, Weinman and Ober, 2008). Due to the formation of a hydration layer (hydrogen bonding of functional groups of material and water molecules), which plays a role of physical barrier, hydrophilic surfaces are resistant to the adhesion of fouling agents (Chen et al., 2010). Therefore, the antibiofouling effectiveness of coatings based on hydrophilic compounds is strongly related to the physiochemical properties of the material such as molecular weights and the chains conformation of the polymer, which have an impact on the strength of the hydration layer. Poly(2-hydroxyethyl methacrylate) (PEG), polyamides, polyurethanes, and naturally occurring polysaccharides, including chitosan and dextran, are examples of hydrophilic polymers, which additionally are electrically neutral, and hydrogen-bond acceptors/donor, characterized by good antifouling properties. Among them, PEG and PEGylated polymers are currently the best as well as mostly studied and widely used protein-resistant surfaces. For example, titanium surfaces coated with the methoxy-terminated poly(ethylene glycol) conjugated to the adhesive amino acid L-3,4-dihydroxyphenylalanine, presented in Figure 14.9, exhibit good antifouling and fouling release properties [marine biofouling assays against green algal spores (*Ulva linzaand*) and diatoms (*Navicula perminuta*)] (Statz et al., 2006) and good resistance to protein and mammalian cell fouling (Dalsin and Messersmith, 2005). Similarly, application of block copolymer with PEGylated fluoroalkyl side chains (Figure 14.9) results in low settlement density of barnacle larvae (Weinman et al., 2007), weak adhesion of both *Ulva* and *Navicula* (Krishnan et al., 2006) and very good protein resistance.

Additionally, it is worth noting that for all hydrophilic antifouling coatings, prolonged exposure to biological environment results in the surface oxidation, damaging the hydration layer, and leading to inhibition of antifouling properties.

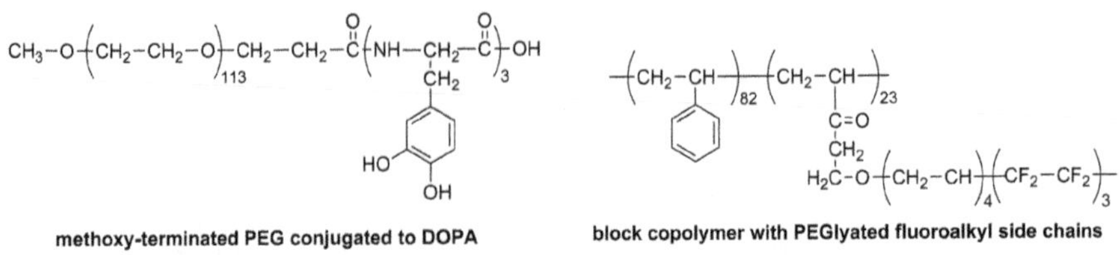

methoxy-terminated PEG conjugated to DOPA   block copolymer with PEGlyated fluoroalkyl side chains

FIGURE 14.9 Examples of PEGylated polymer antifouling coatings. (Sufficiently adapted, modified, and redrawn based on Krishnan, Weinman and Ober, 2008 and Statz et al., 2006.)

There are also some reports on the application of hydrophobic antifouling polymer coatings. One of the mostly applied is poly(dimethylsiloxane) (PDMS), which is not resistant to protein adsorption, but because of its nonpolar nature, is nonadhesive to cells and organisms (except diatoms, which adhere strongly to hydrophobic surfaces). For example, a triblock copolymer might be obtained in the ring-opening polymerization of ε-caprolactone (PCL) with PDMS as initiator (Azemar et al., 2015). The external PCL blocks are biodegradable, and PDMS maintains a core block. This copolymer might be used as binder in hydrophobic and biodegradable paints with a reduced environmental impact and applied to prevent microorganisms' settlement and proliferation. The biocides release is obtained by hydration, degradation, and erosion of the binder (PCL blocks) and the fouling release through a pronounced hydrophobicity of PDMS blocks.

Polyzwitterionic compounds are the next-generation, very promising antifouling and antimicrobial biomaterials. Examples of antifouling zwitterionic polymers are presented in Figure 14.10.

They are positive- and negative-charged polybetaines (the same monomer unit, e.g., 2-methacryloyloxylethyl phosphorylcholine, sulfobetaine methacrylate or carboxybetaine methacrylate), and polyampholytes (1:1 positive and negative charges on two different monomer units such as natural amino acids). The most studied zwitterionic materials are poly(sulfobetaine) and poly(carboxybetaine) adhered to methacrylate or acrylamide backbones. They preserve charge-neutral surface and form a hydration layer on the surface, which is bound through electrostatic interactions (Harding and Reynolds, 2014).

Recently, other interesting approaches toward development of eco-friendly and nontoxic antifoulants have been introduced. It is based on the incorporation of antifouling properties of bacteriocin produced by marine biofilm bacteria into epoxy paint (Palanichamy and Subramanian, 2017). The invention uses bacteriocins isolated from marine biofilm-forming bacteria: *Pseudomonas aeruginosa*, *Bacillus* sp., *Micrococcus* sp., and *Flavobacterium* sp., which exhibit antibacterial activities against each other. Additionally, it was found that the bacteriocin isolated from *P. aeruginosa* contains two compounds: 1,2-benzenedicarboxylic acid, mono(2-ethylhexyl) ester and 1,2-benzene dicarboxylic acid, di-isooctyl ester, which are responsible for the antibacterial and antifouling activities. Incorporation of 2% (w/v) of bacteriocin in epoxy paint resulted in 15% times less fouling load, compared to the reference PVC samples, in the period of 6 months. Both life expectancy and antifouling properties of coatings can be enhanced by increasing the biocide concentration.

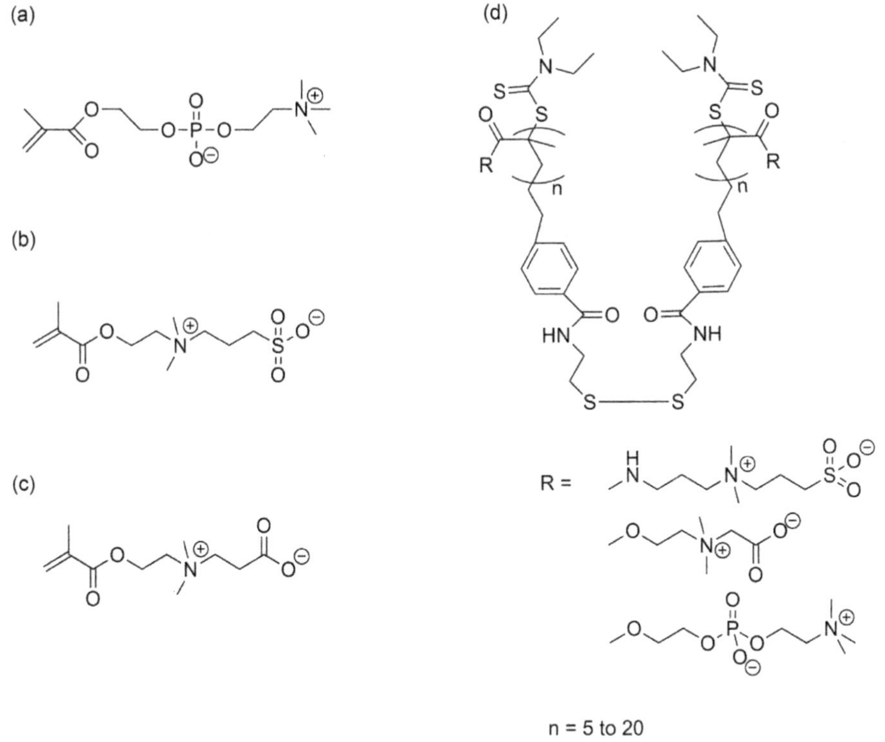

(a)

(d)

(b)

(c)

R =

n = 5 to 20

FIGURE 14.10 Chemical structures of three types of zwitterionic monomer: (a) 2-methacryloyloxyethyl phorylcholine, (b) sulfobetaine methacrylate, and (c) carboxybetaine methacrylate as well as an example of antifouling polymers with zwitterionic side chains and a disulfide group for attachment to gold substrates. (Sufficiently adapted, modified, and redrawn based on Krishnan, Weinman and Ober, 2008 and Zhang et al. 2015.)

14.5　ANTIFOULING BIOACTIVE SURFACES

Numerous publications show that the studies conducted in this field go a step further, focusing on designing surfaces that are characterized by both antifouling and bioactive properties.

A relatively large group of antifouling bioactive materials are based on aforementioned poly(ethylene glycol). In order to gain additional bioactive properties, it is necessary to introduce PEG appropriate functional groups (e.g., $-NH_2$, $-COOH$, $-NHS$, $-NSC$, or $-OH$) for attachment of the bioactive molecule and, at the same time, creating an effective antifouling/spacer polymer for the preparation of bioactive surfaces based on, for example, gold, glass, SiO_2, PDMS, PLG, or PU (Yu et al., 2011). The interesting example of such process is the immobilization of heparin, a well-known and widely used anticoagulant, on polydimethylsiloxane elastomer surfaces through a heterobifunctional PEG spacer (Chen et al., 2005). An asymmetric PEG with an allyl group and an N-succinimidyl carbonate (NSC) group was attached by hydrosilylation onto a Si–H-functionalized PDMS surface. The obtained results indicate that heparin immobilization to PDMS *via* a heterobifunctional PEG spacer leads to improved thrombogenic properties of materials by increasing the density of active heparin on the surface.

An interesting solution of coatings meeting both antifouling and bioactive require-ments is based on dual-functional allyl group terminated polyethylene glycol-poly-hexamethylene biguanide (PEG-PHMB) block dual-functional copolymers, obtained in aqueous phase, using a plasma/UV-assisted surface graft polymerization (Zhi et al., 2017). PHMB, an example of polymeric guanidines, is known for its broad-spectrum effectiveness toward Gram-negative bacteria, Gram-positive bacteria, fungi, some viruses, and other pathogenic microorganisms (Hübner and Kramer, 2010). In order to obtain dual-functional coatings, PHMB was functionalized with (1) monoallyl-functionalized PEG (allyloxy polyethylene glycol, APEG, $\overline{M_w}=1200$ and 2400) or (2) monoallyl groups without PEG block (A-PHMB). Figure 14.11 presents the process of the synthesis of copolymers and then grafting them onto a silicone rubber surface (polydimethylsiloxane, PDMS) as a bottlebrush-like coating, in aqueous phase under UV irradiation.

Both obtained polymer coatings (A-PHMB and APEG$_{1200/2400}$-PHMB) exhibit broad-spectrum antimicrobial properties against Gram-negative/positive bacteria and fungi. Additionally, compared with A-PHMB and APEG$_{1200}$-PHMB, APEG$_{2400}$-PHMB coating shows an improved antibiofilm-antifouling properties and a long reusable cycle.

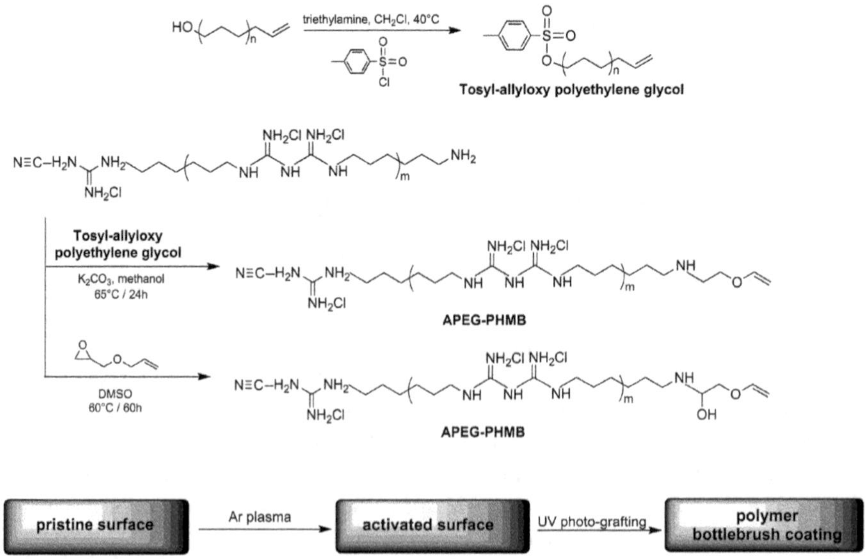

FIGURE 14.11 Syntheses of A-PHMB, APEG-PHMB, and A-PHMB/APEG-PHMB bot-tlebrush-like coating formation on polymeric substrate. (Sufficiently adapted, modified, and redrawn based on Zhi et al., 2017.)

REFERENCES

Ayala-Zavala, J. F., Soto-Valdez, H., Gonzalez-Leon, A., Alvarez-Parrilla, E., Martin-Belloso, O., and Gonzalez-Aguilar, G. A. 2008. Microencapsulation of cinnamon leaf (Cinnamomum zeylanicum) and garlic (Allium sativum) oils in β-cyclodextrin. *Journal of Inclusion Phenomena and Macrocyclic Chemistry*, 60(3–4): 359–368.

Ayala-Zavala, J., Vega-Vega, V., Rosas-Domínguez, C., Palafox-Carlos, H., Villa-Rodriguez, J. A., Siddiqui, M. W., Dávila-Avina, J. E., and González-Aguilar, G. A. 2011. Agroindustrial potential of exotic fruit byproducts as a source of food additives. *Food Research International*, 44(7): 1866–1874.

Azemar, F., Faÿ, F., Réhel, K. and Linossier, I. 2015. Development of hybrid antifouling paints. *Progress in Organic Coatings*, 87: 10–19.

Bazaka, K., Jacob, M. V., Truong, V. K., Wang, F., Pushpamali, W. A. A., Wang, J. Y., Ellis A. V., Berndt, C. C. Crawford, R. J., and Ivanova, E. P. 2010. Plasma-enhanced synthesis of bioactive polymeric coatings from monoterpene alcohols: A combined experimental and theoretical study. *Biomacromolecules*, 11(8): 2016–2026.

Boccaccini, A. R., Erol, M., Stark, W. J., Mohn, D., Hong, Z., and Mano, J. F. 2010. Polymer/bioactive glass nanocomposites for biomedical applications: A review. *Composites Science and Technology*, 70(13): 1764–1776.

Brzeziński, S., Jasiorski, M., Maruszewski, K., Ornat, M., Malinowska, G., Borak, B., and Karbownik, I. 2007. Bacteriostatic textile-polymeric coat materials modified with nanoparticles. *Polimery*, 52(5).: 362–366.

Campoccia, D., Montanaro, L., and Arciola, C. R. 2013. A review of the biomaterials technologies for infection-resistant surfaces. *Biomaterials*, 34(34): 8533–855.

Chambers, L. D., Stokes, K. R., Walsh, F. C., and Wood, R. J. 2006. Modern approaches to marine antifouling coatings. *Surface and Coatings Technology* 201(6): 3642–3652.

Chen S., Li L., Zhao C., and Zheng J. 2010. Surface hydration: Principles and applications toward low-fouling/nonfouling biomaterials. *Polymer* 51(23): 5283–5293.

Chen, H., Zhao, C., Zhang, M. Z., Chen, Q., Ma, J., and Zheng, J. 2016. Molecular understanding and structural-based design of polyacrylamides and polyacrylates as antifouling materials. *Langmuir*, 32:3315–3330.

Chen, H., Chen, Y., Sheardown, H., and Brook, M. A. 2005. Immobilization of heparin on a silicone surface through a heterobifunctional PEG spacer. *Biomaterials* 26(35): 7418–7424.

Chiu, P. E., and Lai, L. S. 2010. Antimicrobial activities of tapioca starch/decolorized hsian-tsao leaf gum coatings containing green tea extracts in fruit-based salads, romaine hearts and pork slices. *International Journal of Food Microbiology*, 139(1–2): 23–30.

Chouirfa, H., Bouloussa, H., Migonney, V., and Falentin-Daudré, C. 2019. Review of titanium surface modification techniques and coatings for antibacterial applications. *Acta Biomaterialia*, 83: 37–54.

Chudzik, S. J., Anderson, A. B., Chappa, R. A., and Kloke, T. M. 2005. *Bioactive Agent Release Coating*. U.S. Patent No. 6,890,583. Washington, DC: U.S. Patent and Trademark Office.

Chung, D., Papadakis, S. E. and Yam, K. L. 2003. Evaluation of a polymer coating containing triclosan as the antimicrobial layer for packaging materials. *International Journal of Food Science & Technology*, 38(2): 165–169.

Cloutier, M., Mantovani, D., and Rosei, F. 2015. Antibacterial coatings: Challenges, perspectives, and opportunities. *Trends in Biotechnology*, 33(11): 637–652.

Couto, D. S., Alves, N. M., and Mano, J. F. 2009. Nanostructured multilayer coatings combining chitosan with bioactive glass nanoparticles. *Journal of Nanoscience and Nanotechnology*, 9(3): 1741–1748.

Dalsin, J. L., and Messersmith, P. B. 2005. Bioinspired antifouling polymers. *Materials Today* 8(9): 38–46.

Davies, D. (2003). Understanding biofilm resistance to antibacterial agents. *Nature Reviews Drug Discovery*, 2(2): 114–122.

Del Toro-Sánchez, C. L., Ayala-Zavala, J. F., Machi, L., Santacruz, H., Villegas-Ochoa, M. A., Alvarez-Parrilla, E., and González-Aguilar, G. A. 2010. Controlled release of antifungal volatiles of thyme essential oil from β-cyclodextrin capsules. *Journal of Inclusion Phenomena and Macrocyclic Chemistry*, 67(3–4): 431–441.

Dobretsov, S. V., and Qian, P. Y. 2002. Effect of bacteria associated with the green alga Ulva reticulata on marine micro-and macrofouling. *Biofouling* 18(3): 217–228.

Durand, L., Habran, N., Henschel, V., and Amighi, K. (2010). Encapsulation of ethylhexyl methoxycinnamate, a light-sensitive UV filter, in lipid nanoparticles. *Journal of Microencapsulation*, 27(8): 714–725.

Elena, P., and Miri, K. 2018. Formation of contact active antimicrobial surfaces by covalent grafting of quaternary ammonium compounds. *Colloids and Surfaces B: Biointerfaces*, 169: 195–205.

Fabra, M. J., Chambin, O., Voilley, A., Gay, J. P., and Debeaufort, F. 2012. Influence of temperature and NaCl on the release in aqueous liquid media of aroma compounds encapsulated in edible films. *Journal of Food Engineering*, 108(1): 30–36.

Glinel, K., Thebault, P., Humblot, V., Pradier, C. M., and Jouenne, T. 2012. Antibacterial surfaces developed from bio-inspired approaches. *Acta Biomaterialia*, 8(5): 1670–1684.

Gorustovich, A. A., Roether, J. A., and Boccaccini, A. R. 2009. Effect of bioactive glasses on angiogenesis: A review of in vitro and in vivo evidences. *Tissue Engineering Part B: Reviews*, 16(2), 199–207.

Green, J. B. D., Fulghum, T., and Nordhaus, M. A. 2011. A review of immobilized antimicrobial agents and methods for testing. Biointerphases, 6(4): MR13–MR28.

Harding, J. L., and Reynolds, M. M. 2014. Combating medical device fouling. *Trends in Biotechnology*, 32(3): 140–146.

Hoque, J., Akkapeddi, P., Yadav, V., Manjunath, G. B., Uppu, D. S., Konai, M. M., Yarlagadda, V., Sanyal, K., and Haldar, J. 2015. Broad spectrum antibacterial and antifungal polymeric paint materials: Synthesis, structure–activity relationship, and membrane-active mode of action. *ACS Applied Materials & Interfaces*, 7(3): 1804–1815.

Hübner, N. O., and Kramer, A. 2010. Review on the efficacy, safety and clinical applications of polihexanide, a modern wound antiseptic. *Skin Pharmacology and Physiology* 23(Suppl. 1): 17–27.

Kayaci, F., and Uyar, T. 2012. Encapsulation of vanillin/cyclodextrin inclusion complex in electrospun polyvinyl alcohol (PVA) nanowebs: Prolonged shelf-life and high temperature stability of vanillin. *Food Chemistry*, 133(3): 641–649.

Krishnan, S., Ayothi, R., Hexemer, A., Finlay, J. A., Sohn, K. E., Perry, R., Ober, C. K., Kramer, E. J., Callow, M. E., Callow, J. A., and Fischer, D. A. 2006. Anti-biofouling properties of comblike block copolymers with amphiphilic side chains. *Langmuir* 22(11): 5075–5086.

Krishnan, S., Weinman, C. J., and Ober, C. K. 2008. Advances in polymers for anti-biofouling surfaces. *Journal of Materials Chemistry* 18(29): 3405–3413.

Lee, S. B., Koepsel, R. R., Morley, S. W., Matyjaszewski, K., Sun, Y., and Russell, A. J. 2004. Permanent, nonleaching antibacterial surfaces. 1. Synthesis by atom transfer radical polymerization. *Biomacromolecules*, 5(3): 877–882.

Lee, S. J., Heo, D. N., Lee, H. R., Lee, D., Yu, S. J., Park, S. A., Ko, W.K., Park, S.W., Im, S. G., Moon J.H., and Kwon, I. K. 2015. Biofunctionalized titanium with anti-fouling resistance by grafting thermo-responsive polymer brushes for the prevention of peri-implantitis. *Journal of Materials Chemistry B*, 3(26): 5161–5165.

Li, Z., Lee, D., Sheng, X., Cohen, R. E., and Rubner, M. F. 2006. Two-level antibacterial coating with both release-killing and contact-killing capabilities. *Langmuir*, 22(24): 9820–9823.

Lichter, J. A., Van Vliet, K. J., and Rubner, M. F. 2009. Design of antibacterial surfaces and interfaces: Polyelectrolyte multilayers as a multifunctional platform. *Macromolecules*, 42(22): 8573–8586.

Lin, J., Chen, X., Chen, C., Hu, J., Zhou, C., Cai, X., Wang, W., Zheng, C., Zhang, P.-P., Cheng J., and Guo, Z.H. 2018. Durably antibacterial and bacterially antiadhesive cotton fabrics coated by cationic fluorinated polymers. *ACS Applied Materials & Interfaces*, 10(7): 6124–6136.

Muñoz-Bonilla, A., and Fernández-García, M. 2012. Polymeric materials with antimicrobial activity. *Progress in Polymer Science*, 37(2): 281–339.

Palanichamy, S., and Subramanian, G. 2017. Antifouling properties of marine bacteriocin incorporated epoxy based paint. *Progress in Organic Coatings*, 103: 33–39.

Ponce, A. G., Roura, S. I., del Valle, C. E., and Moreira, M. R. 2008. Antimicrobial and antioxidant activities of edible coatings enriched with natural plant extracts: In vitro and in vivo studies. *Postharvest Biology and Technology*, 49(2): 294–300.

Popa, A., Davidescu, C. M., Trif, R., Ilia, G., Iliescu, S., and Dehelean, G. 2003. Study of quaternary 'onium' salts grafted on polymers: Antibacterial activity of quaternary phosphonium salts grafted on 'gel-type' styrene–divinylbenzene copolymers. *Reactive and Functional Polymers*, 55(2):151–158.

Quirós-Sauceda, A. E., Ayala-Zavala, J. F., Olivas, G. I., and González-Aguilar, G. A. 2014. Edible coatings as encapsulating matrices for bioactive compounds: A review. *Journal of Food Science and Technology*, 51(9): 1674–1685.

Shirazi, S., Lin, C. J., and Chen, D. 2010. Inorganic fouling of pressure-driven membrane processes - a critical review. *Desalination*, 250(1): 236–248.

Statz, A., Finlay, J., Dalsin, J., Callow, M., Callow, J. A., and Messersmith, P. B. 2006. Algal antifouling and fouling-release properties of metal surfaces coated with a polymer inspired by marine mussels. *Biofouling* 22(6): 391–399.

Stigter, M., Bezemer, J., De Groot, K., and Layrolle, P. 2004. Incorporation of different antibiotics into carbonated hydroxyapatite coatings on titanium implants, release and antibiotic efficacy. *Journal of Controlled Release*, 99(1): 127–137.

Stoodley, P., Sauer, K., Davies, D. G., and Costerton, J. W. 2002. Biofilms as complex differentiated communities. *Annual Reviews in Microbiology* 56(1): 187–209.

Venkateswaran, S., Wu, M., Gwynne, P.J., Hardman, A., Lilienkampf, A., Pernagallo, S., Blakely, G., Swann, D. G., Gallagher M. P., and Bradley, M. 2014. Bacteria repelling poly (methylmethacrylate-co-dimethylacrylamide) coatings for biomedical devices. *Journal of Materials Chemistry B*, 2(39):6723–6729.

Weinman, C. J., Krishnan, S., Park, D., Paik, M. Y., Wong, K., Fischer, D. A., Handlin D. L., Kowalke, G. L., Wendt, D. E., Sohn, K. E., and Kramer, E. J. 2007. Antifouling block copolymer surfaces that resist settlement of barnacle larvae. *PMSE Preprints* 96: 597–598.

Wu, J., Ye, J. J., Zhu, J. J., Xiao, Z. C., He, C. C., Shi, H. X., Wang, Y. D., Lin, C., Zhang, H. Y., Zhao, Y. Z., Fu, X. B., Cheng, H., Li, X. K., Li, L., Zheng, J., and Xiao, J. 2016. Heparin-based coacervate of FGF2 improves dermal regeneration by asserting a synergistic role with cell proliferation and endogenous facilitated VEGF for cutaneous wound healing. *Biomacromolecules*, 17:2168–2177.

Xynos I. D., Hukkanen M. V. J., Batten J. J., Buttery L. D., Hench L. L., and Polak J. M. 2000. Bioglass 45S5 stimulates osteoblast turnover and enhances bone formation in vitro: Implications and applications for bone tissue engineering. *Calcified Tissue International*, 67(4):321–9.

Yu, Q., Wu, Z., and Chen, H. 2015. Dual-function antibacterial surfaces for biomedical applications. *Acta Biomaterialia*, 16: 1–13.

Yu, Q., Zhang, Y., Wang, H., Brash, J., and Chen, H. 2011. Anti-fouling bioactive surfaces. *Acta Biomaterialia* 7(4): 1550–1557.

Zhang, B., Myers, D., Wallace, G., Brandt, M., and Choong, P. 2014. Bioactive coatings for orthopaedic implants—recent trends in development of implant coatings. *International Journal of Molecular Sciences*, 15(7): 11878–11921.

Zhi, Z., Su, Y., Xi, Y., Tian, L., Xu, M., Wang, Q., Padidan S., Li, P., and Huang, W. 2017. Dual-functional polyethylene glycol-b-polyhexanide surface coating with in vitro and in vivo antimicrobial and antifouling activities. *ACS Applied Materials and Interfaces* 9(12): 10383–10397.

Zhou, Z., and Meyerhoff, M. E. 2005. Preparation and characterization of polymeric coatings with combined nitric oxide release and immobilized active heparin. *Biomaterials*, 26(33): 6506–6517.

15 Adhesion of Polymer Coatings

S. Verma and S.K. Nayak
CIPET, Chennai, India

S. Mohanty
CIPET: SARP-LARPM, Bhubaneswar, India

CONTENTS

15.1 INTRODUCTION

Adhesion Science and Engineering is based on the principle of physics, chemistry, and rheology of the polymer coating system to be casted onto any substrate or adherend. The interdependence of these disciplines of science on the phenomenon of adhesion is shown in Figure 15.1. This multidisciplinary field of adhesion science requires the adaption of the disparate academic disciplines as evident from the flowchart (Petrie 2000). Polymer materials have one unique fundamental property of adhesion

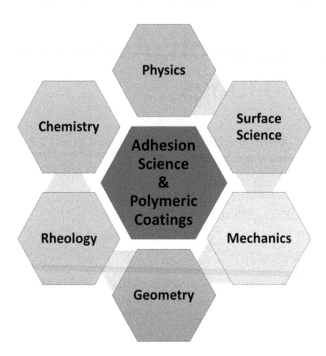

FIGURE 15.1 Equally essential parts related to adhesion science of polymer coatings.

which varies from one polymer to another depending upon the individual chemical composition (Petrie 2000). Apart from this fact, polymers possess characteristics of non-Newtonian fluid and can be made to flow in molten state depending upon the viscosity of the solution, which also promotes adhesion. In addition, the adhesive polymer chemistries respond in different ways to different types of substrates such as wood, metal, or plastic. Due to this fact, the chemistry of a viscous polymer also influences its proper binding with the substrate. After proper binding of the polymer with the substrate, the geometry of the polymer coating governs the direction and intensity of stress, which in turn is estimated with the help of the shape of the bond between the polymer coating and the substrate (3M 2020).

Adhesion is the term used to denote the joining of two or more different substances/layers or a combination of both. Similarly, polymeric coatings are the adhesives that join two surfaces and resist their separation by forming an adhesive bonding (Watts 2005; Whitehouse 1987). That being said, the current chapter focuses fundamentally on the modes of adhesion bonding, the issues associated with proper bonding, the physical and chemical factors affecting the adhesion, ways to the adhesion performance, and the characterization techniques.

15.2 WORK OF ADHESION

The adhesion of polymer coatings refers to the attraction between the coating and the substrate, which when brought together in contact remains inseparable until any work is done to separate them. The work done to separate the polymer coating from its respective substrate is termed as mechanical work of adhesion and is denoted by W_A, as expressed in Eq. (15.1) (Petrie 2000):

$$W_A = \gamma_C + \gamma_S - \gamma_{CS}, \tag{15.1}$$

where γ_C and γ_S represent the interfacial energies of the polymer coating and the substrate, respectively. The interfacial energy existing between the two surfaces in contact is denoted by γ_{CS}.

Contrastingly, the viscoelastic behavior of polymers is evaluated in terms of factors responsible for viscoelasticity (ζ) reported mathematically to estimate the practical adhesion (W_P) with the help of calculated work of adhesion (W_A). Hence, the relationship between the two works of adhesion is reported in the findings of Ahagon and Gent, and represented in Eq. (15.2) (Ahagon and Gent 1975):

$$W_P = W_A + f\left(W_A\right)\zeta. \tag{15.2}$$

This equation is significant in establishing a connection between the viscoelastic properties of the polymer coating and the mechanical behavior of adhesion in the form of absorption energy or analogous terms. Apart from analyzing such structure–property relationship of polymer coatings, an insight on various modes of adhesion is also provided in the next section.

15.3 MODES OF ADHESION

15.3.1 Physical Interactions

The physical adsorption takes place with the help of van der Waals' <forces> that evolve around all the atoms and molecules at the interface between the polymeric coating and the substrate. The van der Waals forces are basically the intermolecular forces of attraction categorized into three types: the forces acting between (1) molecules and permanent dipoles, (2) permanent dipole and non-polar molecule, and (3) non-polar molecules (Comyn 2005). The van der Waals forces are weak forces of attraction that are unable to sustain the adhesive bonding between the coating and the substrate in the long run, and thus tend to fail eventually under stress. Hence, the physical adsorption is not the effective mode of adhesion in the case of polymer coatings used for marine or underwater applications since the high surface tension of water is capable of displacing the adhesive polymeric layer from the metallic substrate (Comyn 2005). Therefore, in the direction of establishing stronger adhesive bonding between the coating and the substrate, a more feasible mode of adhesion that consists of chemical interactions will be discussed further.

15.3.2 Chemical Interactions

The polymers can be polymerized by a set of chemical reactions in which the substrate surface acts as the adherend and the polymer coating as the adhesive forming a strong solid bond which leads to long-term stability comprising two types of chemical interactions, namely, primary and secondary chemical bonds. The primary chemical bonds are known for chemical bonding with bond energy of the order of 300 kJ/mol, which includes ionic and covalent bonds effective within the distance of 0.15 nm (Whitehouse 1987; Weldon 2009). The example of covalent bonding (C–O bonds) can be observed from the reaction between hydroxyl (–OH) groups present on the surface of substrate (wood, metal) and isocyanate-based polyurethane (PU) adhesives resulting in adhesion due to carbon–oxygen bonds. Another strong evidence of covalent bonding is observed in the case of silane coupling agents having the tendency to react with both substrate and the coating, which results in network formation of strong and durable covalent bonds across the substrate–coating interface (Comyn 2005). The secondary chemical bonds that denote polar interactions include London dispersion, hydrogen bonding, or Lewis acid–base interactions. For example, the formation of hydrogen bonds is also observed in the case of wood substrates because of its rich cellulosic content, which tends to react with hydroxyl (–OH) groups present in isocyanate group of polyurethane adhesives. Similarly, interfacial hydrogen bonding is also observed as the mode of adhesion between polyurethane coating and float glass substrates (Comyn 2005; Agrawal and Drzal 1996). Against this background, the electrostatic theory of adhesion reveals the presence of electrons on the surface of metal substrate, which also take part in reaction with the polymer material applied onto the substrate and contributes to adhesion. This mode of adhesion occurring between the coating and substrate is termed as static electrification. For example, in the case of solvent-borne coatings, the metal substrate containing electrons on its surface gets wet by a layer of adhesive polymeric coating cast onto a pretreated substrate and creates an interfacial surface which then disappears as both the surfaces have adhered to each other with the help of interfacial bonds.

In a research study, the reaction mechanism of adhesion has been reported through a set of chemical equations (Dey et al. 2018). Figure 15.2 depicts the adhesion mechanism of a superhydrophobic coating of methylhydrogen polysiloxane polymeric resin over phosphated steel substrate.

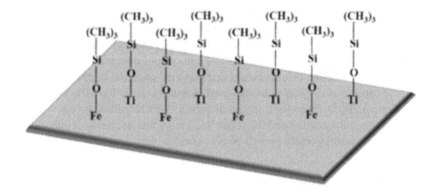

FIGURE 15.2 Probable mechanism of adhesion depicting chemical interactions between mild steel substrate, TiO_2 particles, and siloxane moieties (Dey et al. 2018). (Reproduced with kind permission from Elsevier, Copyright 2018.)

The authors have reported that the principle of strong adhesion and film formation is chemical bond formation between organic methyl groups (CH_3) and inorganic siloxane bonds (Si-O-Si) on phosphate steel substrate in the *<immediate>* presence of methylhydrogen polysiloxane adhesion promoter (Dey et al. 2018). It was also found that the proposed reaction mechanism is based on the reactivity of several silanol units formed after the hydrolysis of methylhydrogen polysiloxane and its rapid adsorption onto the phosphated metal substrate to form Si-O-M chemical bonds (Dey et al. 2018). Apart from promoting good adhesion, the current reaction mechanism is responsible for enhancing the corrosion resistance property of superhydrophobic material coated substrates. This being the fact, silanol groups of polysiloxane resin also take part in condensation reaction to form covalent bonds and actively block the corrosion spots on the substrate surface (Dey et al. 2018).

15.3.3 MECHANICAL INTERLOCKING

The third component of adhesion (apart from physical and chemical interactions) is the mechanical interlocking depending on the mechanical force that exists between polymer coating and crevices at the substrate surface originating from roughening operations (Weldon 2009; Kinloch 1982). It has been found that coating adheres better to a rough surface than to smooth one, favored by the production of chemically active sites. The coating flows into the pores and crevices of the rough substrate giving rise to mechanical interlocking. The irregular surface in which the polymer coating enters undergoes polymerization and solidifies on account of mechanical interlocking.

In this direction, Kim et al. (2010) reported mechanical interlocking as the factor responsible for enhancement in the adhesion strength at polymer–metal interface and investigated the effect of surface topography of an epoxy–steel bond toward adhesion performance by systematically assessing varying micropattern dimensions fabricated over metal surfaces. Figure 15.3 depicts the mechanical interlocking at metal polymer interface followed by the microline pattern created at steel interface and its corresponding surface and cross-sectional scanning electron microscopy (SEM) micrographs.

In this work, the authors have highlighted that the adhesion between a surface-treated specific metal and a particular polymer is achieved by adsorption of the polymer over the substrate and is dependent upon the topography and mechanical properties of the polymer which in turn promotes mechanical interlocking (Kim et al. 2010). In addition, the combined effect of stress distribution, fracture characteristics, and surface morphology between the polymer and the substrate is also considered while optimizing the design to transfer the load within the adhesive bond as shown in Figure 15.3b. On the other hand, from the SEM micrographs, it is evident that the prepared steel surface will lead to the uniform distribution of epoxy by completely filling the microgrooves on the micropatterned surface. Hence, on creating microline patterned steel–polymer interface, the complete mechanical interlocking of the polymer into the steel surface cavities can be obtained (Kim et al. 2010).

Reasoning from the aforementioned discussion, in order to understand the mechanism of adhesion completely, the factors affecting the adhesion performance of the

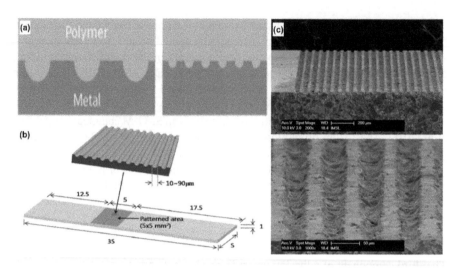

FIGURE 15.3 Microline patterned (a) metal–polymer interface, (b) steel specimen, and (c) scanning electron micrographs of a microline-patterned steel surface (Kim et al. 2010). (Reproduced with kind permission from Elsevier, Copyright 2010.)

polymer coatings need to be taken into account such as substrate preparation. Each of these physical and chemical factors has been discussed in detail in the next section.

15.4   FACTORS AFFECTING ADHESION PERFORMANCE OF POLYMER COATINGS

The physical and chemical factors responsible for the successful formation of the adhesive film are listed in Table 15.1. The major physical factors affecting the adhesion performance are surface smoothness, film thickness, viscosity of the polymer film, etc. Besides, the difference in thermal expansion denotes the tendency of material to change its shape, area, and volume in response to temperature. But, when this difference is high the bonding strength weakens under temperature variations. In contrast, the chemical factors affecting the adhesion performance such as degree of polarization denote the extent of polymerization. Similarly, the complexity of adhesive molecule represents the chain length and molecular weight of the polymer coating. Effect of pH shows the influence of strong acid/alkali, and the presence of electric dipole indicates the separation of charges in the polymer molecule.

The detailed findings on a combination of factors affecting the adhesion performance such as substrate preparation, surface roughness, and wettability of the substrate are discussed in the following.

15.4.1   Substrate Preparation

Adhesive bonding of the polymer coating over the substrate is governed by intermolecular forces established between the two to confer required bond strength and durability to the polymer coating. In order to ensure significant penetration of polymer

TABLE 15.1

Physical and Chemical Factors Affecting the Adhesion Mechanism

S. No.	Physical Characters of Polymer Adhesive/Coating Layer	Chemical Characters of Polymer Adhesive/Coating Layer
1.	Smoothness of the substrate surface	Polarity of the polymer coating
2.	Surface tension and thickness of the film	Degree of polymerization
3.	Viscosity and adhesive strength of the polymer coating	Complexity of the polymer adhesive molecule
4.	Difference in thermal expansion of coating and substrate	Effect of pH

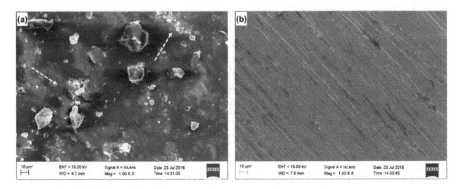

FIGURE 15.4 FESEM micrograph of mild steel substrate: (a) without surface treatment and (b) with surface treatment.

coating onto substrate, it is essential to pretreat the substrate with simple and sophisticated chemical methods such as abrasion treatment and corona discharge (Watts 2005). These substrate preparation techniques remove any foreign materials present on the substrate surface like oil, dirt, moisture, grease, etc. These contaminants give rise to a layer that is cohesively weak, whereas clean surfaces can give strong bonding to the polymer coatings. Figure 15.4a shows the Field Emission Scanning Electron Microscopy (FESEM) image of untreated mild steel substrate depicting the presence of corrosion products and other impurities (indicated by dotted arrow) on the surface of mild steel substrate. On the other hand, Figure 15.4b illustrates the FESEM image of surface-treated MS plate showing better morphology free from any impurities.

In this regard, the selection of appropriate method for surface preparation depends upon the polymer resin, material and nature of substrate before application of coating, desired adhesive strength, and time available (Petrie 2000). The nature of substrate plays a predominant role in improving the adhesion performance of polymer coating. In case of metallic substrates, these are the most commonly protected substrates from the environmental hazards of moisture/humidity leading to corrosion and other problems. Sand blasting is the most effective technique of substrate treatment pertaining to ISO 8501-1 standard in which a standard abrasive (SA) profile is created depending upon the type of metal finish required at the surface such as gray metal visible, shades of gray, near white metal, and white metal finish. According to British Standards Institute (BSI), Swedish Standards Association (SSA), National Association of Corrosion Engineers (NACE), and Steel Structure Painting Council (SSPC), out of the four standards of blast cleaning, the Brush-off Blast (SSPC-SP 10) is the most effective method that denotes surface area (Sa 2.5) surface profile. In this procedure, foreign materials such as mill scale, rust, dirt, grease, corrosion products and its oxides are removed completely by using abrasive blasting technique. The criteria should fulfil 95% of each square inch of surface area of visible residues; thus, this is probably the best quality of surface preparation (Talbert 2008).

15.4.2 SURFACE ROUGHNESS

Surface roughness is quite different from surface cleanliness as discussed earlier: the former provides a suitable method for cleaning of the substrate, whereas the latter provides teeth to the substrate, which promotes mechanical interlocking of polymer coating layer with the roughened substrate consisting of peaks and valleys. As per ISO-8503-1 to ISO-8503-5, surface roughness (R_a) profile determines the measure of peak to trough amplitude of blast profile created during sand blasting and it is expressed in microns. It is also found that the surface roughness and dry film thickness (DFT) of any polymer coating are proportional to each other. For example, to obtain a DFT of 20 µ, a surface roughness profile of 70 µ is not suitable since it will result in high paint consumption. Therefore, before selecting the type of surface preparation of the substrate to obtain an appropriate surface roughness profile, the desired DFT should always be considered. In this regard, Balabanava et al. (2007) reported the effect of surface roughness on various properties such as adhesion, wear, friction behavior, and optical, which are very important in microdomain. The authors claimed that the surface roughness aids in controlling the adhesive properties of microstructures by modifying the working surface by varying its surface roughness.

In order to measure the surface roughness or study the surface topography of the treated substrate, Figure 15.5 shows the various types of testing methodology adopted for characterizing the surface profile obtained after the surface treatment. Generally, the methods such as surface replica tape, surface comparators, or surface profile needle gauges are used. On the other hand, the more precise and sophisticated methods for the evaluation of surface topography are Stylus profilometry, SEM, and contact angle analysis for wetting of liquids on the pretreated solid substrates, which are the physical methods. In addition, the chemical methods of analysis include X-ray photoelectron spectroscopy, electron microscopy, mass spectroscopy, etc., which aim to determine the specific chemistry at the coating–substrate interface so as to provide prolonged adhesive bonding between the two (Watts 2005).

In consideration of the foregoing discussion, stylus profiler (Nano Map, AEP Technology, New York, USA) was utilized in a research study by (Dey et al. 2018) to quantitatively measure the surface roughness of the substrates coated with neat TiO_2 particles and methylhydrogen polysiloxane-treated TiO_2 particles. The surface

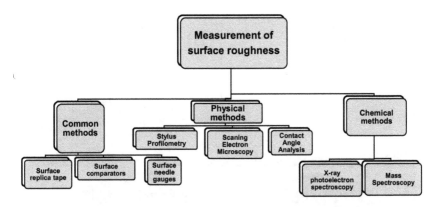

FIGURE 15.5 Methods for the evaluation of surface topography of the substrate after surface treatment.

topography image revealed an elevation in surface roughness values from 7.94 to 13.11 μm on increasing the concentration of adhesion promoter.

15.4.3 Wettability and Spreadability

Wetting is defined as the spreading of liquid over the solid surface and establishing a close contact with it. To ensure complete wettability, the grooves, roughness, or crevices created over the substrate surface should be filled up by the polymer coating before the entrapment of air leading to the formation of air pockets. Subsequently, this air entrapment results into the formation of weak boundary layer over the substrate surface hindering the adhesion phenomenon. Besides wetting the substrate, polymer coating in its liquid state must be stable to flow and reach the substrate surface at micro- and nanolevel to avoid gas pocket occlusion (Petrie 2000). Apart from wettability, to fulfil the criteria of proper spreadability of the polymer coating over the substrate, these basic universal requirements should always be considered prior to casting/spraying (Petrie 2000):

1. Cleanliness of the substrate
2. Wetting of the substrate
3. Solidification of the polymer coating over the substrate forming an adhesive joint
4. Geometry of polymer coating after solidifying should be free from environmental stresses

In this regard, a research work emphasized on the measurement of wettability of the substrate to study the surface characteristics (hydrophobic or hydrophilic) of the bare and coated samples using a contact angle measurement instrument (Data Physics, OCA-15EC, Filderstadt, Germany) (Dey et al. 2018). Figure 15.6 displays the optical image of the water droplets acquired for bare as well as coated substrates at different compositions. It is clearly evident from Figure 15.6 that the bare mild steel substrate

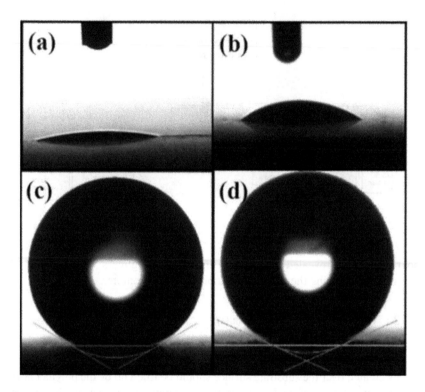

FIGURE 15.6 Water contact angle optical images of (a) surface-treated bare mild steel substrate, (b) neat TiO$_2$-coated substrate, and (c,d) modified TiO$_2$-coated substrate at different compositions (Dey et al. 2018). (Reproduced with kind permission from Elsevier, Copyright 2018.)

after phosphate treatment exhibited water contact angle (WCA) of 11° denoting super-hydrophilic characteristics. Furthermore, as apparent from Figure 15.6b–d, the contact angle values were found to increase for TiO$_2$-coated substrate revealing hydrophilic character (WCA=41.7°) and the hydrophobic characteristics (WCA=151.5° and 160.7°) for siloxane modified TiO$_2$-coated substrates, respectively. The authors claimed that hydrophobicity is dependent on two factors: (1) surface roughness and (2) chemical composition. Having said that, an increase in the surface roughness values on account of incorporation of methyl groups of siloxane rendered superhydrophobic nature to the coatings (Dey et al. 2018).

After achieving a solidified adhesive bond over the substrate surface, certain stresses arise from solidification or complete curing of polymer coatings, which in turn affects the adhesion performance, which will be discussed in detail in the next section.

15.4.4 STRESSES

Fracture mechanics is a branch of science that deals with the study of load transferred from one coating layer to the other or a single coating to the substrate. Stress is defined as the force per unit area or the force carried per unit area by the polymer coating material acting as an adhesive. The stress fields generated within the polymer coating layer are quite complex and highly nonuniform in nature, which involves both normal and shear stresses acting in several directions (Dillard 2002). The average normal stress (σ) acting over unit area is given by normal component of force (F_n) divided by the increment of area (ΔA) as represented in Eq. (15.3) (Dillard 2002):

$$\sigma = \lim_{\Delta A \to 0} \frac{\Delta F_n}{\Delta A}. \tag{15.3}$$

Similarly, the average shear stress (τ) acting over unit area is given by tangential component of force (F_t) acting perpendicular to the surface divided by the area as shown in Eq. (15.4) (Dillard 2002):

$$\tau = \lim_{\Delta A \to 0} \frac{\Delta F_t}{\Delta A}. \tag{15.4}$$

The singular stresses are perhaps acknowledged by fracture mechanics by providing a framework for quantitative evaluation of severity of cracks. Furthermore, the resistance of polymer materials towards crack propagation and designing of damage-tolerant protective coatings are also studied using fracture mechanics (Liechti 2002). It has been observed that the polymer coatings adhesively bonded to the substrate are quite susceptible to an array of potential fracture mechanisms such as cohesive cracks growing within the polymer coating layer. Another mode of failure describes the growth of cracks along the coating-substrate interface better known as adhesive failure. Further, this crack branching leads to crack nucleation, which subsequently results in cohesive or adhesive failure, or sometimes a combination of both (Liechti 2002). Consequently, delamination, fracture, erosive wear, and shearing of the coating occurs detaching it from the substrate caused due to the residual stress, thermal mismatch and these inbuilt contact stresses that are sustained by the coating prior to adhesive failure (Petrie 2000).

From another point of view, apart from mechanical modes of coating failure, the environmental stress factors impacting the adhesive performance of coatings consists of the stresses that originate after subjecting the coating to atmospheric conditions and are not inbuilt such as weathering stress, blistering, etc., which are developed due to the gradient in osmotic pressure originated at the interface (Weldon 2009). A detailed analysis on the causes and consequences of these environmental stresses is provided in the next section.

15.4.5 BLISTERING

Steel substrate in marine environment is often accompanied by osmotic as well as electroosmotic blistering due to their direct contact with salt water or salt-laden moisture in the air (Weldon 2009). Other possible cause of blistering of polymer coating is the usage of partially water-soluble solvents in the coating formulation. As a consequence, electric potential gradient arises resulting in the passage of water through blisters that leads to the migration of ions within the coating layers. The entrapment

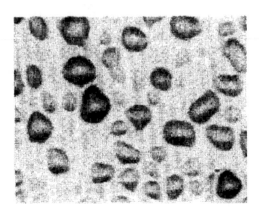

FIGURE 15.7 Microscopy results of water blister formation at the glass-coating interface (Krongauz, Schmid, and Vandeberg 1995). (Reproduced with kind permission from Elsevier, Copyright 1995.)

and subsequent migration of these ions entering from the coating surface migrating up to the metal substrate results in coating delamination and poor adhesion performance. In this regard, Krongauz, Schmid, and Vandeberg (1995) reported reduction in adhesion of urethane acrylate-based polymer coatings on glass substrate after immersing it in water. Furthermore, it has been observed that the water that accumulated beneath the coating surface in droplet form resulted into the formation of blisters at the glass–polymer interface. The authors claimed appearance of foggy droplets on examining the coated glass slides, microscopically which is ascribed to the accumulation of water drops within the interfacial region between the urethane coating and the glass slide surface. Figure 15.7 reports the result of microscopic imaging denoting the amount of water accumulation in the form of blisters.

Therefore, the effect of such environmental factors on the adhesive performance of designed polymer coating should be analyzed and forecasted well in advance with the help of modelling and simulation before subjecting the coatings to real field applications.

15.5 WAYS TO IMPROVE ADHESION PERFORMANCE

In order to overcome the adhesion losses which tend to occur due to various types of mechanical and environmental stresses sustained by the bonded polymer coating system, few ways to improve the adhesion performance will be discussed. With the aim of providing protection against moisture-induced de bonding, it is equally important to maintain good adhesion between the different polymeric layers coated onto the substrate. In view of this, the aforementioned criteria can be fulfilled either by using suitable adhesion promoters, or, by incorporating potential fillers within the polymer coating system.

15.5.1 ADHESION PROMOTERS

The adhesion strength of the coating can be increased by means of adhesion promoters such as organosilane compounds, which can provide strong interfacial interactions between the coating and the substrate (Esfandeh et al. 2007). In the field of adhesives, paints, coatings, and ink, adhesion promoters are utilized in two ways: (1) either they can be coated as primer layer onto the substrate prior to the application of topcoat or (2) they can be mixed with the existing coating system and applied as single layer onto the substrate. The procedure of using silane compounds as primer involves the dilution of the silane with the help of solvents such as methanol, ethanol, propanol, or mixture of solvents. These alcohol-based solvents promote the hydrolysis and/or condensation reactions of silane adhesion promoter while adhering to the substrate by forming siloxane bridges (–Si–O–Si) (Pape 2018). The optimum quantity of solvent usage is between 2% and 5% of silane compound followed by post curing of the polymer coating (<125°C), which helps in complete evaporation of the solvent. However, heating above the temperature of 125°C induces excessive crosslinking of the silane primer layer and hinders diffusion occurring between the polymer coating and the silane primer interphase region (Pape 2018). Apart from maintaining correct amount of solvent and temperature of curing, it is important to control the thickness of the silane primer layer, which leads to poor adhesion on account of high thickness and also results in insufficient bond strength and water resistance because of low thickness (Pape 2018). For example, as per the findings of Comyn (2005), epoxysilane and aminosilane of appropriate thickness were used as adhesion promotors on to surface-treated aluminum substrates followed by a topcoat of epoxy and polyurethane paints, respectively. Similarly, from the findings of (Esfandeh et al. 2007), it has been observed that the adhesion of topcoat over an epoxy-coated aluminum substrate has been improved by using an organosilane as the tie layer between the two layers. Further, the effect of tie layer composition on adhesion strength was investigated by changing the silane compound and its percentage. Figure 15.8a shows the SEM micrograph taken from the interface of the primer layer (epoxy) and the intermediate layer (silicone) in equal ratio (50:50) without tie-coat (silane layer). It is evident from the micrograph that both the layers are clearly separated thereof, the adhesion strength was found to be quite low. Figure 15.8b represents the SEM image of the two coating layers consisting of silane tie coat (1 wt.%) embedded in between the primer and intermediate layer. It can be seen from the image that on addition of silane layer, a good interfacial bonding between the epoxy primer and the intermediate layer has

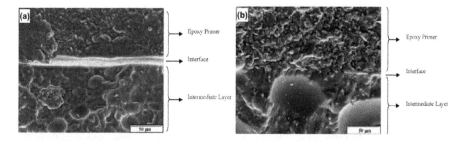

FIGURE 15.8 SEM micrograph of the primer and the intermediate layer (a) silicone/epoxy, 50/50 wt.% and (b) silicone/epoxy, 50/50, 1 wt.% silane (Esfandeh et al. 2007). (Reproduced with kind permission of Elsevier, Copyright 2007.)

been obtained, which leads to improved adhesion strength within the multilayered coating system (Esfandeh et al. 2007).

Alternatively, silane compounds can also be used in bulk for improving the adhesion performance of the polymer coatings as explained by the following mechanism. The organofunctional and alkoxy group present in silane coupling agents are capable of forming strong bonds between the polymer coating and the filler particles (Pape 2018). In other words, the silane compounds consist of silicon bonded to oxygen and carbon groups and are capable to react in multiple ways to form cross-linked networks within the polymer matrix. They can act as material medium to enhance the wettability of filler particles within the polymer matrix and promote adhesion. In this regard, a research study reported a tremendous increase in the properties upon treating the TiO_2 nanoparticles with the adhesion promoter (methylhydrogen polysiloxane) coated over mild steel substrates (Dey et al. 2018). The silanol units from the siloxane compound were absorbed on the exposed metal surface forming TieOeSi/FeeOeSi bonds (Dey et al. 2018). Upon increasing the concentration of adhesion promoter beyond 10 wt.%, the silane film layer was found to be formed on the top of TiO_2 layer (Dey et al. 2018).

15.5.2 FILLERS

Fillers are the complex chemical compounds used within the polymer matrix to impart either better performance or cost effectiveness to the polymer coating. In consideration of the foregoing discussion on adhesion promoters, the filler material fulfils the present criteria by reacting with the matrix material and bestows good adhesion between the different phases (Subramanian 2013). In view of this, the incorporated filler (organic, inorganic, or mineral) promote adhesion between the various components of the polymer-based paint, which is used as end-product for casting/spraying over different substrates. However, before selection of an appropriate filler the compatibility and chemical reactivity of the filler with the immiscible polymers should be considered. Apart from the degree of reactivity, the interfacial polymerization also influences the adhesion phenomenon occurring at the phase boundary between the two immiscible polymer phases (Subramanian 2013). The fillers usually contain high amount of chemically combined water within their molecular structure, having said that the usage of such fillers should be avoided for polymer coating to be cast on metal substrates. For the present application of adhesion promotion out of the different variety of fillers available, mineral fillers are specifically used to provide desired characteristics to the adhesive bond without sacrificing on the other properties. The mineral fillers used at minimal loadings exhibit better performance on account of better dispersion forming a dilute particle suspension in particular solvents (Subramanian 2013). According to a research study, the incorporation of fillers at nanoscale results in improvement of adhesion performance by incorporating nanoclay (Mohamadpour, Pourabbas, and Fabbri 2011). The authors reported that adhesion is directly related to the quality of the coating-substrate interface, the type of clay and dispersion state of the clay inside the polymer matrix. The roughness induced by clay at the coating–substrate interface resulted in change of adhesion magnitude (Mohamadpour, Pourabbas, and Fabbri 2011).

15.6   EVALUATION OF ADHESION PERFORMANCE

The complex, nonuniform, multidirectional stresses as discussed in Section 15.4.4 portray the need for testing the adhesion of polymer coating thoroughly. In view of this, even a single polymer coating layer of adhesive is found to be relatively complex structure with nonuniform stresses (Dillard 2002). Hence, prior to testing, an understanding of these complexities is vital in selection, conduction, and interpretation of adhesion performance tests of polymer coatings.

The current practice to determine the failure modes are more focused on testing adhesion phenomenon and designing of adhesive material based on the load it can sustain. The basic modes of failure in which an adhesive material fails is classified as adhesive or cohesive failure, or a fusion of both. In case of single-layered polymer coating cast onto a different material substrate, if the cracks grow entirely within the coating layer, it is termed as cohesive cracking. It means that the bond between the polymer coating and the substrate is working well. On the other hand, if the crack is found to grow along the interface between the coating and the substrate then it is considered to be an adhesive cracking (Liechti 2002).

In other words, adhesive failure is an interfacial bond failure between the coating and the substrate and it occurs at the coating–substrate interface, whereas cohesive failure occurs when the layer of the polymer coating remains adhered on the substrate without any signs of detaching/debonding and it occurs in bulk (Petrie 2000). The methods used to determine the adhesive or cohesive failure modes of a polymeric coating are discussed in detail in the following.

15.6.1   PULL-OFF ADHESION

The pull-off adhesion strength of coatings on metal substrates is measured using ASTM D 4541-09 reporting the results in the easiest form as "pass with a pull-off strength" (the value obtained) (American Society for Testing and Materials 2009). The test results can be reported in the following format for a particular coating system: strength (MPa or psi), failure zone (B, B/C or C/D) followed by its description (within primer, between primer and mid-coat, or between mid-coat and topcoat) after minute observations and expertise in testing (Francisco and Claudino 2016). The nomenclature of various layers is done as per ASTM D 4541-09 standard: the specimen as substrate A, upon which successive coating layers B, C, D, etc., have been applied, including the adhesive, Y, that secures the fixture or dolly, Z, to the top coat (American Society for Testing and Materials 2009; Verma, Mohanty, and Nayak 2020). Soer et al. (2009) reported that polyester-based coatings at different compositions applied on to aluminum alloy substrate revealed a large difference in the stress at delamination. The schematic representation of the testing scheme used for pull-off test is shown in Figure 15.9a, which consists of an assembly of stud affixed onto the substrate using a glue supported by the holders and Figure 15.9b represents the details of various coating layers including the glue layer up to stud as designated by ASTM D 4541-09. After fixing of the stud, tensile stress was measured as a function of displacement of the fixed stud, whereas, the final force required to pull off the coating from the substrate was used to determine the final adhesion force at a pulling

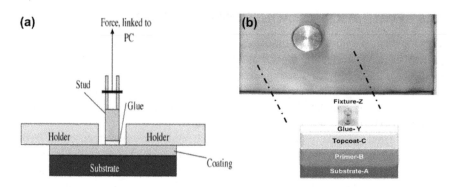

FIGURE 15.9 (a) Methodology of pull-off test (Soer et al. 2009). (Reproduced with kind permission from Elsevier, Copyright 2009). (b) Details of coating layers as per ASTM D 4541-09 (Verma, Mohanty, and Nayak 2020). (Reproduced by permission of The Royal Society of Chemistry.)

rate of 1 mm/min. In addition, before initiating the test, the point of zero displacement was set at the point where the measured tensile stress was 1 N. The reason of failure of poly(styrene-alt-maleic anhydride) coating being the shielding effect of long aliphatic tails of the copolymer, which prevented the interaction between the acid groups present in the polymer with the substrate thereby resulted in poor adhesion (Soer et al. 2009).

According to a report on waterborne coatings, Francisco et al. (Francisco and Claudino 2016) have carried out life prediction analysis of polymer coatings in terms of adhesion strength *via* Pull-off test. Prior to the analysis, the exposed samples were cured at 77°C with relative humidity maintained at 65% RH. The adhesion strength of the coated samples was measured twice before and after subjecting the samples to rigorous chemical tests (salt-spray and cyclic weathering) for a minimum duration of 1344 hours and the results were reported in terms of any loss of adhesion post exposure (Francisco and Claudino 2016). On another note, the adhesion behavior of elastomeric polymer coatings was measured in terms of modelling parameters and peeling behavior as proposed by Kendall and team (Kendall 1971; Verma, Mohanty, and Nayak 2019). The siloxane duplex coatings were studied using modified or extended Kendall's model to report the pull-off force required to detach a rigid cylinder bonded with commercial epoxy glue from a dual-layer elastomeric coating system (Kohl and Singer 1999). It was found that the pull-off force is inversely proportional to the thickness of each coating layer. Contrastingly, in another research study by Huang et al. (2001) on polyether-polydimethylsiloxane copolymer it was observed that the static coefficient of friction (μ_s) affects the adhesion performance of the polymer coating with respect to the different substrates it is cast onto. Hence, it is suggested to measure and optimize the μ_s values prior to casting and evaluating the adhesion strength of the polymer coating system (Huang et al. 2001). In another research on epoxy-polydimethylsiloxane nanocomposite coatings over mild steel substrates, the authors reported the digital images post conducting the Pull-off adhesion test as per ASTM D 4541-09 using a portable adhesion tester (Verma, Mohanty, and Nayak 2020). Figure 15.10 represents the Pull-off adhesion strength test results represented

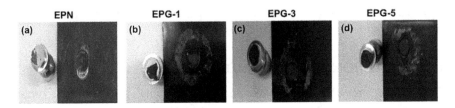

FIGURE 15.10 The digital images of the coated substrates after pull-off adhesion test at various compositions (Verma, Mohanty, and Nayak 2020). (Reproduced by permission of The Royal Society of Chemistry.)

as digital images captured after the occurrence of failure (cohesive or adhesive) on various coating systems. The authors revealed that on increasing the loading of graphene oxide nanosheets beyond the critical filler concentration, an increase in the extraction of topcoat layer was observed. On comparing the properties of neat and nanocomposite coatings, the nanocomposite at optimized concentration showed better adhesion strength. This might be due to the reaction of organosilicon adhesion promoter (3-aminopropyltriethoxysilane) used in the study, which reacts with various functional groups of graphene oxide nanosheets and promotes interlayer adhesion in the designated multilayer coating system.

15.6.2 T-Peel Test

The peel test is one of the most commonly used test to report the adhesion performance in terms of fracture energy of the adhesive coat and the substrate system (Crocombe 2005). The test method known as 180° peel test as per ASTM D903 is suitable for one rigid (substrate) and one flexible polymeric film (paint coating), which can be easily bent up to 180° without cracking as reported by (Rudawska 2019). The parameters that needs to be optimized are: rate of peeling, coating thickness, dimension of specimen, and preconditioning for at least 72 hours prior to testing in standard laboratory conditions (23°C±2°C, 50%±5% RH) (Martin-Martinez 2005). The adhesion strength or bond strength (Newton per meter) calculated from an average of minimum five samples per composition along with the type of failure occurred is reported: (1) adhesive failure (A), (2) cohesive failure (C), (3) noncoalescence failure (N), (4) surface cohesive failure (S) and (5) cohesive failure of substrate (Martin-Martinez 2005). This test can also be utilized for durability analysis or life-cycle assessment of coatings after subjecting them to aging experiments. In another research study, the adhesion strength of urethane coatings on glass substrate was measured by a 180° peel method (Krongauz, Schmid, and Vandeberg 1995). After preconditioning of the cured films (2.54 cm width) deposited onto glass substrate, it was taken out of the humidity chamber and the peel resistance was evaluated using UTM-Instron machine. The peel-test test results were reported in terms of peel resistance force for (1) dry adhesion, (2) wet adhesion and (3) % transmission (ratio of attached area to the total film area). This force required to peel the cured coating from the glass varied at different concentrations of various alkylsilane compounds used as adhesion promoters. The authors concluded that the incorporation of alkylsilane reduced the loss of adhesion as compared to the nonsilane-based coatings (Krongauz, Schmid, and Vandeberg 1995).

15.7 CRITERION FOR GOOD ADHESION

The adhesion strength of the polymer coating is basically attributed to the surface forces acting between the atoms of the two surfaces, viz. polymer coating and the substrate on grounds of different interactions among them. These interactions should evolve sufficiently between the two surfaces in order to promote good adhesion. On the contrary, the formation of boundary layer between the polymer coating and the substrate results in the poor adhesion performance discussed as follows. The boundary layer theory by Bikerman (1967) proposed that the occurrence of a finite boundary layer deposited at the interface consisting of adsorbed molecules from surrounding environment such as foreign materials is the major cause of poor adhesion. In this regard, Figure 15.11 discusses the effects of formation of weak boundary layer over the surface of polymer coating, which can penetrate through the adhesive layer from the top surface as well as from cross section as depicted.

As evident from Figure 15.11, the external stresses such as moisture, humidity, or atmospheric oxides are capable of forming a weak boundary layer over the surface of the polymer coating. These weak boundary layers are formed due to several reasons and the exact composition is unpredictable and difficult to determine. The common examples of weak boundary layer are air entrapment, low molecular weight impurities, oxide layers, catalytic air oxidation and corrosion products (Petrie 2000). Hence, the polymer coating should be able to withstand these external stresses by disrupting any contact between the absorbed molecules of external agents and the bulk material through its outstanding surface properties. If any coating is able to fulfil the aforementioned condition by withstanding the stresses and displays outstanding surface properties, then it is supposed to obey the criteria for good adhesion. Otherwise, in case of any failure, the rupture of adhesion bond proceeds through a single material initiated at a point where the external stress exceeds the adhesive bond strength of the coating (Petrie 2000; Bikerman 1967).

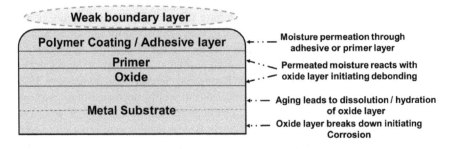

FIGURE 15.11 Mechanism of operation of weak boundary layer within a polymer coating system.

15.8 CONCLUSIONS

The adhesion of polymeric coatings is governed by the chemical nature of the polymer matrix, fillers, or modifiers added to alter its properties; is the outcome of different strategies used for its polymerization; and relies on the resultant cured geometry. A good adhesion is not only restricted to better and improved adhesion strength but is also interrelated to other properties of polymer coatings such as corrosion resistance and hydrophobicity. There exists a strong synergy between the adhesion performance of polymer coatings and the aforementioned properties. This structure–property relationship is affected right from the beginning of fabrication and designing of the bonded structures such as surface topography, loading modes, fracture toughness, and other mechanical properties of the polymer coating. There have been recent developments on these structure–property relationships and the factors controlling the adhesive performance and the ways to improve it, are emphasized in the current study.

For the future outlook of work, more sophisticated analysis on certain advanced characterization techniques needs to be introduced for evaluation of adhesive performance of polymer coatings using various modeling and simulation techniques such as Monte Carlo simulation. Alternatively, an in-depth analysis to study the effect of stresses on debonding under various conditions such as quasi-static, creep, impact and fatigue loading should also be conducted. This will result in determining novel adhesive characteristic properties of paints and coatings, and will reciprocate in appropriate selection of polymer coating system to bond properly with the respective substrate.

REFERENCES

3M. 2020. "Adhesion Basics: Introduction to the Science of Adhesion." https://www.3m.com/3M/en_US/bonding-and-assembly-us/resources/adhesion-basics/.

Agrawal, R. K., and L. T. Drzal. 1996. "Adhesion mechanisms of polyurethanes to glass surfaces. III. Investigation of possible physico-chemical interactions at the interphase." *Journal of Adhesion* 55 (3–4): 221–43. doi:10.1080/00218469608009949.

Ahagon, A., and A. N. Gent. 1975. "Effect of interfacial bonding on the strength of adhesion." *Journal of Polymer Science: Polymer Physics Edition* 13 (7): 1285–1300.

American Society for Testing and Materials. 2009. *"ASTM D4541–09: Standard Test Method for Pull-Off Strength of Coatings Using Portable Adhesion Testers."* West Conshohocken, PA: ASTM D4541-09

Balabanava, N., R. Wierzbicki, M. Zielecka, and Z. Rymuza. 2007. "Effect of roughness on adhesion of polymeric coatings used for microgrippers." *Microelectronic Engineering* 84 (5–8): 1227–30. doi:10.1016/j.mee.2007.01.183.

Bikerman, J. J. 1967. "Causes of poor adhesion: Weak boundary layers." *Industrial & Engineering Chemistry* 59 (9): 40–44. doi:10.1021/ie51403a010.

Comyn, J. 2005. "What are adhesives and sealants and how do they work?" In *Adhesive Bonding Science, Technology and Applications*, edited by R. D. Adams, 1st ed., 23–50. Cambridge: Woodhead Publishing Limited.

Crocombe, A. 2005. "Stress analysis." In *Adhesive Bonding Science, Technology and Applications*, edited by R. D. Adams, 1st ed., 104–5. England: Woodhead Publishing Limited.

Dey, S., S. Chatterjee, B. P. Singh, S. Bhattacharjee, T. K. Rout, D. K. Sengupta, and L. Besra. 2018. "Development of superhydrophobic corrosion resistance coating on mild steel by electrophoretic deposition." *Surface and Coatings Technology* 341: 24–30. doi:10.1016/j.surfcoat.2018.01.005.

Dillard, D. A. 2002. "Fundamentals of stress transfer in bonded systems." In *Adhesion Science & Engineering -I The Mechanics of Adhesion*, edited by A.V. Pocius and D.A. Dillard, 4–50. New York: Elsevier.

Esfandeh, M., S. M. Mirabedini, S. Pazokifard, and M. Tari. 2007. "Study of silicone coating adhesion to an epoxy undercoat using silane compounds effect of silane type and application method." *Colloids and Surfaces A: Physicochemical and Engineering Aspects* 302: 11–16. doi:10.1016/j.colsurfa.2007.01.031.

Francisco, J. S., and T. F. Claudino. 2016. "Waterborne coatings for heavy duty applications in M&PC market." In SSPC 2011 Conference Proceedings (SSPC: The Society for Protective Coatings), ASM International, 1–16.

Huang, W., Y. Yao, Y. Huang, and Y. Yu. 2001. "Surface modification of epoxy resin by polyether – polydimethylsiloxanes – polyether triblock copolymers." *Polymer* 42: 1763–66.

Kendall, K. 1971. "The adhesion and surface energy of elastic solids." *Journal of Physics D: Applied Physics* 4 (8): 1186–95. doi:10.1088/0022–3727/4/8/320.

Kim, W.-s., I.-h. Yun, J.-j. Lee, and H.-t. Jung. 2010. "Evaluation of mechanical interlock effect on adhesion strength of polymer – metal interfaces using micro-patterned surface topography." *International Journal of Adhesion and Adhesives* 30. Elsevier: 408–17. doi:10.1016/j.ijadhadh.2010.05.004.

Kinloch, A. J. 1982. "The science of adhesion - part 2 mechanics and mechanisms of failure." *Journal of Materials Science* 17 (3): 617–51. doi:10.1007/BF00540361.

Kohl, J. G, and I. L. Singer. 1999. "Pull-off behavior of epoxy bonded to silicone duplex coatings." *Progress in Organic Coatings* 36: 15–20.

Krongauz, V. V., S. R. Schmid, and J. T. Vandeberg. 1995. "Imaging in evaluation of polymer coatings adhesion to glass at high humidity." *Progress in Organic Coatings* 26: 145–62. doi:10.1016/0300-9440(95)00565-X.

Liechti, K. M. 2002. "Fracture mechanics and singularities in bonded systems." In *The Mechanics of Adhesion*, edited by D.A. Dillard and A.V. Pocius, 46–48. Amsterdam: Elsevier.

Martin-Martinez, J M. 2005. "Shoe industry." In *Adhesive Bonding Science, Technology and Applications*, edited by R. D. Adams, 1st ed., 441–42. England: Woodhead Publishing Limited.

Mohamadpour, S., B. Pourabbas, and P. Fabbri. 2011. "Anti-scratch and adhesion properties of photo-curable polymer/clay nanocomposite coatings based on methacrylate monomers." *Scientia Iranica* 18 (3 F). Elsevier B.V: 765–71. doi:10.1016/j.scient.2011.06.001.

Pape, P. G. 2018. *"Silane - an Overview."* *Elsevier.* https://www.sciencedirect.com/toipcs/chemistry/silane.

Petrie, E. M. 2000. *Handbook of Adhesives and Sealants.* 2nd ed. New York: McGraw-Hill Publishing.

Rudawska, A. 2019. *Peel Test.* Elsevier. sciencedirect.com/topic/engineering/peel-test.

Soer, W. J., W. Ming, C. E. Koning, R. A. T. M. van Benthem, J. M. C. Mol, and H. Terryn. 2009. "Barrier and adhesion properties of anti-corrosion coatings based on surfactant-free latexes from anhydride-containing polymers." *Progress in Organic Coatings* 65: 94–103. doi:10.1016/j.porgcoat.2008.10.003.

Subramanian, M. N. 2013. "Chapter 3: Types of additives." In *Plastics Additives and Testing*, 1st ed., 53–56. Salem: Scrivener Publishing.

Talbert R. 2008. *Paint Technology Handbook.* Boca Raton, FL: CRC Press.

Verma, S., S. Mohanty, and S. K. Nayak. 2019. "A review on protective polymeric coatings for marine applications." *Journal of Coatings Technology and Research* 16 (2). Springer US: 307–38. doi:10.1007/s11998-018-00174-2.

Verma, S., S. Mohanty, and S. K. Nayak. 2020. "Preparation of hydrophobic epoxy – polydimethylsiloxane – graphene oxide nanocomposite coatings for antifouling application." *Soft Matter* 16. Royal Society of Chemistry: 1211–26. doi:10.1039/C9SM01952A.

Watts, J F. 2005. "Surfaces: How to assess." In *Adhesive Bonding Science, Technology and Applications*, edited by R. D. Adams, 84–104. Cambridge: Woodhead Publishing Limited.

Weldon, D. G. 2009. *Failure Analysis of Paints and Coatings*. Revised. Imperial, PA.

Whitehouse, R.S. 1987. "Silicon adhesives, sealants and coupling agents." In *Synthetic Adhesives and Sealants* Edited by W.C. Wake, 98–99. 1st ed. Chichester: John Wiley & Sons.

16 Modeling and Simulation in Polymer Coatings

Akarsh Verma, Avinash Parashar,
and Sandeep Kumar Singh
Indian Institute of Technology, Roorkee, India

Naman Jain
Meerut Institute of Engineering and
Technology, Meerut, India

Sanjay Mavinkere Rangappa
and Suchart Siengchin
King Mongkut's University of Technology North Bangkok

CONTENTS

16.1 INTRODUCTION

Polymer coatings are designed for various demands in engineering sector and, therefore, have to fulfill different requirements according to the demand such as in chemical sensing, corrosion protection, and nanocomposites [1–10]. Hence, it includes a variety of synthetic materials according to the applications. Polymer coatings can be applied to a metal surface to provide anticorrosion protection from the corrosive causing elements. On the contrary, the polymer coatings develop very often microdefects (void sites) in structural applications, reducing the lifespan of the material. Therefore, an early sensing, diagnosis, and repair of these microcracks occurred by polymer coatings are important [11].

Molecular modeling and simulation techniques are pretty imperative tools for polymer/materials science and engineering [12–18]. These computational methods allow predictions and offer explanations to experimentally observed structure of the

309

polymers and various macroscopic/microscopic, mechanical, fracture, and thermal properties. With recent advancements in the computing power and reduction in the computational cost, polymer coating simulations can guide the research scientists to molecular materials design. To ensure that simulation and modeling are giving a correct representation of data, one must care to ensure the validity, accuracy, and reproducibility of the simulation results [19–21].

16.2 WHICH MODELING AND SIMULATION TECHNIQUE SHOULD I USE?

It is essential for those who are thinking of using modeling and simulation as a potential solution tool for their research problem related to the polymer coating system, "why and how do I want the modeling and simulation to accomplish and solve the problem?" Before proceeding further, one must know the answer to the question mentioned. The "best model" depends on the problem on which one wants to work and the answer to question. Broadly, for classical molecular dynamics simulations, most technical problems come under two categories: the first one, atomistic or coarse-grained model, the latter containing either generic models or models that are parameterized/optimized for specific chemistries [22,23].

16.2.1 Molecular Dynamics Simulations

Molecular dynamics (MD)-based technique is an emerging substitute to perform the atomistic simulations for various applications. In the past decade, the applications of MD-based simulations have been extended in the field of materials science and polymer science to predict the properties of conventional as well as nonconventional materials [24–31]. In MD-based simulations, atoms are treated as the classical particles, and interaction between them is estimated with the help of empirically derived (or from high-fidelity quantum mechanics simulations) interatomic potentials. The fundamental equation solved in MD-based simulation is mathematically given by Eq. (16.1):

$$m_\alpha a_\alpha = F_\alpha = -\left(\partial E/\partial r_\alpha\right)$$
$$\text{where } \alpha = 1, 2, ..., N \tag{16.1}$$

The term potential energy "E" is composed of an internal part (E_{int}) and an external part (E_{ext}) that deal with the interaction between atoms and external fields, respectively. For updating position and velocity vectors after each time step, certain numerical integration algorithms are necessary; for example, the Verlet, velocity-Verlet (an extended version of the Verlet algorithm), leapfrog, and Beeman's algorithm come in handy. Among these algorithms, velocity-Verlet is the most commonly utilized scheme in the MD simulations. The velocity-Verlet algorithm utilizes the fundamental Taylor's expansion for the iteration process to continue. Expressions for the position vector "r" and velocity vector "v" at time "t," subsequent to an increment t are given by Eqs. (16.2) and (16.3), respectively:

$$r(t + \Delta t) \approx r(t) + \dot{r}(t)\Delta t + (\ddot{r}(t)\Delta t^2) / 2 + ..., \qquad (16.2)$$

$$v(t + \Delta t) = v(t) + (F(t + \Delta t) + F(t) / 2m)\Delta t. \qquad (16.3)$$

16.2.2 DENSITY FUNCTIONAL THEORY

Quantum mechanics-based density functional theory (DFT) is considered as the most accurate computational tools amongst all the other computational techniques. The fundamental equation (Schrödinger equation) for the electronic wave functions in presence of a potential field generated by stationary atoms is solved in quantum mechanics framework [32–40].

The total energy $E[\rho]$, as a functional of the electron energy density "ρ", is expressed mathematically through Eq. (16.4):

$$E[\rho] = T[\rho] + E_H[\rho] + E_{xc}[\rho] + E_{ext}[\rho] + E_{zz}. \qquad (16.4)$$

Solving the Schrödinger equation without approximations is a tedious and highly computationally intensive task; hence, Kohn–Sham (KS) has proposed a simplified form of the equation for performing DFT calculations, which provides relatively good approximation to the energy calculations [41]. KS equations for this new system model are given by Eqs. (16.5) and (16.6):

$$\left[\frac{1}{2}\nabla^2 + V_{eff}(r) \right] \phi_i(r) = \epsilon_i \, \phi_i, \qquad (16.5)$$

where

$$V_{eff} = \frac{\delta(E_H[\rho] + E_{xc}[\rho] + E_{ext}[\rho]}{\delta\rho} = V_H[\rho] + V_{xc}[\rho] + V_{ext}[\rho] \qquad (16.6)$$

is called the KS or effective potential, ϕ_i are the KS one-electron orbitals, whereas the electron density $\rho(r)$ is given by Eq. (16.7):

$$\rho(r) = \sum_{i=1}^{N} |\varphi_i|^2. \qquad (16.7)$$

In order to solve the KS equations, an initial guess of electron density "ρ" is required. In a solid-state system, the pseudopotentials (PPs) are used for defining interactions between the core and valence electrons. In PP approach, the key is dividing electrons into two categories depending on their contributing significances. The first category belongs to core electrons, which do not participate significantly in the bonding process and can be ignored, whereas the other category belongs to valence electrons, which are lesser in number and are treated with plane-wave basis sets.

PPs are of two types: empirical PP and ab initio PP. Ab initio PPs are commonly preferred over the empirical PPs, and this is based on the local solutions of wave functions near the atomic nuclei. However, the nonlocal form of PPs is used in more accurate DFT implementations. The PP approach is much faster in all electron calculations because tightly bound core electrons are not taken directly into consideration. The process of solving KS equations for development of the density matrix from the Fock/KS matrix and after that solving for density functional is a costly process, which makes DFT a computationally intensive technique with lots of tedious tasks.

Strain energy (U) obtained from DFT-based simulations is illustrated in Eq. (16.8):

$$U = U_\varepsilon - U_{\varepsilon=0}. \qquad (16.8)$$

This strain energy is again processed to predict Young's modulus (Y in N/m^2) and 2D Young's modulus (Y_s in N/m) as demonstrated in Eqs. (16.9) and (16.10), respectively:

$$Y = (\partial^2 U / \partial \varepsilon^2) / V. \qquad (16.9)$$

$$Y_s = (\partial^2 U / \partial \varepsilon^2) / S. \qquad (16.10)$$

Here, S and V are the surface area and volume of the atomistic system, respectively.

16.3 CASE STUDIES

In 2009, Griffin [42] developed a mathematical model to grasp the temperature profile through a polymer coating when a radiant heat source is exposed to their surface. He included the effects of many parameters that include the exothermic and endothermic reactions, radiation and convection heat transfer, and the thermal resistance that enlarge because of the coating expansion and various gases formation. Furthermore, the results were validated with the experimental observations in conjunction with a sensitivity analysis.

Earlier in 2000, Price and Cairncross [43] described mathematically a drying model that would find the optimal drying conditions for the polymer solution coating (refer Figure 16.1 for the model description). They pondered over the air temperature and heat transfer coefficients on the substrate and coating sides, and concluded that the coating side heat transfer coefficient is always greater than or equal to the substrate side.

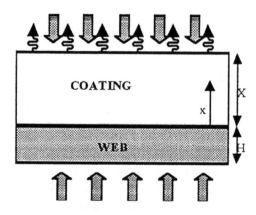

FIGURE 16.1 Configuration of a typical coating/substrate system. The detailed information can be found from the respective article [43].

Jiang et al. [44] utilized the three-dimensional (3D) finite element simulations to evaluate the stress–strain responses, failure modes, and deformation mechanisms of polymer coatings scratches. As an example, they calculated the von-Mises stress field variation for the acrylic-steel system (refer Figure 16.2) as well as the scratch mechanisms (refer Figure 16.3).

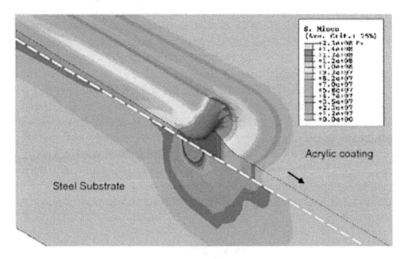

FIGURE 16.2   The von-Mises stress field for the acrylic–steel coating system [44].

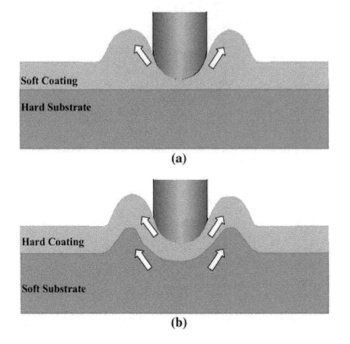

FIGURE 16.3   Illustration of the coating thinning and pile-up under scratch. (a) Soft coating on a hard substrate and (b) hard coating on a soft substrate [44].

Kondoh et al. [45] simulated the surface acoustic wave sensors response with different polymer coating films (glassy, rubbery, glassy-rubbery). Hickey et al. [46] conducted the MD simulations for analyzing quenched electro-osmotic flow through dynamic uncharged polymer coatings (refer Figure 16.4 for the simulation setup). The same group then investigated the grafted and charged polymer coatings for the electro-osmotic flow mobility case [47]. They also used the MD simulations to see the adsorbed polymer coatings effect on the electro-osmotic flow inside a capillary [48].

Kyrychenko et al. [49] predicted the effect of poly(vinylpyrrolidone) oligomer chain length on coating silver nanoparticles with the help of atomistic MD simulations (refer Figure 16.5 for the configuration). Handge et al. [50] analyzed a shear lag model (interfacial shear stress) to investigate the sequential cracking in a polymer coating (using nonlinear elastic stress transfer). Earlier, Tabor et al. [51] and Krzak et al. [52] conducted finite difference-based simulations to investigate the water diffusion through the polymer coatings and sandwich-structured films; and then verified their mathematical model with the experimental techniques. Hopkins et al. [53] provided insights into the mechanism of polymer coatings delamination during stent deployment through computational investigation. The same team combined the experimental and computational approaches to evaluate the stent polymer coatings

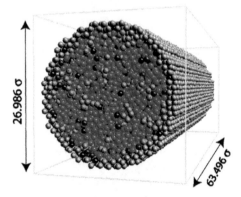

FIGURE 16.4 Snapshot of the system under consideration. The detailed information can be found from the respective article [46].

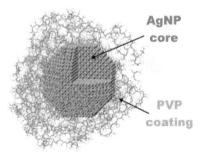

FIGURE 16.5 Snapshot of the system under consideration. The detailed information can be found from the respective article [49].

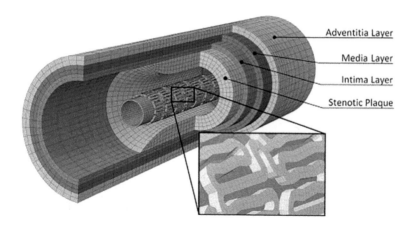

Adventitia Layer
Media Layer
Intima Layer
Stenotic Plaque

FIGURE 16.6 Finite element mesh for balloon, Xience stent, and stenotic artery. The detailed information can be found from the respective article [55].

adhesion properties [54]. Schiavone et al. [55] also checked out the effects of coatings material and design on stent deployment inside a stenotic artery through the finite element simulations (Figure 16.6).

Recently in 2016, Kumar and Manik [56] performed MD simulations to investigate the surface properties of polyvinyl acetate-perfluorooctane-based antistain coatings. Hinderliter and Croll [57] examined the changes in macroscopic properties on coatings due to the weathering, with the help of Monte Carlo simulations. Peters et al. [58] recorded the response of silica nanoparticle coatings to alkane solvents (decane and squalene) through the atomistic simulations. Lyakhova et al. [59] evaluated the interfaces role in self-replenishing composite coatings through the coarse-grained simulations. Sakhnenko et al. [60] predicted and simulated a metal–polymer coating system in order to determine the service life.

Rahmani et al. [61] utilized the reactive molecular simulations to observe the damage mitigation efficacy of polyimide-based coatings when they were exposed to atomic oxygen bombardment (see Figure 16.7 for the simulation setup). Mitsoulis [62] investigated the viscoelastic effects through the blade-over-roll coatings by employing the numerical simulations. In an interesting study done by Rossi et al. [63], the MARTINI force field (MD simulations) was used to describe the coarse-grained particles of thermoset polyester coating. With the help of coupled finite element and meshfree method, Gong et al. [64] simulated the impact on a ductile metal pipe with a coating of polymer over it. Bhatti et al. [65] numerically modeled (finite element technique) a fiber-optic phase modulator using a piezoelectric polymer coating. Bahlakeh et al. [66] did MD and quantum mechanics simulations to reveal the interfacial adhesion mechanism on steel (having cerium oxide film) with epoxy/polyamide coating. Al-Muslimawi et al. [67] used a numerical model to solve various flows for tube-tooling die-extrusion coating. Hsu et al. [68] conducted investigations through finite element analysis on thin polymer substrates when subjected to fatigue bending loadings. Rottler and Robbins [69] explored the mechanical and failure behavior in the glassy polymer bonds through MD simulations. Adema et al.

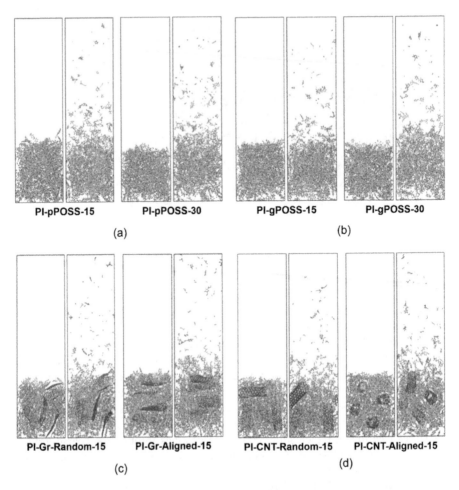

FIGURE 16.7 Simulation setup for atomic oxygen bombardment. The detailed information can be found from the respective article [61].

[70] used a numerical method in order to simulate (by Monte Carlo simulations) the photodegradation process in polyester-urethane polymer coatings.

Jarray et al. [71] conducted a multiscale investigation to check out the plasticizer-polymer compatibility during coating formulation (see Figure 16.8 for the setup snapshot). Paul et al. [72] used 3D finite element simulations to see the consequence of polymer coatings adhesion. Ren et al. [73] explored the electrochemical corrosion characteristics of conducting polymer coatings when subjected to a proton exchange membrane fuel cells. Maksimov and Maksimova et al. [74] simulated the polymer coatings adhesion on the steel sheets surface. Michels and Moons [75] conducted numerical simulations for phase separation in a solution-processed polymer blend. Theoretically, Andrei et al. [76] studied about the surface leveling of thermosetting powder coatings. Wredenberg and Larsson [77] presented the numerical results for the delamination of thin coatings during scratching. Zhu and Braatz [78] developed a

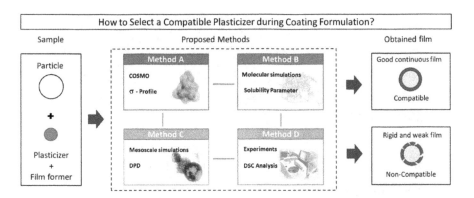

FIGURE 16.8 Selection criteria of a compatible plasticizer during coating formulation. The detailed information can be found from the respective article [71].

mechanistic mathematical model in a polymer stent coating for drug-release application. Zhang et al. [79] used theoretical simulations (using drop-tower technique) for polyvinyl alcohol coating. Arya and Bhargava [80] analyzed the drying of polymer (ternary) coatings through simulation study. Makki et al. [81] employed a novel simulation approach by combining dissipative particle dynamics with the kinetic Monte Carlo method to study the photodegradation (a type of chemical degradation) of polymer coatings. Jayachandran et al. [82] designed a multilayer polymeric coating in order to capture the resistance offered during the indentation process and the results were given by the constitutive models. Iliopoulos and Scriven [83] employed the finite element/Galerkin simulation technique to study a blade-coating design. Lin et al. [84] used the lattice Monte Carlo simulation technique to analyze the conducting polymer coatings morphology. Van der Aa [85] performed a numerical study (based on Lagrange Euler method) to explore the wall ironing process of polymer-coated sheet metal. In 2016, Dey et al. [86] developed a hydrodynamic model to explain the polymer solution in a dip coating in the evaporative regime. Burch et al. [87] did a computer analysis on a polymer coating when exposed to the realistic weather conditions.

Moore et al. [88] used the MD simulations to analyze the coating growth on nanofibers in an axisymmetric manner. Kaunisto et al. [89] discussed various mechanistic models from polymer coated matrix systems for the drug-release application. Monfared and Sharif [90] used a mathematical model and simulated to provide design guidelines for creating the antifouling and self-polishing coatings. Flack et al. [91] developed a mathematical model to derive the mechanisms involved in spin coating of polymer resists. Wang et al. [92] utilized the dissipative particle dynamics simulations technique to explore the mechanism of disassembly and self-assembly of pH-sensitive polymeric micelle with coating repair agent. In 2017, Bahlakeh and Ramezanzadeh [93] conducted both the experimental investigation and MD simulations to investigate the interfacial bonding mechanisms operating between carbon steel sheets and an epoxy adhesive (refer Figure 16.9 for the simulation setup). Li et al. [94] also conducted MD simulations to investigate the absorption process of coating and isolating of olefin group polymer. Xiao et al. [95] predicted a computational

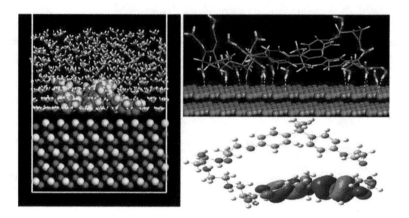

FIGURE 16.9 Snapshots of the simulation setup. The detailed information can be found from the respective article [93].

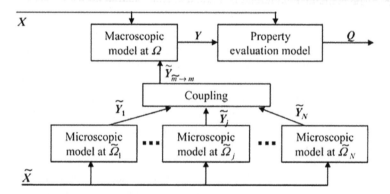

FIGURE 16.10 Multiscale hierarchical modeling framework. The detailed information can be found from the respective article [93].

design of polymer nanocomposite coatings (as seen in the Figure 16.10). Croll and Hinderliter [96] performed the Monte Carlo simulations to approximate the coating service lifetime during natural weathering conditions. Sokolov et al. [97] used a scalar model to examine the crack dynamics of a polymer coating domain.

ACKNOWLEDGMENTS

The authors would like to acknowledge the financial support received by the United States Department of State (DOS), United States-India Educational Foundation (USIEF), and Institute of International Education (IIE). In addition, academic support from the Pennsylvania State University, USA and Indian Institute of Technology Roorkee, India is highly appreciable.

CONFLICTS OF INTEREST

There are no conflicts of interest to declare by the authors.

REFERENCES

1. Cho, S.H., White, S.R. and Braun, P.V., 2009. Self-healing polymer coatings. *Advanced Materials*, *21*(6), pp. 645–649.
2. Verma, A., Baurai, K., Sanjay, M.R. and Siengchin, S., 2019. Mechanical, microstructural, and thermal characterization insights of pyrolyzed carbon black from waste tires reinforced epoxy nanocomposites for coating application. *Polymer Composites*, *41*(1), pp. 338-349.
3. McGill, R.A., McGill, R.A., Abraham, M.H., Abraham, M.H. and Grate, J.W., 1994. Choosing polymer coatings for chemical sensors. *Chemtech*, *24*(PNNL-SA-24862).
4. Verma, A., Budiyal, L., Sanjay, M.R. and Siengchin, S., 2019. Processing and characterization analysis of pyrolyzed oil rubber (from waste tires)-epoxy polymer blend composite for lightweight structures and coatings applications. *Polymer Engineering & Science*, *59*(10), pp. 2041–2051.
5. Tan, C.K. and Blackwood, D.J., 2003. Corrosion protection by multilayered conducting polymer coatings. *Corrosion Science*, *45*(3), pp. 545–557.
6. Bertrand, P., Jonas, A., Laschewsky, A. and Legras, R., 2000. Ultrathin polymer coatings by complexation of polyelectrolytes at interfaces: Suitable materials, structure and properties. *Macromolecular Rapid Communications*, *21*(7), pp. 319–348.
7. Mansfeld, F., 1995. Use of electrochemical impedance spectroscopy for the study of corrosion protection by polymer coatings. *Journal of Applied Electrochemistry*, *25*(3), pp. 187–202.
8. Levkin, P.A., Svec, F. and Fréchet, J.M., 2009. Porous polymer coatings: A versatile approach to superhydrophobic surfaces. *Advanced Functional Materials*, *19*(12), pp. 1993–1998.
9. Nandivada, H., Chen, H.Y., Bondarenko, L. and Lahann, J., 2006. Reactive polymer coatings that "click". *Angewandte Chemie International Edition*, *45*(20), pp. 3360–3363.
10. Wei, Q., Becherer, T., Angioletti-Uberti, S., Dzubiella, J., Wischke, C., Neffe, A.T., Lendlein, A., Ballauff, M. and Haag, R., 2014. Protein interactions with polymer coatings and biomaterials. *Angewandte Chemie International Edition*, *53*(31), pp. 8004–8031.
11. Mphahlele, K., Ray, S.S. and Kolesnikov, A., 2017. Self-healing polymeric composite material design, failure analysis and future outlook: A review. *Polymers*, *9*(10), p. 535.
12. Verma, A. and Parashar, A., 2017. The effect of STW defects on the mechanical properties and fracture toughness of pristine and hydrogenated graphene. *Physical Chemistry Chemical Physics*, *19*(24), pp. 16023–16037.
13. Verma, A., Parashar, A. and Packirisamy, M., 2019. Effect of grain boundaries on the interfacial behaviour of graphene-polyethylene nanocomposite. *Applied Surface Science*, *470*, pp. 1085–1092.
14. Verma, A., Kumar, R. and Parashar, A., 2019. Enhanced thermal transport across a bi-crystalline graphene–polymer interface: An atomistic approach. *Physical Chemistry Chemical Physics*, *21*(11), pp. 6229–6237.
15. Chaurasia, A., Verma, A., Parashar, A. and Mulik, R.S., 2019. Experimental and computational studies to analyze the effect of h-BN nanosheets on mechanical behavior of h-BN/polyethylene nanocomposites. *The Journal of Physical Chemistry C*, *123*(32), pp. 20059–20070.
16. Molenaar, J. and Koopmans, R.J., 1994. Modeling polymer melt-flow instabilities. *Journal of Rheology*, *38*(1), pp. 99–109.
17. Yashin, V.V. and Balazs, A.C., 2006. Modeling polymer gels exhibiting self-oscillations due to the Belousov–Zhabotinsky reaction. *Macromolecules*, *39*(6), pp. 2024–2026.
18. Odegard, G.M., Gates, T.S., Wise, K.E., Park, C. and Siochi, E.J., 2003. Constitutive modeling of nanotube–reinforced polymer composites. *Composites Science and Technology*, *63*(11), pp. 1671–1687.

19. Makki, H., Adema, K.N., Peters, E.A., Laven, J., van der Ven, L.G., van Benthem, R.A. and de With, G., 2016. Multi-scale simulation of degradation of polymer coatings: Thermo-mechanical simulations. *Polymer Degradation and Stability, 123*, pp. 1–12.

20. Hinderliter, B. and Croll, S., 2005. Monte Carlo approach to estimating the photodegradation of polymer coatings. *Journal of Coatings Technology and Research, 2*(6), pp. 483–491.

21. Fermeglia, M. and Pricl, S., 2007. Multiscale modeling for polymer systems of industrial interest. *Progress in Organic Coatings, 58*(2–3), pp. 187–199.

22. Gartner III, T.E. and Jayaraman, A., 2019. Modeling and simulations of polymers: A Roadmap. *Macromolecules, 52*(3), pp. 755–786.

23. Allen, M.P. and Tildesley, D.J., 2017. *Computer Simulation of Liquids.* Oxford: Oxford University Press.

24. Verma, A. and Parashar, A., 2018. Molecular dynamics based simulations to study failure morphology of hydroxyl and epoxide functionalised graphene. *Computational Materials Science, 143*, pp. 15–26.

25. Verma, A. and Parashar, A., 2018. Molecular dynamics based simulations to study the fracture strength of monolayer graphene oxide. *Nanotechnology, 29*(11), p. 115706.

26. Verma, A., Parashar, A. and Packirisamy, M., 2018. Tailoring the failure morphology of 2D bicrystalline graphene oxide. *Journal of Applied Physics, 124*(1), p. 015102.

27. Verma, A. and Parashar, A., 2018. Reactive force field based atomistic simulations to study fracture toughness of bicrystalline graphene functionalised with oxide groups. *Diamond and Related Materials, 88*, pp. 193–203.

28. Verma, A. and Parashar, A., 2018. Structural and chemical insights into thermal transport for strained functionalised graphene: A molecular dynamics study. *Materials Research Express, 5*(11), p. 115605.

29. Singla, V., Verma, A. and Parashar, A., 2018. A molecular dynamics based study to estimate the point defects formation energies in graphene containing STW defects. *Materials Research Express, 6*(1), p. 015606.

30. Zeng, Q.H., Yu, A.B. and Lu, G.Q., 2008. Multiscale modeling and simulation of polymer nanocomposites. *Progress in Polymer Science, 33*(2), pp. 191–269.

31. Neyts, E.C. and Bogaerts, A., 2014. Combining molecular dynamics with Monte Carlo simulations: implementations and applications. In *Theoretical Chemistry in Belgium* (pp. 277–288). Berlin, Heidelberg: Springer.

32. Runge, E. and Gross, E.K., 1984. Density-functional theory for time-dependent systems. *Physical Review Letters, 52*(12), p. 997.

33. Sham, L.J. and Schlüter, M., 1983. Density-functional theory of the energy gap. *Physical Review Letters, 51*(20), p. 1888.

34. Petersilka, M.G.U.J., Gossmann, U.J. and Gross, E.K.U., 1996. Excitation energies from time-dependent density-functional theory. *Physical Review Letters, 76*(8), p. 1212.

35. Tannor, D.J., 2007. *Introduction to Quantum Mechanics: A Time-Dependent Perspective.* Sausalito, CA: University Science Books.

36. Cooper, F., Khare, A. and Sukhatme, U., 1995. Supersymmetry and quantum mechanics. *Physics Reports, 251*(5–6), pp. 267–385.

37. Bender, C.M., Brody, D.C. and Jones, H.F., 2002. Complex extension of quantum mechanics. *Physical Review Letters, 89*(27), p. 270401.

38. Parr, R.G., 1980. Density functional theory of atoms and molecules. In *Horizons of Quantum Chemistry* (pp. 5–15). Dordrecht: Springer.

39. Yang, W., 1991. Direct calculation of electron density in density-functional theory. *Physical Review Letters, 66*(11), p. 1438.

40. Gross, E.K.U. and Kohn, W., 1990. Time-dependent density-functional theory. In *Advances in Quantum Chemistry* (Vol. 21, pp. 255–291). Boston: Academic Press.

41. Verma, A., Parashar, A. and Packirisamy, M., 2018. Atomistic modeling of graphene/hexagonal boron nitride polymer nanocomposites: A review. *Wiley Interdisciplinary Reviews: Computational Molecular Science*, 8(3), p.e1346.
42. Griffin, G.J., 2010. The modeling of heat transfer across intumescent polymer coatings. *Journal of Fire Sciences*, 28(3), pp. 249–277.
43. Price Jr, P.E. and Cairncross, R.A., 2000. Optimization of single-zone drying of polymer solution coatings using mathematical modeling. *Journal of Applied Polymer Science*, 78(1), pp. 149–165.
44. Jiang, H., Browning, R., Whitcomb, J.D., Ito, M., Shimouse, M., Chang, T.A. and Sue, H.J., 2010. Mechanical modeling of scratch behavior of polymeric coatings on hard and soft substrates. *Tribology Letters*, 37(2), pp. 159–167.
45. Kondoh, J., Shiokawa, S., Rapp, M. and Stier, S., 1998. Simulation of viscoelastic effects of polymer coatings on surface acoustic wave gas sensor under consideration of film thickness. *Japanese Journal of Applied Physics*, 37(5S), p. 2842.
46. Hickey, O.A., Harden, J.L. and Slater, G.W., 2009. Molecular dynamics simulations of optimal dynamic uncharged polymer coatings for quenching electro-osmotic flow. *Physical Review Letters*, 102(10), p. 108304.
47. Hickey, O.A., Holm, C., Harden, J.L. and Slater, G.W., 2011. Influence of charged polymer coatings on electro-osmotic flow: molecular dynamics simulations. *Macromolecules*, 44(23), pp. 9455–9463.
48. Hickey, O. A., Harden, J.L. and Slater, G.W., 2012. Computer simulations of time-dependent suppression of EOF by polymer coatings. *Microfluidics and Nanofluidics*, 13(1), pp. 91–97.
49. Kyrychenko, A., Korsun, O.M., Gubin, I.I., Kovalenko, S.M. and Kalugin, O.N., 2015. Atomistic simulations of coating of silver nanoparticles with poly (vinylpyrrolidone) oligomers: Effect of oligomer chain length. *The Journal of Physical Chemistry C*, 119(14), pp. 7888–7899.
50. Handge, U.A., 2002. Analysis of a shear-lag model with nonlinear elastic stress transfer for sequential cracking of polymer coatings. *Journal of Materials Science*, 37(22), pp. 4775–4782.
51. Tabor, Z., Krzak, M., Nowak, P. and Warszyński, P., 2012. Water diffusion in polymer coatings containing water-trapping particles. Part 1. Finite difference-based simulations. *Progress in Organic Coatings*, 75(3), pp. 200–206.
52. Krzak, M., Tabor, Z., Nowak, P., Warszyński, P., Karatzas, A., Kartsonakis, I.A. and Kordas, G.C., 2012. Water diffusion in polymer coatings containing water-trapping particles. Part 2. Experimental verification of the mathematical model. *Progress in Organic Coatings*, 75(3), pp. 207–214.
53. Hopkins, C.G., McHugh, P.E. and McGarry, J.P., 2010. Computational investigation of the delamination of polymer coatings during stent deployment. *Annals of Biomedical Engineering*, 38(7), pp. 2263–2273.
54. Hopkins, C., McHugh, P.E., O'dowd, N.P., Rochev, Y. and McGarry, J.P., 2013. A combined computational and experimental methodology to determine the adhesion properties of stent polymer coatings. *Computational Materials Science*, 80, pp. 104–112.
55. Schiavone, A., Zhao, L.G. and Abdel-Wahab, A.A., 2014. Effects of material, coating, design and plaque composition on stent deployment inside a stenotic artery—Finite element simulation. *Materials Science and Engineering: C*, 42, pp. 479–488.
56. Kumar, N. and Manik, G., 2016. Molecular dynamics simulations of polyvinyl acetate-perfluorooctane based anti-stain coatings. *Polymer*, 100, pp. 194–205.
57. Hinderliter, B.R. and Croll, S.G., 2006. Simulations of nanoscale and macroscopic property changes on coatings with weathering. *Journal of Coatings Technology and Research*, 3(3), pp. 203–212.

58. Peters, B.L., Lane, J.M.D., Ismail, A.E. and Grest, G.S., 2012. Fully atomistic simulations of the response of silica nanoparticle coatings to alkane solvents. *Langmuir*, *28*(50), pp. 17443–17449.

59. Lyakhova, K., Esteves, A.C.C., van de Put, M.W., van der Ven, L.G., van Benthem, R.A. and de With, G., 2014. Simulation-experimental approach to investigate the role of interfaces in self-replenishing composite coatings. *Advanced Materials Interfaces*, *1*(3), p. 1400053.

60. Sakhnenko, N.D., Ved, M.V. and Nikiforov, K.V., 1998. Simulation of metal–polymer coating system for service life prediction. *Simulation Practice and Theory*, *6*(7), pp. 647–656.

61. Rahmani, F., Nouranian, S., Li, X. and Al-Ostaz, A., 2017. Reactive molecular simulation of the damage mitigation efficacy of POSS-, graphene-, and carbon nanotube-loaded polyimide coatings exposed to atomic oxygen bombardment. *ACS Applied Materials & Interfaces*, *9*(14), pp. 12802–12811.

62. Mitsoulis, E.V.A.N., 2010. Numerical simulation of viscoelastic effects in blade-over-roll coating forming flows. *Computer Methods in Materials Science*, *10*(3), pp. 156–166.

63. Rossi, G., Giannakopoulos, I., Monticelli, L., Rostedt, N.K., Puisto, S.R., Lowe, C., Taylor, A.C., Vattulainen, I. and Ala-Nissila, T., 2011. A MARTINI coarse-grained model of a thermoset polyester coating. *Macromolecules*, *44*(15), pp. 6198–6208.

64. Gong, Y., Yang, Z.G. and Wang, Y.F., 2012. Impact simulation on ductile metal pipe with polymer coating by a coupled finite element and meshfree method. *Journal of Failure Analysis and Prevention*, *12*(3), pp. 267–272.

65. Bhatti, A., Al-Raweshidy, H.S. and Murtaza, G., 1999. Numerical modeling of a fiber-optic phase modulator using piezoelectric polymer coating. *IEEE Photonics Technology Letters*, *11*(7), pp. 812–814.

66. Bahlakeh, G., Ramezanzadeh, B., Saeb, M.R., Terryn, H. and Ghaffari, M., 2017. Corrosion protection properties and interfacial adhesion mechanism of an epoxy/polyamide coating applied on the steel surface decorated with cerium oxide nanofilm: complementary experimental, molecular dynamics (MD) and first principle quantum mechanics (QM) simulation methods. *Applied Surface Science*, *419*, pp. 650–669.

67. Al-Muslimawi, A., Tamaddon-Jahromi, H.R. and Webster, M.F., 2013. Numerical simulation of tube-tooling cable-coating with polymer melts. *Korea-Australia Rheology Journal*, *25*(4), pp. 197–216.

68. Hsu, J.S., Lee, C.C., Wen, B.J., Huang, P.C. and Xie, C.K., 2016. Experimental and simulated investigations of thin polymer substrates with an indium tin oxide coating under fatigue bending loadings. *Materials*, *9*(9), p. 720.

69. Rottler, J. and Robbins, M.O., 2003. Molecular simulations of deformation and failure in bonds formed by glassy polymer adhesives. *Journal of Adhesion Science and Technology*, *17*(3), pp. 369–381.

70. Adema, K.N., Makki, H., Peters, E.A., Laven, J., van der Ven, L.G. and van Benthem, R.A., 2015. Kinetic Monte Carlo simulation of the photodegradation process of polyester-urethane coatings. *Physical Chemistry Chemical Physics*, *17*(30), pp. 19962–19976.

71. Jarray, A., Gerbaud, V. and Hemati, M., 2016. Polymer-plasticizer compatibility during coating formulation: A multi-scale investigation. *Progress in Organic Coatings*, *101*, pp. 195–206.

72. Paul, J., Liping, Z., Ngoi, B. and Zhong Ping, F., 2004. Bragg grating temperature sensors: modeling the effect of adhesion of polymeric coatings. *Sensor Review*, *24*(4), pp. 364–369.

73. Ren, Y.J., Chen, J., Zeng, C.L., Li, C. and He, J.J., 2016. Electrochemical corrosion characteristics of conducting polypyrrole/polyaniline coatings in simulated environments of a proton exchange membrane fuel cell. *International Journal of Hydrogen Energy*, *41*(20), pp. 8542–8549.

74. Maksimov, A. and Maksimova, O., 2014. Simulation of adhesion of polymer coatings on the surface of steel sheets. *Journal of Chemical Engineering and Chemistry Research, 1*(2), pp. 94–100.
75. Michels, J.J. and Moons, E., 2013. Simulation of surface-directed phase separation in a solution-processed polymer/PCBM Blend. *Macromolecules, 46*(21), pp. 8693–8701.
76. Andrei, D.C., Hay, J.N., Keddie, J.L., Sear, R.P. and Yeates, S.G., 2000. Surface levelling of thermosetting powder coatings: Theory and experiment. *Journal of Physics D: Applied Physics, 33*(16), p. 1975.
77. Wredenberg, F. and Larsson, P.L., 2009. Delamination of thin coatings at scratching: experiments and numerics. *Journal of Mechanics of Materials and Structures, 4*(6), pp. 1041–1062.
78. Zhu, X. and Braatz, R.D., 2015. A mechanistic model for drug release in PLGA biodegradable stent coatings coupled with polymer degradation and erosion. *Journal of Biomedical Materials Research Part A, 103*(7), pp. 2269–2279.
79. Zhang, Z., Qi, X., Tang, Y., Li, B., Wang, C. and Huang, Y., 2006. Theoretical simulation of PVA coating using drop-tower technique. *High Power Laser and Particle Beams, 18*(11), pp. 1837–1840.
80. Arya, R.K. and Bhargava, C.K., 2015. Simulation analysis of drying of ternary polymeric solution coatings. *Progress in Organic Coatings, 78*, pp. 155–167.
81. Makki, H., Adema, K.N., Peters, E.A., Laven, J., van der Ven, L.G., van Benthem, R.A. and de With, G., 2014. A simulation approach to study photo-degradation processes of polymeric coatings. *Polymer Degradation and Stability, 105*, pp. 68–79.
82. Jayachandran, R., Boyce, M.C. and Argon, A.S., 1995. Design of multilayer polymeric coatings for indentation resistance. *Journal of Computer-Aided Materials Design, 2*(2), pp. 151–166.
83. Iliopoulos, I. and Scriven, L.E., 2005. A blade-coating study using a finite-element simulation. *Physics of Fluids, 17*(12), p. 127101.
84. Lin, B., Sureshkumar, R. and Kardos, J.L., 2003. Morphology of conducting polymeric coatings: quantitative comparison between theory and experiment. *Chemical Engineering Science, 58*(11), pp. 2445–2447.
85. Van der Aa, H.C.E., Van der Aa, M.A.H., Schreurs, P.J.G., Baaijens, F.P.T. and Van Veenen, W.J., 2000. An experimental and numerical study of the wall ironing process of polymer coated sheet metal. *Mechanics of Materials, 32*(7), pp. 423–443.
86. Dey, M., Doumenc, F. and Guerrier, B., 2016. Numerical simulation of dip-coating in the evaporative regime. *The European Physical Journal E, 39*(2), p. 19.
87. Burch, D., Martin, J.W. and VanLandingham, M.R., 2002. Computer analysis of a polymer coating exposed to field weather conditions. *Journal of Coatings Technology, 74*(924), pp. 75–86.
88. Moore, K., Clemons, C.B., Kreider, K.L. and Young, G.W., 2007. Modeling and simulation of axisymmetric coating growth on nanofibers. *Journal of Applied Physics, 101*(6), p. 064305.
89. Kaunisto, E., Marucci, M., Borgquist, P. and Axelsson, A., 2011. Mechanistic modelling of drug release from polymer-coated and swelling and dissolving polymer matrix systems. *International Journal of Pharmaceutics, 418*(1), pp. 54–77.
90. Monfared, H. and Sharif, F., 2008. Design guidelines for development of tin-free antifouling self-polishing coatings using simulation. *Progress in Organic Coatings, 63*(1), pp. 79–86.
91. Flack, W.W., Soong, D.S., Bell, A.T. and Hess, D.W., 1984. A mathematical model for spin coating of polymer resists. *Journal of Applied Physics, 56*(4), pp. 1199–1206.
92. Wang, X., Gao, J., Wang, Z., Xu, J., Li, C., Sun, S. and Hu, S., 2017. Dissipative particle dynamics simulation on the self-assembly and disassembly of pH-sensitive polymeric micelle with coating repair agent. *Chemical Physics Letters, 685*, pp. 328–337.

93. Bahlakeh, G. and Ramezanzadeh, B., 2017. A detailed molecular dynamics simulation and experimental investigation on the interfacial bonding mechanism of an epoxy adhesive on carbon steel sheets decorated with a novel cerium–lanthanum nanofilm. *ACS Applied Materials & Interfaces*, 9(20), pp. 17536–17551.

94. Li, B., Sheng, X., Xing, W.G., Dong, G.L., Liu, Y.J., Zhang, C.Q., Chen, X.J., Zhou, N.N. and Qin, Z.B., 2010. Molecular dynamic simulation on the absorbing process of isolating and coating of α-olefin drag reducing polymer. *Chinese Journal of Chemical Physics*, 23(6), p. 630.

95. Xiao, J., Huang, Y. and Manke, C.W., 2010. Computational design of polymer nano-composite coatings: A multiscale hierarchical approach for barrier property prediction. *Industrial & Engineering Chemistry Research*, 49(17), pp. 7718–7727.

96. Croll, S.G. and Hinderliter, B.R., 2005. Monte Carlo approach to estimating coating service lifetime during weathering. *Surface Coatings International Part B: Coatings Transactions*, 88(3), pp. 177–183.

97. Sokolov, I.M., Morgenstern, O. and Blumen, A., 1994, April. Models for cracking of polymer coatings. In *Macromolecular Symposia* (Vol. 81, No. 1, pp. 235–248). Basel: Hüthig & Wepf Verlag.

17 Future Challenges and Applications of Polymer Coatings

Mohit H, Sanjay Mavinkere Rangappa, and Suchart Siengchin
King Mongkut's University of Technology North Bangkok

Hemath Kumar G.
Madanapalle Institute of Technology and Science

Arul Mozhi Selvan V. and Ruban R.
National Institute of Technology, Tiruchirappalli, India

CONTENTS

17.1 INTRODUCTION

The concrete have a broad range of applications for construction, roads, bridges, etc., and it also has higher durability and strength when compared with metal. Presently, the concrete is examined as a noncorrosive component. For instance, chemical and physical reactions from the environment affect the structure after a short period of service, which leads to failures [1,2]. Epoxy polymer coating from neat epoxy as major component can produce dense layer at concrete surface after the process, which restricts from erosive media and also improves mechanical durability [3,4]. From previous literature, it is observed that the protection from the epoxy polymer coating on concrete is not efficient, because the effects of erosive solutions like Cl⁻ and OH⁻ slowly erode the coating and react with material [5].

The corrosion properties of polymer coatings is an essential factor for improving the mechanical durability of the product. Hence, it is very important to develop the epoxy polymer coatings by reinforcing adequate nanoparticles for the improvement of corrosion resistance [6,7]. Moreover, the nanofillers can be easily accumulated inside the polymer matrix because of their higher surface area and surface energy. In the same manner, the presence of nanofillers will form cracks or damages within the polymer coating, which degrade the performance of coating [8]. For an illustration, Li et al. [9] observed that the addition of nanofillers on polymer filmed-coated concrete blocks has reduced the water absorption capacity and increased the contact angles of water. On the contrary, incorporation of nano organoclay can efficiently enhance the shielding effect of coatings on concrete, hence improving the restriction of chloride ion and moisture content on concrete. Currently, the graphene and its derivative components have been examined as nanoparticles for the polymer coatings because of their superior ion-diffusion barrier characteristics, identical nanostructures, excellent compatibility with polymers, and higher specific surface areas [10–12]. Zhang et al. [13] investigated the influence of graphene oxide/polyvinylpyrrolidone at epoxy polymer coatings that have significantly enhanced the corrosion and erosion resistance of the coating. Li et al. [14] fabricated a series of modified graphene reinforced water-based polyurethane hybrid and showed that graphene has good dispersion in hybrid coating. Javidparvar et al. [15] observed that graphene oxide with epoxy polymer coating would be applied in storing cation inhibitors for the steel corrosion sectors.

Zheng et al. [16] prepared epoxy polymer coating modified with graphene oxide using solution blending process and found that at 0.3 wt.% of graphene oxide content reduced the water sorption capacity and chloride ion diffusion of polymer-coated concrete. The improvement principle for chloride penetration is assigned to production of cross-linking in polymer coating, protection effects of the graphene oxides, and enhanced hydrophobicity. Chi et al. [17] fabricated a new epoxy polymer coating from cross-linked solution using the Diels-alder reaction process. It was found that the coating has lower viscosity, which permeates into damages of the concrete till 1.5 mm, which leads to formation of a robust coating/concrete composites. It also acquires a lower concentration of volatile organic components, so as it shows a less negative impact on the environment even though there are some investigations on corrosion resistance for metallic materials by nanoparticles polymer coatings. There are some investigations on improving mechanical durability of blocks coated from ceramic or metal or whisker nanoparticles-based polymer coatings.

From recent literature, it was found that there are no metal matrix-based nanoparticles that have been used in polymer coating for concrete block applications. In this chapter, the epoxy polymer coatings are fabricated using aluminum silicon carbide (Al-SiC) as filler reinforcement. Epoxy polymer coating is sprayed on the surfaces of concrete blocks. Analyses of internal relative humidity, adhesion, tensile properties, corrosion resistance, chloride ion permeation, and techno-economic are carried out.

17.2 MATERIALS AND METHODS

17.2.1 PREPARATION OF AL-SIC NANOPARTICLES

In this work, Al 6061 aluminum nanoparticles (100 nm) have been selected and procured from Ganapathy Colours, Chennai, India. The silicon carbide (SiC-10 μm) microparticles were supplied from Carborundum Universal, Kochi, India. The mechanical alloying process is conducted in high-energy ball mill by mixing 75 wt.% of Al6061 aluminum powder and 25 wt.% of SiC particles. 3 wt.% of silica gel is poured into the mixture to avoid severe cold welding. Balls to powder weight ratio is 10:1, and different diameters of ceramic balls from 8 to 20 mm are employed. The ball-milling time is fixed as 3 hours, and an interim period of 30 minutes for every 1 hour, to prevent from overheating, and a speed of 200±5 rpm are selected. After the mechanical alloying process, the prepared Al-SiC is separated by a 50 nm nanosieve.

17.2.2 AL-SIC REINFORCED EPOXY POLYMER COATINGS FABRICATION

The epoxy resin (LY 556) and hardener (HY 951) were purchased from Sakthi Fiber Glass, Inc., India. Ultrasonicator probe machine (UP 400ST, Hielscher) was utilized to disperse Al-SiC nanofillers within the epoxy polymer. The Al-SiC nanoparticles were weighed and combined in 500 ml of beaker. After ultrasonication, hardener (HY 951) was mixed by a mechanical agitator for 300 seconds [18]. The coating was poured on the surface of concrete blocks and cured for 240 minutes at 80°C±5°C to reduce the residual stress and enhance the cross-linking density between the concrete block and Al-SiC polymer coating.

17.2.3 AL-SIC NANOPARTICLES REINFORCED POLYMER COATINGS CHARACTERIZATION

Perkin Elmer spectrum 2 is utilized to examine the Fourier transform infrared (FTIR) spectrum of Al-SiC epoxy polymer coating. The spectra were measured between 4000 and 400 cm^{-1} with a resolution of 1 cm^{-1} and a scan rate of 32 per minute.

The internal relative humidity of Al-SiC epoxy polymer-coated concrete block was measured at a time duration of 1 year exposed into atmospheric conditions by using the thermo-hygrometric sensor, which was kept in a small hole of the sample at a depth of 15 mm [19]. The Al-SiC epoxy polymer coating viscosity was measured from the AR-G2 rheometer under the temperature of 25°C and 50 s^{-1} shear rate.

Posi-Test ATA determined the adhesion of the Al-SiC epoxy polymer coating as per the ASTM D7234 standard [17]. The Al-SiC epoxy polymer coating was poured on the concrete blocks and cured at room temperature for 3 hours. Then, stainless steel studs were placed on coating surface with the help of glue. Finally, these specimens were placed in the furnace at 100°C for 24 hours to enhance the cross-linking density within the coating and stud. The measurement of epoxy coating was done with 0.25 MPa/s of pulling rate. The five samples were measured to get a mean value.

The tensile test was conducted on the Tinius Olsen Universal testing machine attached with a load cell of 15 kN at 1 mm/min crosshead speed. Five specimens (100×10×2 mm) were experimented for each type of specimen to get a mean quantity. Test was conducted under room temperature (28°C±2°C) and 60%±2% of relative humidity.

The standard concrete cube specimen (150×50×30 mm) was fabricated by mixing cement, sand, and gravel in the ratio of 1:3:5 with an adequate amount of water. Al-SiC epoxy polymer coating coated on the surface of the cube concrete sample and cured at an atmospheric temperature of 30°C±2°C and relative humidity of 60%±2% for 2 hours. Then, the coated concrete cube was placed in a furnace at 120°C for 24 hours to remove the residual moisture content and enhance the cross-linking between epoxy coatings and concrete block. To avoid leaking of solution from the specimen, the paraffin sealant was coated on the other surfaces of the sample, which is before the test. The sample was immersed in 5 wt.% of the NaCl solution at room temperature for 36 hours. The corrosion resistance of the sample was measured using ACM instruments contained with a three-electrode cell arrangement. The concrete specimen, saturated calomel, and platinum mesh were acted as working, reference, and auxiliary electrodes, respectively. Tests performed in frequency between 100 mHz and 100 kHz from an amplitude perturbation of 100 mV.

The flux test applied to measure anticorrosion characteristics of the Al-SiC epoxy polymer coating. First, the Al-SiC epoxy polymer coating of 0.6 kg/m^2 coated on one side of the concrete sample (30 mm × φ 70 mm) and cured at room temperature for 1 hour. Then, it was shifted to the furnace at 100°C for 24 hours. In this testing condition, the sealant was applied to the curved surface to prevent from leaking of ions and water content. The test was conducted according to ASTM C 1202 standard, and five samples were measured to collect a mean value.

17.2.4 TECHNO-ECONOMIC ANALYSIS

The techno-economic analysis was carried out by employing an economic evaluation tool as SuperPro software. This method utilizes the discounted cash flow scheme for the complete life cycle of the project. The equipment costs were estimated from built-in models and other elements of investments were calculated based on the multiplication parameters taken from previous literature and SuperPro software [20]. The quantities are considered as 5% of electrical facilities, 2% of insulation, 30% of auxiliary facilities, 20% of engineering supervision (from a database of SuperPro), 5% of piping, 40% of instrumentation, and 40% of construction [21]. The cost of the equipment was estimated using SuperPro software and depreciation rate measured obtained from all undepreciated equipment prices as a contribution from the individual. The rate of inflation was calculated as 3% and the interest rate at 8.5%. The other miscellaneous prices such as local taxes, insurance, and expenses from the factory were employed in the economic evaluation process from default quantities.

17.3 RESULTS AND DISCUSSION

17.3.1 FTIR SPECTRA

Figure 17.1 shows the FTIR spectrum of Al-SiC epoxy polymer coating on a concrete block. The reinforcement of Al-SiC nanoparticles in the epoxy polymer depicts a new peak at 1150 cm^{-1}, related to Si-O stretch bonding of aliphatic ether, which

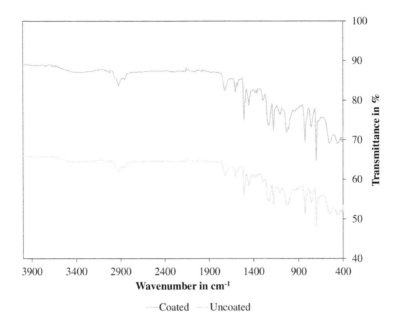

FIGURE 17.1 FTIR spectrum of both uncoated and coated Al-SiC epoxy polymers on concrete block.

indicates that the aliphatic ether found on the surface of Al-SiC epoxy polymer coating. This aliphatic ether connects the epoxy polymer coating with the concrete block [18,22,23]. A general peak at $1780\,cm^{-1}$ assigned to the Al-SiC epoxy polymer chemical compound presented in the spectrum of polymer-coated concrete block cured at 120°C [24–26]. For instance, the coating based on Al-SiC nanoparticles is capable of cross-linking via the aliphatic ether.

17.3.2 RELATIVE HUMIDITY, VISCOSITY, AND ION PERMEABILITY

The internal relative humidities of both Al-SiC epoxy polymer-coated concrete and conventional concrete were measured for 1 year using the hygrometric sensor. Figure 17.2 exhibits a wide variability, which shows the average value as evaluated from epoxy-coated samples. There is no significant difference observed between the coated and uncoated samples [19].

The coating viscosity is important for its permeability. The viscosity of Al-SiC polymer coating as a function of time is shown in Figure 17.3. Initial viscosity of Al-SiC epoxy polymer coating was 130 and 350 under glycidyl-ether (GE) and D230-melamine (D230-M) solutions, respectively. Al-SiC epoxy polymer coating under GE reduces the viscosity. However, the viscosity of the coating rises with time due to the presence of glycidyl ether which chemically reacts with the curing agent and connects the network of epoxy polymer coating without cross-linking [17].

The viscosity of Al-SiC epoxy polymer coating is comparably higher under D230-M solvent, due to its basic reaction within the matrix. The time of spreading

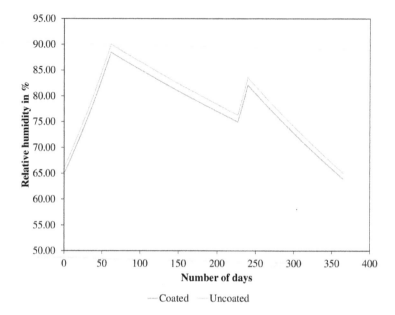

FIGURE 17.2 Internal relative humidity of concrete both coated and uncoated samples exposed to atmospheric condition.

of Al-SiC epoxy polymer on the surface of concrete block and the corresponding depth of permeation at the concrete are displayed in Figure 17.3a and b, respectively. Al–SiC epoxy polymer coating poured on the concrete block, which permeates up to 2 mm of depth with a period of 240 seconds. The lower initial viscosity and slightly slower increment in viscosity are important parameters for the excellent permeability of Al-SiC epoxy polymer coating. Thus, it can fully permeate in the concrete block within 3600 seconds, which showed that lower initial viscosity and slight increment in viscosity are advantageous for coatings permeability [17,19].

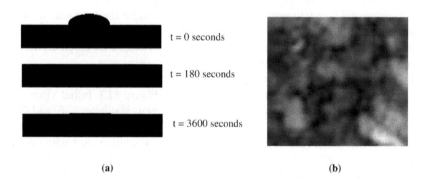

FIGURE 17.3 Spreading of Al-SiC polymer coating on the concrete surface (a) and photographic view of depth of permeation of Al-SiC polymer coating in concrete (b).

17.3.3 Mechanical Properties

The stress-strain curve of Al-SiC epoxy polymer coating on the concrete block is shown in Figure 17.4. Table 17.1 summarizes the elongation at break, tensile strength, and modulus of Al-SiC epoxy polymer coating on concrete block. The epoxy coating showed higher elongation at break, tensile strength, and modulus of about 14.5%, 4.8 MPa, and 3.2 GPa, respectively, when compared with conventional concrete block. The reason may be attributed to the fact that the epoxy group connects with the curing agent to produce several polar groups in the network of epoxy polymer without cross-linking [17,27]. As mentioned earlier, the Al-SiC epoxy polymer coating at lower viscosity could permeate and restrict gaps of concrete blocks. Therefore, cross-linking mechanism within the concrete block and Al-SiC epoxy polymer coating under cured temperature showed higher mechanical strength when compared with conventional concrete block.

The bonding strength between the Al-SiC epoxy polymer coating and concrete block observed by measuring the failure mode and pull-off strength are shown in Table 17.1. Al-SiC epoxy polymer-coated concrete block showed higher pull-off strength when compared with uncoated one. The reason is assigned to the coating that permeates into the concrete block and in-situ cross-linking, which tends to the production of robust concrete structure [17,19]. Therefore, both the polymerization process and coating permeability acts as an essential role in the bonding strength. Failure exhibited in concrete block for specimens, signifying that block coated from Al-SiC polymer, has an excellent bond strength.

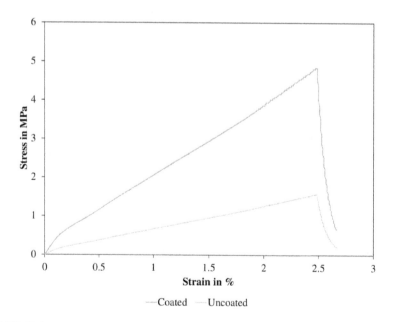

FIGURE 17.4 Stress-strain curves of both uncoated and coated concrete blocks.

TABLE 17.1

Properties of Concrete Block Uncoated and Coated with Al-SiC Epoxy

Parameter	Uncoated Concrete Brick (%)	Al-SiC Epoxy Polymer-Coated Concrete Brick (%)
Tensile strength in MPa	1.56±0.84	4.80±0.75
Elongation at break in %	10.25±1.37	14.5±2.35
Tensile modulus in GPa	0.75±0.54	3.2±0.96
Bonding strength in MPa	0.27±1.04	0.65±1.33
Anticorrosion charge in C	3600±2.47	1400±1.02
Corrosion rate in mm/year	0.015±1.56	$1.57 \times 10^{-4} \pm 1.75$
Volatile organic component in wt.%	3.02±0.96	1.02±0.85

17.3.4 CORROSION RESISTANCE

Chloride-embedded corrosion is an important problem that leads to degradation of concrete blocks. Resistance of Al-SiC epoxy polymer coatings under chlorine environment is measured for anticorrosion property, and the results are listed in Table 17.1. The conventional concrete block has 3600 C of total charge passed, which signifies vulnerability for corrosion. The block coated with Al-SiC epoxy polymer showed a total charge of 1400 °C, which signifies its resistance from chlorine. The reason may be assigned to higher cross-linking between the epoxy polymer coating and concrete block [17,28,29]. As per the previous literature [28], the pull-off strength primarily depends on the coating permeation depth, whereas the property of anticorrosion is measured by the polymer coating cross-linking density. The Al-SiC epoxy polymer coating has shown a relatively lower total charge when compared with conventional one, because it can block the pores of the concrete surface, which significantly increases the polymer coating thickness. Generally, the corrosive ions should strike the coating thickness. Hence, the chloride-embedded corrosion is remarkably exhibited by the influence of cross-linking and good permeability within the polymer coating and concrete block. Electrochemical impedance spectroscopy is applied to measure the property of anticorrosion in real schemes for Al-SiC epoxy polymer (Figure 17.5). It is also found that the radius of the arc correlates with resistance characteristics.

Figure 17.5 Nyquist plot exhibits that the conventional concrete block showed a smaller radius, which signifies the weak resistance from corrosion. However, Al-SiC epoxy polymer-coated concrete blocks showed a larger arc radius, which signifies the higher corrosion resistance due to preliminary reaction between the epoxy polymer and concrete block with the absence of pores. The two parameters that also contribute for the enhancement of corrosion resistance are excellent permeation and higher cross-linking network within the concrete block and polymer coating. Table 17.1 also shows the volatile organic compounds (VOC) content of Al-SiC epoxy polymer coating. The VOC content (~ 1.2 wt.%) is comparably lower than the published literature [17] due to the preliminary reaction between the Al-SiC, epoxy resin, and concrete block. Therefore, the Al-SiC epoxy polymer coating has a lower content of VOC and has a positive impact on the environment.

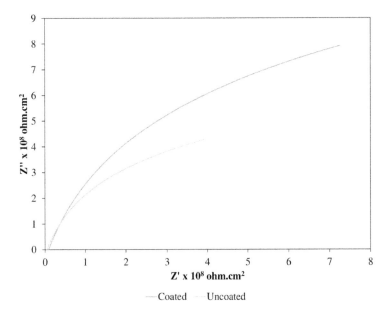

FIGURE 17.5 Electrochemical impedance spectroscopic Nyquist curve for uncoated and coated concrete blocks.

17.3.5 TECHNO-ECONOMIC ANALYSIS

In the present investigation, a novel technique for the fabrication of epoxy polymer coatings on concrete bricks derived from metal matrix composites has been displayed from the perspective of techno-economic studies of the process. The results estimated from SuperPro software and the major components are shown in Table 17.2. In this primary scenario, the price of fabricating 1 ton of epoxy polymer coating material is $5000. Assuming a gain of 8.4% (product cost of $4000 per ton) results in 12.6% of return on investment with 10 years of the payback period.

The total capital cost was estimated at $12.4 million, with $2.9 million in plant indirect and $4.1 million in direct plant costs. There is a total amount of $664,000 between operating costs and revenues. The total equipment prices were $1865,000, a reasonable quantity, considering the account that the technology itself is the combination of simple and complex designs that requires adequate sophisticated equipment [21]. The benefits of this investment are driven by the principle that there is a consequential void between the operating prices and revenues under the given case. The operating prices contain utilities, labor, materials, laboratory, facility-dependent cost, quality control, and assurance. The operating breakdown costs are shown in Table 17.3. On the other hand, most of the operating costs are combined with the prices of materials utilized in this project. The electricity for running the ultrasonicator probe constitutes lower than 2% of the total operating prices. This reason may be assigned to the processing pathway which contains sonication operations that would acquire significant dispersion of nanoparticles within the epoxy matrix.

TABLE 17.2

Major Economic Factors Estimated for the Process

Parameters	Price in USD
Total capital investment	12,400,000
Total plant indirect price	2,900,000
Total plant direct price	4,100,000
Direct fixed capital price	8,658,000
Operating price	6,450,000
Revenues	9,424,000
Unit production price per ton	4490
Unit production revenue per ton	5000
Return on investment	12.6
Payback period	10 years

TABLE 17.3

Detailed Breakdown of the Operating Costs

Cost Unit	Value	Percentage
Utilities	$76,860	1.03
Materials	$2,663,500	37.36
Facility development	$1,976,800	32.26
Labor	$2,294,880	19.87
Laboratory/quality control/quality assurance	$344,400	9.48

Hence, the major prices are associated with the mixing of two elements and uniformly dispersed within the matrix to get higher performance. The materials contribute to a huge amount of all operating prices. Hence, variations in costs and required amounts will act as the most important role in determining the feasibility of the process. The epoxy resin is one of the possible thermoset polymers to be utilized in this process since the contribution of price related to the cost of the monomer is so significant. It will show a higher interest in the future to consider lower price alternatives such as vinyl ester and polyester, which could reduce the processing cost to a greater extent. Presently, the short payback period is an outcome of relatively low investment when compared with operating costs and revenues. The surplus of above $664,000 between annual prices and plant incomes provides a reasonably quick return on investment. The plant scale is also within the reasonable range, which regards both output products and input concentrations. The annual fabrication of Al-SiC epoxy polymer-coated concrete bricks of 100 tons of Al-SiC is also a reasonable amount when compared with other polymer coatings worldwide. It can be affirmed that such products will not affect the prices of the market in a significant condition, and as much as the conventional polymer coating market is increasing, it is ascribed that the demand lack is not a threat to this project work.

17.4 OTHER APPLICATIONS

In recent years, the addition of various nanoparticles has influenced the optoelectronics and electrical properties of polymer coatings drastically. The nanocomposite polymer coatings can be applied as promising materials in different fields such as biological or chemical sensors, electronic nanodevices, biomedicine, Electro Magnetic Interference (EMI) shielding, microwave absorption, and energy storage applications. These types of metal-based polymer coatings showed higher performance in highly aggressive environments, which acts as a corrosion inhibitor to metals. When compared with previous literature [30–32], it is found that Al-SiC polymer coating has superior anticorrosive properties. It can also be applied in heat exchanger applications for the enhancement of heat transfer rate and thermal stability. The heat transfer rate of these polymer coatings has higher thermal conductivity than antiscale effect coating and can be applied for heavy-duty industrial applications [33]. These types of Al-SiC epoxy polymer coatings can be applied in different machine elements due to its superior tribological properties such as higher wear resistance and a lower coefficient of friction under different mediums and higher contact pressures. It has also been proved that under higher contact pressures, the debris of wear trapped and generated at the sliding motion, which played as an important role for further resistance. These polymer coatings have higher resistance from wear, coefficient of friction, and any other environmental effect when compared with conventional coatings [34].

17.5 CONCLUSION AND FUTURE CHALLENGES

The basic and latest investigations on the application of Al-SiC epoxy polymer coating to improve the corrosion resistance, mechanical properties, and permeability properties of concrete brick have been summarized in this chapter. The developed novel Al-SiC epoxy polymer coating has higher mechanical strength, excellent permeability, and lower VOC content. The Al-SiC epoxy polymer coating can permeate into concrete blocks under lower viscosity. However, it could cross-link at higher curing temperatures by producing a robust concrete structure. The unfulfilled expectations of Al-SiC epoxy polymer coatings must be addressed to enable these nanobased composites to attain their complete possibility. The challenges of dispersing Al-SiC-based nanoparticles within the polymer coatings and necessity to avoid the poor interfacial adhesion in Al-SiC/polymer nanocomposites with the effect of covalent bond interactions. In addition, the development of new fabrication techniques should be done to reduce solvent waste and explosion risk. The performance of the coatings mainly depends on the ability to limit the destruction of coated metal.

Further investigations are necessary to enhance Al-SiC epoxy polymer coatings as new automatic solutions that comprise multifunctionalities of coatings such as reduced wear, decrement in drag, biofouling, corrosion protection, heat transfer enhancement, higher mechanical properties, and superhydrophobicity. Apart from these challenges, polymer-based nanoparticles reinforced composite coatings are applied at the industrial level, and their effect is expected to increase significantly in the future. A significant impact on the expansion of Al-SiC nanoparticle fabrication has been elevated by considering that Al-SiC. The main benefit of reinforcing Al-SiC nanofillers over graphene and carbon-based in polymer nanocomposites, because it is non-toxic and able to withstand under a higher range of temperatures.

REFERENCES

1. Pan, X., Shi, Z., Shi, C., Ling, T., Li, N. 2017. A review on surface treatment for concrete part 2: performance. *Construction Building Materials* 133:81–90.
2. Pan, X., Shi, Z., Shi, C., Ling, T., Li, N. 2017. A review on concrete surface treatment Part I:types and mechanisms. *Construction and Building Materials* 132:578–590.
3. Bauer, R.S. 1979. *Epoxy Resin Chemistry.* Washington, DC: American Chemical Society.
4. Ibrahim, M., Al-Gahtani, A.S., Maslehuddin, M., Dhakil, F.H. 1999. Use of surface treatment materials to improve concrete durability. *Journal of Materials in Civil Engineering* 11:36–40.
5. Merah, A., Khenfer, M.M., Korichi, Y. 2015. The effect of industrial coating type acrylic and epoxy resins on the durability of concrete subjected to accelerated carbonation. *Journal of Adhesives Science and Technology* 29(22):2446–2460.
6. Hooda, A., Goyat, M.S., Gupta, R., Prateek, M., Agarwal, M., Biswas, A. 2017. Synthesis of nano-textured polystyrene/ZnO coatings with excellent transparency and superhydrophobicity. *Materials Chemistry and Physics* 193:447–452.
7. Zhou, J., Tan, Z., Liu, Z., Jing, M., Liu, W., Fu, W. 2017. Preparation of transparent fluorocarbon/TiO2-SiO2 composite coating with improved self-cleaning performance and antiaging property. *Applied Surface Science* 396:161–168.
8. Sahnesarayi, M.K., Sarpoolaky, H., Rastegari, S. 2014. Effect of heat treatment temperature on the performance of nano-TiO$_2$ coating in protecting 316L stainless steel against corrosion under UV illumination and dark conditions. *Surface Coatings and Technology* 258:861–870.
9. Li, G., Yue, J., Guo, C.H., Ji, Y.S. 2018. Influences of modified nanoparticles on hydrophobicity of concrete with organic film coating. *Construction and Building Materials* 169:1–7.
10. Meyer, J.C., Geim, A.K., Katsnelson, M.I., Novoselov, K.S., Booth, T.J., Roth, S. 2007. The structure of suspended graphene sheets. *Nature* 446(7131):60–63.
11. Zhu, Y., Murali, S., Cai, W., Li, X., Suk, J.W., Potts, J. 2010. Graphene and graphene oxide: synthesis, properties, and applications. *Advanced Materials* 22(46):3906–3939.
12. Zhou, S.G., Wu, Y.M., Zhao, W.J., Yu, J.J., Jiang, F.W., Ma, L.Q. 2018. Comparative corrosion resistance of graphene sheets with different structures in waterborne epoxy coatings. *Colloid Surfaces Part A-Physicochemical Engineering Aspects* 556(5): 273–283.
13. Zhang, Z., Zhang, W., Li, D., Sun, Y., Wang, Z., Hou, C., Chen, L., Cao, Y., Liu, Y. 2015. Mechanical and anticorrosive properties of graphene/epoxy resin composites coating prepared by in-situ method. *International Journal of Molecular Science* 16:2239–22.
14. Li, P., Ren, H., Qiu, F., Xu, J., Yang, D. 2014. Preparation and properties of graphene oxidemodified waterborne polyurethane-acrylate hybrids. *Polymers-Plastics Technology and Engineering* 53(13):1408–1416.
15. Javidparvar, A.A., Naderi, R., Ramezanzadeh, B. 2019. Epoxy-polyamide nanocomposite coating with graphene oxide as cerium nanocontainer generating effective dual active/barrier corrosion protection. *Composites Part B: Engineering* 172:363–375.

16. Zheng, W., Chen, W.G., Feng, T., Li, W.Q., Liu, X.T., Dong, L.L., Fu, Y.Q. 2020. Enhancing chloride ion penetration resistance into concrete by using graphene oxide reinforced waterborne epoxy coating. *Progress in Organic Coatings* 138:105389.

17. Chi, J., Zhang, G., Xie, Q., Ma, C., Zhang, G. 2020. High performance epoxy coating with cross-linkable solvent via Diels-Alder reaction for anti-corrosion of concrete. *Progress in Organic Coatings* 139:05473.

18. Mohit, H., Selvan, V.A.M. 2019. Physical and thermomechanical characterization of the novel aluminum silicon carbide-reinforced polymer nanocomposites. *Iranian Polymer Journal* 28:823–837.

19. Diamanti, M.V., Brenna, A., Bolzoni, F., Berra, M., Pastore, T., Ormellese, M. 2013. Effect of polymer modified cementitious coatings on water and chloride permeability in concrete. *Construction and Building Materials* 49:720–728.

20. Peters, M.S., Timmerhaus, K.D. 1991. *Plant Design and Economics for Chemical Engineers*. 4th edition, McGraw-Hill, Inc.

21. Bochenski, T., Chan, W.Y., Olsen, B.D., Schmidt, J.E. 2018. Techno-economic analysis for the production of novel, bio-derived elastomers with modified algal proteins as a reinforcing agent. *Algal Research* 33:337–344.

22. Park, S., Dikin, D.A., Nguyen, S.B.T., Ruoff. R.S. 2009. Graphene oxide sheets chemically cross-linked by polyallylamine. *Journal of Physical Chemistry Part C* 113:15801–15804.

23. Zhang, Y., Huang, X., Duan, B., Wu, L., Li, S., Yuan, X, 2007. Preparation of electro-spun chitosan/poly(vinyl alcohol) membranes. *Colloid Polymer Science* 285:855–863.

24. Tarducci, C., Badyal, J.P.S., Brewer, S.A., Willis, C. 2005. Diels-Alder chemistry at furan ring functionalized solid surfaces, *Chemical Communications* 3:406–408.

25. Min, Y.Q., Huang, S.Y., Wang, Y.X., Zhang, Z.J., Du, B.Y., Zhang, X.H., Fan, Z.Q. 2015. Sonochemical transformation of epoxy-amine thermoset into soluble and reusable polymers. *Macromolecules* 48:316–322.

26. Imai, Y., Itoh, H., Naka, K., Chujo, Y. 2000. Thermally reversible IPN organic-inorganic polymer hybrids utilizing the Diels-Alder reaction. *Macromolecules* 33:4343–4346.

27. Ma, S.Q., Liu, X.Q., Fan, L.B., Jiang, Y.H., Cao, L.J., Tang, Z.B., Zhu, J. 2014. Synthesis and properties of a bio-based epoxy resin with high epoxy value and low viscosity, *ChemSusChem* 7:555–562.

28. Liu, B., Fang, Z.G., Wang, H.B., Wang, T. 2013. Effect of cross linking degree and adhesion force on the anti-corrosion performance of epoxy coatings under simulated deep sea environment. *Progress in Organic Coatings* 76:1814–1818.

29. Zhang, G.L., Xie, Q.Y., Ma, C.F., Zhang, G.Z. 2018. Permeable epoxy coating with reactive solvent for anticorrosion of concrete. *Progress in Organic Coatings* 117:29–34.

30. Ozkazanc, H., Zor, S. 2013. Electrochemical synthesis of polypyrrole (PPy) and PPy/metal composites on copper electrode and investigation of their anticorrosive properties. *Progress in Organic Coatings* 76(4):720–728.

31. Olad, A., Rashidzadeh, A. 2008. Preparation and characterization of polyaniline/$CaCO_3$ composite and its applications as anticorrosive coating on iron. *Iranian Journal of Chemical Engineering* 5(2):45–54.

32. Iqbal, S., Ahmad, S. 2018. Recent development in hybrid conducting polymers: synthesis, applications and future prospects. *Journal of Industrial and Engineering Chemistry* 60:53–84.

33. Friis, J.E., Subbiahdoss, G., Gerved, G., Holm, A.H., Santos, O., Blichfield, A.B., Moghaddam, S.Z., Thormann, E., Daasbjerg, K., Iruthayaraj, J., Meyer, R.L. 2019. Evaluation of surface-initiated polymer brush as anti-scaling coating for plate heat exchangers. *Progress in Organic Coatings* 136(105196): 1–12.

34. Nunez, E.E., Gheisari, R., Polycarpou, A.A. 2019. Tribology review of blended bulk polymers and their coatings for high-load bearing applications. *Tribology International* 129:92–111.

18 Incorporation of CeMo Nanocontainers Loaded with Inhibitors, Water, and Chlorine Microtraps into Anticorrosion Coatings onto ZK30

George Kordas
Institute of Nanoscience and Nanotechnology
(INN), NCSR "Demokritos"

CONTENTS

18.1 INTRODUCTION

Nowadays, one of the most important technological problems to be addressed is the degradation of metals and their alloys due to corrosion, which ultimately determines the lifetime of metal constructions. The corrosion of metals, in particular ZK30 alloys, which finds increasing demand in the construction of airplanes, bicycles, etc., has a huge impact on the global economy.

The new generation of smart corrosion-protection systems will be nanostructured coatings via modulating the intimate coating structure by nanocontainers. Upon triggering due to corrosion, the inhibitors leach out the active chemicals stored in the nanocontainers incorporated in the coating. Corrosion protection systems based on such nanocontainer-structured materials have important advantages over conventional approaches, and our group is the first to develop this technology [1–29]. The first is the generation of an active corrosion protection or self-healing even if damages by mechanical or chemical action occurred. The second is the reservoir into the coatings assuring a sustainable long-term protection. The third is the added functionalities such as the abrasion resistance; chemical resistance; or entire microbial, fouling, and adhesion properties that can be tailored by the nanoparticles themselves. The fourth is the chromium-free corrosion protection by their design.

In the present chapter, we report the development of new-generation, smart, corrosion protection systems for ZK30, which are environmentally friendly by tailoring macroscopic material properties at the nanoscale level. This radical paradigm changes by evading chromium-based systems offers the unique possibility to establish a new class of knowledge-based materials for environmentally friendly corrosion protection and move fare beyond the horizon of the present state of the art. The present work offers an innovative corrosion protection system replacing the Cr (VI)-based coatings.

18.2 EXPERIMENTAL

18.2.1 CHARACTERIZATION METHODS

Scanning electron microscopy (SEM) using a PHILIPS Quanta Inspect (FEI Company) microscope with W (tungsten) filament 25 kV equipped with EDAX GENESIS (Ametek Process and Analytical Instruments) was employed for structural characterization. The dynamic light scattering (DLS) technique using a Malvern Instruments Zetasizer DTS 1060 was used for measuring size and zeta potential of the nanocontainers. Volumetric static sorption apparatus BET (Autosorb-1 MP, Quantachrome Instruments) was employed for pore size determination of the nanocontainers. Electrochemical impedance spectroscopy (EIS) apparatus of Solartron (ModuLab XM MTS – Materials Test System) was used for the corrosion-resistance evaluation. The equipment was controlled by the ModuLab software inside a Faraday cage where the open circuit potential was measured by an electrochemical cell from Plexiglas filled with 0.5 M NaCl aqueous solution with a saturated calomel reference electrode, a counter electrode made from a rectangular platinum plate (~1*1 cm^2) and coated metallic substrate as working electrode (area of about 3.15 cm^2). The encapsulation capacity of chemicals into the nanocontainers was assessed by thermogravimetric analysis via a Perkin Elmer (Pyris Diamand S II) apparatus (heating rate = 10°C/minutes[1] in air). The release of the inhibitor from nanocontainers was determined by a HITACHI model 100-40 spectrophotometer apparatus connected with a stirrer and an automatic sampler. All measurements were carried out at room temperature.

18.2.2 PRODUCTION OF NANO-/MICROMATERIALS

The preparation of nano-/microcontainers was done via a three-stage procedure. The first stage involves the production of the nucleus mainly on the polystyrene bases using the chemistry of Table 18.1. Figure 18.1 shows the resulting SEM micrograph

TABLE 18.1

The Conditions Used for the Manufacture of Polystyrene Nanospheres at 80°C

Reagents	Quantity
Styrene	5.5 ml
Potassium persulfate (KPS)	0.65 g
Sodium dodecyl sulfate (SDS)	0.21 g
Water	450 ml

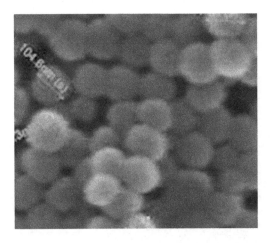

FIGURE 18.1   Polystyrene nanospheres.

of the polystyrene nanospheres. Depending on the amount of chemicals used, the size of the spheres can range between 40 and 500 nm.

The second stage required the employment of the sol-gel method to produce a coating, for example, CeMo. Negatively charged polystyrene balls react with the positively charged products of the Ce $(ACAC)_3$ hydrolysis and with the anion molybdenum. Ce $(ACAC)_3$ condenses on the surface of polystyrene with hydrolysis together with molybdenum anion and forms molecules of cerium-molybdenum oxide. Polyvinylpyrrolidone is added to the mixture to prevent the aggregation of particles. Table 18.2 gives the contentions used for the production of the CeMo shell. In the end, the cerium-molybdenum oxide nanocontainers arise after the heat treatment of the composite core-shell structure, where the internal polystyrene molds are burned and eliminated.

Initially, the coated nanospheres were placed in a clock glass and dried, first at room temperature and then at 60°C. The composite material was then heat-treated for 4 hours at 500°C. The heating rate was 10°C/minutes. Figure 18.2a shows the SEM micrograph of the CeMo nanocontainers that are hollow.

The XRD analysis (Figure 18.2b) shows the structures of the nanocontainers that arose at a temperature of 500°C. The peaks of the sample diagram coincide with those of the structures: CeO_2 oxide (43–1002 Cerianite, SYN) and trioxide cerium

TABLE 18.2

The Conditions Used to Manufacture Coated Polystyrene Nanospheres

Reagents	Quantity
Polystyrene	5 g
Polyvinylpyrrolidone (PVP)	5 g
Cerium (III) acetylacetonate (Ce(acac)$_3$)	2.5 g
Sodium molybdate (Na$_2$MoO$_4$)	0.25 g
Water	500 ml

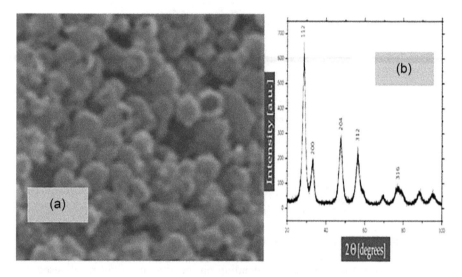

FIGURE 18.2   (a) SEM of CeMo nanocontainers and (b) XRD of CeMo nanocontainers.

molybdenum with the molecular formula Ce (MoO$_4$)$_2$ (33–330 cerium molybdenum oxide) (Library: JCPDS-ICDD 1997 International Center for Diffraction Data).

As for the production of water nanotraps, we proceeded as follows. The methacrylic acid monomer together with ethylene dimethacrylate (EGDMA) is dissolved in acetonitrile in a three-cylinder spherical flask, which is equipped with a freezer, a thermometer, and a nitrogen gas supply. The experiment was conducted at 80°C. After an hour of stirring, the 2,2′-azobis(2-methylpropionitrile) (AIBN) is added, which has previously been dissolved in a quantity of solvent. Stirring continues at the same temperature for 1 more hour. In order to make solvent distillation, the temperature of the experiment increases to the boiling point of the acetonitrile. After a specific amount of solvent is collected, the reaction is terminated and the resulting final solution is in the form of an emulsion. The solution is left at ambient temperature, followed repeatedly by centrifugation (10,000 rpm, 4 minutes) and decanting and resuspension in acetonitrile for two times. At 80°C, the AIBN gives free radicals that react randomly to molecules of the two monomers and give a multipart chain.

At this point, the nucleus of the particle is formed. The increase in temperature at the boiling point of the solvent helps in the further development of particles occurring through polymerization within the spheres, after absorption of monomer molecules, whose concentration is now increased. The development of particles stops when all the developing chains are terminated, or when all monomers are consumed or through the compound of the polymer chains. The polymerization of the spheres follows their modification with caustic soda. The solid collected is reshown to acetonitrile with the help of a supersonic, until the creation of a homogeneous suspension. Caustic soda solution 0.1 M is added to the suspension above, and the new solution is left stirring for 30 minutes at ambient temperature. The following are the centrifugation and rinsing of the sample with acetonitrile and its leaching at ambient temperature. Poly(methacrylic acid) (PMAA)-co-ethylene glycol dimethacrylate (EGDMA) spheres acquire the ability to adsorbed water on their surface after converting carboxylic groups to the corresponding sodium salts. Figure 18.3a shows the water microtraps verifying the homogeneity of the sample. The size of microtraps ranges between 560 and 655 nm.

Figure 18.3b shows the hydrodynamic diameter of the water traps on basis of the water content in solution with the main solvent acetonitrile and with a nanosphere concentration of 50 mg/L. There is an increase in the measured hydrodynamic diameter especially in region of 30 v/v water content. This effect is taken as an indication of the interaction and adsorption of water molecules from water traps. The increase is described with a single Langevin function. The process is reversible.

Another goal of this work is production of chlorine nanotraps to slow down the diffusion of chlorine molecules in epoxy resins. The Stober method was composed of silica spheres [30–32]. The method involves hydrolysis of alkyl-silicate compounds and their following compacting, using ammonia as a catalyst. The diameter of the particles produced depends on the relative contribution between the processes of phenylation and development. Hydrolysis and condensation reactions provide the

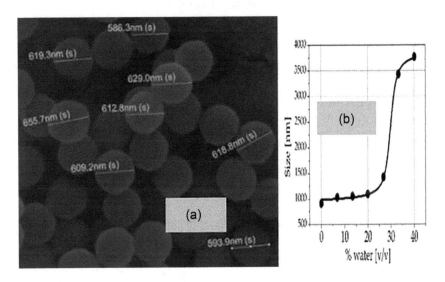

FIGURE 18.3 (a) SEM of water microtraps and (b) water uptake measured by DLS.

molecules of the precursor compound and the necessary hyper saturation for the formation of particles. During the reaction of hydrolysis, the ethoxy group of tetraethyl orthosilicate (TEOS) reacts with the water molecule for the creation of an intermediate product Si $(OC2H5)_{4x}$ $(OH)_x$, with the hydroxyls to substitute the ethoxy groups. Ammonia acts as a key catalyst in this reaction: the reaction of hydrolysis is initiated by the attack of the hydrogels in the molecules of TEOS. The chemical reaction is as follows:

$$Si\,(OC_2H_5)_4 + xH_2O <=> Si(OC_2H_5)_{4-x}(OH)_x + xC_2H_5OH$$

The following reaction of hydrolysis, condensation, occurs instantaneously. The hydroxyls of intermediate products react either with the ethoxy groups of other TEOS molecules or with the product of alcohol molecules where water is produced and as a result we have the formation of Si-O-Si bonds. The solution was released under stirring 18 hours. Then, the sample was centrifuged at 15,000 rpm, and it was repeatedly washed with absolute ethanol and allowed to dry in the air. Table 18.3 lists the detailed conditions.

Then, the silica balls were modified superficially with an organically modified silane (Organically Modified Silicate) under the conditions listed in Table 18.4. The process involves the hydrolysis of 3-glycidoxypropyltrimethoxysilane (GPTMS), that is, the substitution of the methoxy groups of its molecules with hydroxyls. The result of the reaction of the surface hydroxyl of the spheres with the hydroxyls of the hydrolyzed molecule GPTMS, is the creation of silica spheres with free epoxy rings. Figure 18.4 shows the produced chlorine traps.

To fill the nanocontainers with corrosion inhibitor, they were immersed in saturated aqueous solution of ammonium phosphate and were left under stirring for 12 hours. The inhibitor diffuses into the nanocontainers via their porous shell. Then, the sample is centrifuged and allowed to dry in the air.

TABLE 18.3
The Conditions Used for the Silica Nanosphere Production

Reagents	Molecular Weight (g mol⁻¹)	Quantity (ml)
Ethanol trade	46.07	150
TEOS	208.33	3.5
Ammonia solution 32% NH_4OH	35.04	3.5

TABLE 18.4
Conditions Used to Modify the Silica Spheres with GPTMS

Reagents	Molecular Weight (g mol⁻¹)	Quantity (ml)
Ethanol trade	46.07	100
3-Glycolyl-propyl-trimethoxy silane	236.34	0.5

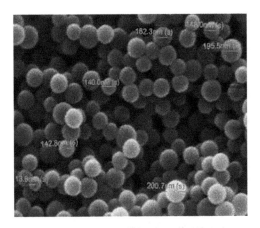

FIGURE 18.4 Chlorine traps by SEM.

18.2.3 SELECTION OF THE CORROSION INHIBITOR FOR THE ZK30 ALLOY

In order to select the appropriate corrosion inhibitor for the ZK30 magnesium alloy, electrochemical studies were carried out in relation to their anticorrosion action chemical compounds, whose choice arose after bibliographical research. The inhibitors were chosen such as zinc nitrate (Zn $(NO_3)_2$), ammonium phosphate dibasic ($(NH_4)2HPO_4$), calcium chloride ($CaCl_2$), sodium fluoroacetate (FCH_2CO_2Na), 8-hydroxycyanolin (8HQ), and sodium dodecylbenzene sulfonate. The procedure for cleaning the test specimens includes the following stages: immersion in ethanol (96% v/v) for 20 minutes at ambient temperature with the help of ultrasound; rinsing with distilled water and insertion of the substrates into an aqueous solution of sodium hydroxide 40 g/L, sodium carbonate (20g/L), and sodium hydroxide-sulfonate (0.1% wt.) for 10 minutes at 70°C; rinsing with distilled water for a few minutes at 40°C; and immersion of the test specimens in a solution of hydrofluoric acid 10% by weight for 20 minutes at room temperature. Finally, the substrates were washed with distilled water and left to air dry. This procedure relieves the substrate from impurities, grease, and grease derived from the manufacturing process of manufacture, and increases the adhesion of the applied coating. In particular, the use of hydrofluoric acid has a decisive effect on the behavior of the alloy in corrosive conditions.

The measurements of electrochemical spectroscopy impedance are shown in Figure 18.5. From the semicircle of a Bode diagram, we get useful information about the impedance of a material consisting of a real and an imaginary part. Each semicircle characterizes a time constant that can be related to the resistance to corrosion of oxide on the metal surface, transport resistance to the metal/electrolyte interface through these pores, or to reduction of corrosion due to diffusion phenomena.

Observing the curves of complex electrochemical resistance, it is evident that the highest value of absolute complex electrochemical resistance corresponds to the solution with the dibasic ammonium phosphate. This indicates that over time it provides the best corrosion protection possibly due to the creation of stable insoluble complexes on the metal surface. The yield of dibasic ammonium phosphate is consistently the best among the remaining compounds in the diagram, increasing the

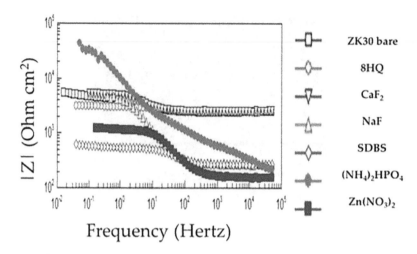

FIGURE 18.5 Bode diagram for the ZK30 magnesium alloy after exposure for 2 hours in a corrosive environment NaCl 5 mM with or without the presence of corrosion inhibitors.

corrosion resistance of the ZK30 alloy in one order of magnitude. It is clarified that the absolute value of complex electrochemical resistance is the sum of the resistance of the solution to high frequencies, the resistance to the medium frequencies corresponding to processes occurring on the surface of the metal and are due to the formation of oxides or complexes of metal inhibitor and the load-carrying resistance.

18.2.4 Synthesis, Application, and Characterization of Hybrid Coating with Embedded Nanocontainers in Magnesium Alloy ZK30

The composition of the coating includes the hydrolysis of 3-aminopropyltriethoxy silane, to absolute ethanol for 1 hour (solution A). At the same time, the epoxy resin is dissolved in absolute ethanol (solution B) and then mixed (solution C). The hardener, HY 943, is dissolved in acetone (solution D) and the final solution of the coating consists of the sub C and D, which remain stirring for 18 hours. The hardener's contribution concerns the improvement of the properties of epoxy resin through the creation of cross-links and the chemical modification with the organically modified silane. Five different solutions were prepared to study the effect of each factor on the efficiency of corrosion protection of coating. The solutions were based on the above hybrid epoxy coating, and each of them included a quantity of the synthesized nanoparticles. The samples were characterized as follows: S1 – Coating, S2 – Coating-A-water traps (WT)-B-chlorine traps (CT), S3 – Coating(inhibitors)-A-water traps (WT)-B-chlorine traps (CT), S4 – Coating(CeMo(empty))-A-water traps (WT)-B-chlorine traps (CT), and S5 – Coating(CeMo(inhibitors))-A-water traps (WT)-B-chlorine traps (CT).

All the samples were studied through EIS in order to assess the effect of nanocontainers and nanotraps on the barrier properties of the coating. Surface equal to 2 cm² of samples was exposed to corrosive environment NaCl concentration 0.05 m for up to 6 days. A three-electrode system was used. Electrode calomel was used as reference electrode, platinum sheet as measuring electrode, and working electrode as the respective sample. All measurements were made on the inside of a faradaic cell to avoid possible electromagnetic interference. At least two samples were measured each time for each of the layers to check the repeatability of the measurements. The figures represent the average of the obtained results. Figure 18.6b shows the FRA results of the samples exposed to salt water for four days. On the top of Figure 18.6a, the FRA of the bare ZK30 sample is shown. The last measurement was done to show the benefits of the coating deposited on the ZK30 substrate (Figure 18.6a).

All coatings exhibit corrosion protection for the ZK30 magnesium alloy, as shown by the absolute values of complex electrochemical resistance. In addition, the hybrid coating without additives shows the best protection as shown by the values of complex electrochemical resistance at low frequencies. As the exposure time of samples in the corrosive environment increases, the values of complex electrochemical resistance are reduced, yet they are still high compared to the noncoated substrate. The phase diagrams we have taken on the basis of the frequency are observed: a time constant at the high frequencies due to the barrier properties of the coating and a time constant

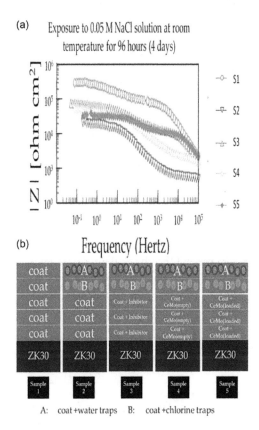

FIGURE 18.6 (a) FRA of the bare ZK30. (b) Bode for the S1, S2, S3, S4, and S5 samples exposed to NaCl concentration 0.05 m for 96 hours.

in the medium frequencies corresponding to the procedures. They are carried out on the metal-coating interface such as the creation of metal oxides, the creation of complexes between the inhibitor and metal, the creation of a films between derived from ions from the nanocontainers, and a third time of relaxation in low frequencies attributed to corrosive processes on the metal surface.

As far as the molybdenum oxide nanocontainers are concerned, these may be involved in increasing the absolute value of complex electrochemical resistance through a mechanism that we explain in the following text. Taking into account the measured pH values that did not exceed the price according to the Pourbaix diagram for the Mo-H_2O system, irrespective of the potential of erosion, the formation of MoO_2 oxide is possible thermodynamic through the reduction of the molybdenum ions:

$$MoO^2{}_4 + 4H^+ + 2e^- \rightarrow MoO_2 + 2H_2O.$$

Alternatively, through the reduction of molybdenum ions in trivalent, tetravalent, or pentavalent molybdenum, oxide is created which has low solubility and is likely to function as a barrier to the diffusion of the harmful ions of the solvent.

The action of molybdenum ions is enhanced by the action of the cerium ion. The suspension mechanism based on them is related to the formation of a highly insoluble oxide/hydroxide precipitation. The produced OH- from the total corrosion reaction of magnesium, consumed by trivalent ions of cerium resulting in the formation of cerium hydroxide ($Ce(OH)_n$).

In order to investigate the possible self-healing of coatings, artificial imperfections were formed in each of the essays of approximately 1 mm length and followed by the exposure of the samples in a corrosive environment NaCl 0.005 m. This small concentration was chosen to reduce the corrosion rate of highly active magnesium alloy. The data were fitted using the model of Figure 18.7.

The model gives the values for R_{coat}, C_{coat}, R_{ct}, and dual layer capacities (C_{dl}), for all coatings as a function of the exposure time. Figure 18.8 shows only the R_{coat} for the samples as a function of the exposure time because of reason of space only this parameter can be discussed here.

Representation of the physical interpretation and the corresponding equivalent circuit

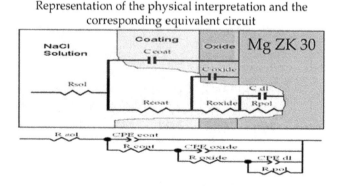

FIGURE 18.7   Model used for the FRA fits.

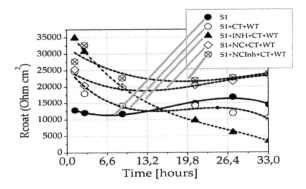

FIGURE 18.8 FRA of samples as a function of immersion time in salt water.

This resistance is related to the rate of corrosion, that is, the flow rate of electrons removed from the metal. Up to 33 hours of exposure, the coatings show a large increase in the load-carrying resistance in the case of hybrid coating without additives. In our case, the increase in the load-carrying resistance is not due to partial self-healing of the coating itself, but to the resistance that occurs due to the reinforcement of the oxides and complexes in the interlayer of metal coating. This can be explained based on the formation of greater thickness of oxides and in their transformation from the less-stable thermodynamic MgO in Mg (OH)$_2$. The second is oxide of a larger molecular size and is responsible for the condensation of the film and the sealing of the pores, hence the inhibition of the diffusion of the solvent on the surface of the metal. In this case, the pH measurements of the solution that gave values higher than 9 in which magnesium hydroxide is stable and insoluble according to the Pourbaix diagram [33,34] are agreed. This is not verified in other cases of coatings where pH measurements gave values 6–7.5. Another possible interpretation of the increase in load transport resistance values, as the coating time increases without additional artificial defects in the corrosive environment, is the existence of groups of primary amines from the phenyl-diamine and epoxy groups from the Aaldite 287 compound, which did not react to each other when creating the coating due to stereochemical barriers and which groups are now reacting when diffusion of the electrolyte. It is important to note here the two coatings: (1) coating with CeMo(empty) @ Coating with water traps @ Coating with chlorine nanotraps and (2) coating with CeMo(inhibitor) @ Coating with water traps @ Coating with chlorine nanotraps show first an decrease of R_{coat} with the time up to 19 hours and then increase with the time. The decrease of R_{coat} is attributed to oxidation reaction while the increase for time greater than 10 hours due to "self-healing."

18.3 CONCLUSIONS

In the present work, we reported the effect of the incorporation of nanocontainers as carriers of corrosion inhibitors as well as traps of chlorine ions and water molecules in the corrosion-protection properties of hybrid organic–inorganic coatings. The objective was to assess the effect of the structures synthesized on the barrier properties of the coating and to investigate possible self-healing. We first evaluated a number of inhibitors via impedance measurements and found that ammonium phosphate performs best. To investigate the possible self-healing of coatings with our nanotechnology, we created artificial defect in each coating and measured the impedance spectrum as function of the time in salt water. The results showed that the coating without additives shows rising values of the current transfer resistance corresponding to the reduced rate of the corrosive process. On the one hand, this may be because of the high resistance values of the formed oxides by increasing the exposure time in the NaCl environment resulting in an increase in their thickness and the transformation of the oxides into the most stable and insoluble hydroxides of magnesium, which may interfere with the diffusion of the electrolyte to the surface of the metal. On the other hand, speculated that groups of primary amines and epoxy groups that did not react during the preparation of the coating, react with the effect of coating exposure on the electrolyte. For the samples with CeMo nanocontainers filled or not with inhibitors, an oxidation is observed for times up to 10 hours in salt water, and then an increase of R_{coat} is detected due to "self-healing."

ACKNOWLEDGMENTS

The work is conducted under generous funding from the European Union under the programs FP6 MULTIPROTECT and FP7 MUST. The author appreciates the collaboration with DAIMLER, FIAT, EADS, Liebherr, Mankiewicz Gebr. & Co., Chemetal GmbH, SIKA, VARNISH Srl, Re-Turn AS, and Wilckens Farben GMBH generously offering samples, chemical, and more important industrial experience to corrosion protection of metals.

REFERENCES

1. Kartsonakis IA, Liatsi P, Danilidis I, Kordas G. Synthesis characterization, and antibacterial action of hollow ceria nanospheres with/without a conductive polymer coating. *Journal of the American Ceramic Society* 2008: 91 (2), 372–378.
2. Kartsonakis IA, Liatsi P, Danilidis I, Bouzarelu D, Kordas G. Synthesis, characterization and antibacterial action of hollow titania spheres. *Journal of Physics and Chemistry of Solids* 2008: 69 (1) 214–221.
3. Tapeinos C, Kartsonakis I, Liatsi P, Kordas G. Synthesis and characterization of magnetic nanocontainers. *Journal of the American Ceramic Society* 2008: 91 (4), 1052–1056.
4. Kartsonakis IA, Danilidis IA, Kordas G. Encapsulation of the corrosion inhibitor 8-hydroxyquinoline into Ceria Nanocontainers. *Journal Sol Gel Science and Technology* 2008: 48 (1–2), 2431.
5. Tsirimpis A, Danilidis I, Liatsi P, Kordas G. Effect of sweep rate and electrolyte concentration on the potentiodynamic deposition of intrinsically conductive polymer coatings onto Al2024-T3 substrates. *Progress in Organic Coatings* 2010: 67, 389–397.
6. Kartsonakis IA, Kordas G. Synthesis and characterization of cerium molybdate nanocontainers and their inhibitor complexes. *Journal of the American Ceramic Society* 2010: 93 (1), 65–73
7. Kartsonakis IA, Danilidis IL, Pappas GS, Kordas G. Encapsulation and release of corrosion inhibitors into titania nanocontainers. *Journal of Nanoscience and Nanotechnology* 2010: 10 (1–9), 5912–5920.

8. Mekeridis E, Kartsonakis IA, Pappas G, Kordas G. Release studies of corrosion inhibitors from cerium titanium oxide nanocontainers. *Journal of Nanoparticle Research* 2011: 13 (2), 541–554.

9. Balaskas AC, Kartsonakis IA, Kordas G, Cabral AM, Morais PJ. Influence of the doping agent on the corrosion protection properties of polypyrrole grown on aluminium alloy 2024-T3. *Progress in Organic Coatings* 2011: 71, 181–187.

10. Balaskas AC, Kartsonakis IA, Snihirova D, Montemor MF, Kordas G. Improving the corrosion protection properties of organically modified silicate–epoxy coatings by incorporation of organic and inorganic inhibitors. *Progress in Organic Coatings* 2011: 72 653– 662.

11. Balaskas AC. Kartsonakis IA, Tziveleka LA, Kordas G. Improvement of anti-corrosive properties of epoxy-coated AA 2024-T3 with TiO_2 nanocontainers loaded with 8-hydroxyquinoline. *Progress in Organic Coatings* 2012: 74 (3) 418–426.

12. Kartsonakis IA, Balaskas AC, Kordas G. Influence of cerium molybdate containers on the corrosion performance of epoxy coated aluminium alloys 2024-T3. *Corrosion Science* 2011: 53, 3771–3779.

13. Balaskas AC, Kartsonakis IA, Bilalis P, Karatzas A, Kordas G. Epoxy coatings containing nanocontainers loaded with corrosion inhibitors for corrosion Protection of AA2024-T3. *Meeting Abstracts* 2011: 1656–1656.

14. Kartsonakis IA, Balaskas AC, Koumoulos EP, Charitidis CA, Kordas G. Incorporation of ceramic nanocontainers into epoxy coatings for the corrosion protection of hot dip galvanized steel. *Corrosion Science* 2012: 57, 50–53.

15. Kartsonakis IA, Koumoulos EP, Balaskas AC, Pappas GS, Charitidis CA, Kordas G. Hybrid organic–inorganic multilayer coatings including nanocontainers for corrosion protection of metal alloys. *Corrosion Science* 2012: 57, 56–66.

16. Montemor MF, Snihirova DV, Taryba MG, Lamaka SV, Kartsonakis IA, Balaskas AC, Kordas G, Tedim J, Kuznetsova A, Zheludkevich ML, Ferreira MGS. Evaluation of self-healing ability in protective coatings modified with combinations of layered double hydroxides and cerium molybdate nanocontainers filled with corrosion inhibitors. *Electrochimica Acta* 2012: 60, 31– 40.

17. Kartsonakis IA, Balaskas AC, Kordas G. Influence of TiO_2 nanocontainers on hybrid organic-inorganic coatings for corrosion protection of magnesium alloy. *International Journal of Structural Integrity* 2013: 4(1), 127–142.

18. Kordas G, Balaskas AC, Kartsonakis IA, Efthimiadou EK. A Raman study of 8-Hydroxyquinoline release from loaded TiO_2 nanocontainer *International Journal of Structural Integrity* 2013: 4(1), 121–126.

19. Karatzas A, Bilalis P, Kartsonakis IA, Kordas G. Reversible spherical organic water microtraps. *Journal of Non-Crystalline Solids* 2012: 358(2) 443–445.

20. Mekeridis ED, Kartsonakis IA, Kordas G. Multilayer organic–inorganic coating incorporating TiO_2 nanocontainers loaded with inhibitors for corrosion protection of AA2024-T3. *Progress in Organic Coatings* 2012: 73 (2–3), 142–148.

21. Kartsonakis IA, Balaskas AC, Koumoulos EP, Charitidis CA, Kordas G. Incorporation of ceramic nanocontainers into epoxy coatings for the corrosion protection of hot dip galvanized steel. *Corrosion Science* 2012: 57, 30–41.

22. Krzak M, Tabor Z, Nowak P, Warszyński P, Karatzas A, Kartsonakis IA, Kordas G. Water diffusion in polymer coatings containing water-trapping particles. Part 2. Experimental verification of the mathematical model. *Progress in Organic Coatings* 2012: 75 (3), 207–214.

23. Kartsonakis IA, Balaskas AC, Koumoulos EP, Charitidis CA, Kordas G. ORMOSIL-epoxy coatings with ceramic containers for corrosion protection of magnesium alloys ZK10. *Progress in Organic Coatings* 2013: 76 (2–3), 459–470.

24. Kartsonakis IA, Athanasopoulou E, Snihirova D, Kordas G. Multifunctional epoxy coatings combining a mixture of traps and inhibitor loaded nanocontainers for corrosion protection of AA2024-T3. *Corrosion Science* 2014: 85, 147–159.

25. Kordas G. Nanotechnology to improve the biofouling and corrosion performance of marine paints: From lab experiments to real tests in sea. *International Journal of Physics Research and Applications* 2019: 2, 053–057.

26. Kordas G. Novel antifouling and self-healing eco-friendly coatings for maritime applications enhancing the performance of commercial maritime paints. *IntechOpen* 2019: 1–9.

27. Kordas G. Incorporation of spherical shaped CuO@SiO$_2$ light micro-traps into CuCoMnO$_x$ spinels to enhance solar absorbance. *Journal of the American Ceramic Society* 2019, https://doi.org/10.1111/jace.16851.

28. Kordas G. CuO (Bromosphaerol) and CeMo (8 Hydroxyquinoline) microcontainers incorporated into commercial marine paints. *Journal of the American Ceramic Society* 2019:1–11, doi:10.1111/jace.16917.

29. Kordas G. Nanotechnology in cancer treatment as a Trojan horse: From the bench to preclinical studies. *Nanostructured polymer composite for biomedical applications.* Eds. Swain SK, Jawaid M. Micro and Nano Technologies. London: Elsevier, 2019: 323–365

30. Stober, W., A. Fink, and E. Bohn, Controlled growth of monodisperse silica spheres in the micron size range. *Journal of Colloid and Interface Science* 1968, 26(1): 62–69.

31. Kaiser, C., et al., Nonporous silica microspheres in the micron and submicron size range. *Fine Particles Science and Technology* 1996, 71–84.

32. Ibrahim, I.A.M., A.A.F. Zikry, M.A. Sharaf, Preparation of spherical silica nanoparticles: Stober silica. *Journal of American Science* 2010;6(11).

33. Lamaka, S.V., et al., Complex anticorrosion coating for ZK30 magnesium alloy. *Electrochimica Acta* 2009. 55(1): 131–141.

34. Persaud-Sharma, D. McGoron A. Biodegradable magnesium alloys: A review of material development and applications. *Journal of Biomimetics Biomaterials Tissue Engineering* 2012, 12: 25–39.

Index

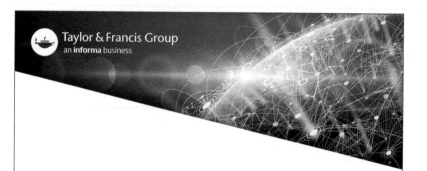